# Global Perspective of Information Technology Management

Felix B. Tan, Ph.D.
University of Auckland, New Zealand

**IRM Press**
**Publisher of innovative scholarly and professional information technology titles in the cyberage**

Hershey • London • Melbourne • Singapore • Beijing

Acquisitions Editor: Mehdi Khosrow-Pour
Managing Editor: Jan Travers
Assistant Managing Editor: Amanda Appicello
Copy Editor: Amanda Appicello
Cover Design: Tedi Wingard
Printed at: Integrated Book Technology

Published in the United States of America by
IRM Press
1331 E. Chocolate Avenue
Hershey PA 17033-1117
Tel: 717-533-8845
Fax: 717-533-8661
E-mail: cust@idea-group.com
Web site: http://www.irm-press.com

and in the United Kingdom by
IRM Press
3 Henrietta Street
Covent Garden
London WC2E 8LU
Tel: 44 20 7240 0856
Fax: 44 20 7379 3313
Web site: http://www.eurospan.co.uk



Library of Congress Cataloguing-in-Publication Data

Global perspective of information technology management / [edited by] Felix B. Tan.
p. cm.
Includes bibliographical references and index.
ISBN 1-931777-11-X (paper)
1. Information technology--Economic aspects. 2. Information technology--Management. I. Tan, Felix B., 1959-

HC79.I55 G5795 2002
658'.05--dc21 2002017217

eISBN: 1-931777-32-2

British Cataloguing-in-Publication Data
A Cataloguing-in-Publication record for this book is available from the British Library.

# Global Perspective of Information Technology Management

# Table of Contents

# Foreword

## THE NATURE OF GLOBAL IT RESEARCH

Managing information technology (IT) on a global scale presents a number of opportunities and challenges. IT can drive the change in global business strategies and improve international coordination. At the same time, IT can be an impediment to achieving globalization. IT as an enabler of and inhibitor to globalization raises interesting questions. To what extent does IT facilitate globalization? In what ways can IT be a constraint to globalization? Is managing IT in a global context largely the same as managing IT in a domestic context? If it is not, then what aspects are different? How are differences in culture important to the implementation of global information systems?

Research into these and other related topics have grown tremendously over the last 10 years. Journals focusing on global aspects of IT (i.e., global IT) were established driven by the need to understand the relationship between IT and globalization. New research tracks on global IT were created at international conferences to generate interest and permit research to be disseminated as soon as they are completed.

What characterizes research into global IT? First, it is a narrow field of study. Global IT research is a subset of information systems (IS) research. It focuses on IS issues that are international in flavor. Global IT research is also an extension of the traditional IS field – beyond a single system in a single location to multiple systems, with diverse users in a global environment. However, global IT research is distinct from general IS research. One way of differentiating is by the research themes examined. Brent Gallupe and I (1999) identified six major themes in global IT research. The following table describes the key global themes. These themes distinguish global IT research from other IS research.

| Research Themes In Global IT Research | Description |
|---|---|
| Global Enterprise Management | Functional management of enterprises across national boundaries using IT. Includes the management of multinational and trans-national corporations. For example, the impact of IT on global supply chains, global distribution or global marketing, including global electronic commerce. |
| Global Information Resources Management | The development, operations, management and use of IT in a global context. Includes the management of information resources in a regional but not domestic context. For example, managing global IT outsourcing; motivating global IT development teams; and managing the introduction of telecommunications technology in Latin America. |
| Comparative Study of Nations | The comparison of IT development, operations, management and use between two or more countries. Culture is not a variable in these studies. For example, comparing the skills of systems analyst in Canada, New Zealand and Singapore. |
| Culture / Socio-economic Issues | The effect of national culture on IT development, operations, management and use. These are regarded as "pure" cross-cultural studies as opposed to those merely comparing nations. For example, exploring the effect of complex cultural dimensions on the level of IT transfer in Arab and East European nations. Also includes socio-economic issues relating to IT like government policy, legislation and economic factors. |
| Research Frameworks and Issues | Conceptual research offering frameworks, theory and research agendas on various aspects of global information management. Key issues studies in IS management from around the globe are included. |
| Single Country Studies | The development, operations, management and use of IT in a domestic context. Does not include the management of information resources in a global context. For example, the adoption of IT in rural China. |

We were surprised to observe that there were fairly few IS research into culture and its effects. We, therefore, encourage more research into the impact of culture on various IT issues. Our understanding of the relationship between culture and global IT is still very limited. There is a need for more theory-based assessment of cultures (Straub, Loch, Evaristo, Karahanna, & Srite, 2002) and the need to have a more dynamic view of culture – one that sees culture as contested, temporal and emergent (Myers & Tan, 2002). Others have warned global IT researchers not to ignore structural conditions present in different geographical locations as the differences in these conditions can exert a strong influence in global e-commerce activities (Markus & Soh, 2002).

Second, global IT research is very diverse. There are a wide variety of variables studied and methodology used. Variables examined were internal to the organization (e.g., characteristics of the development, operations and management environments, the characteristics of the IT being studied and extent of adoption and diffusion of IT), as well as the external environments (e.g., culture, infrastructure, government policy and legal system). Both quantitative and qualitative approaches have been widely employed – primarily, field and case studies (Gallupe & Tan, 1999). Examples include a study of the fit between managerial decision tasks and types of systems and its impact on the amount of IT use between Greek and U.S. managers using field surveys (Ferratt & Vlahos, 1998); an assessment of Arab culture and IT transfer using a combination of qualitative (Hill, Loch, Straub, & El-Sheshai, 1998) and quantitative techniques (Straub, Loch, & Hill, 2001).

Third, the narrow and diverse nature of the field presents difficulties in conducting global IT research. The difficulties intrinsic in global research revolve around resource constraints and methodological problems, for example, language barriers between researchers and research participants, research instrument translation, conceptualizing and measuring culture, biases, financial and time related limitations. Karahanna, Evaristo and Srite (2002) provide a comprehensive discussion on the methodological issues in conducting cross-cultural research. The authors present an overview of the key methodological concerns and review ways of preventing or detecting these problems.

Finally, the study of IT management in the global context can be thought to have considerable overlaps with the field of international business. However, surprisingly few global IT publications have used international business theories for appropriating variables and measures or for putting discussion issues into a broader context (Niederman, Boggs, & Kundu, 2002). There is, therefore, a huge opportunity for researchers to consider the connections between international business and global IT.

## REFERENCES


Ferratt, T. W., & Vlahos, G. E. (1998). An investigation of task technology fit for managers in Greece and the US. *European Journal of Information Systems*, 7(2): pp. 123-136.

Gallupe, R. B., & Tan, F. B. (1999). A research manifesto for global information management. *Journal of Global Information Management*, 7(3): pp. 5-18.

Hill, C. E., Loch, K. D., Straub, D. W., & El-Sheshai, K. (1998). A qualitative assessment of Arab culture and information technology transfer. *Journal of Global Information Management*, 6(3): pp. 29-38.

Karahanna, E., Evaristo, R., & Srite, M. (2002). Methodological issues in MIS cross-cultural research. *Journal of Global Information Management*, 10(1): pp. 48-55.

Markus, M. L., & Soh, C. (2002). Structural Influences on Global E-Commerce Activity. *Journal of Global Information Management*, 10(1): pp. 6-14.

Myers, M. D., & Tan, F. B. (2002). Beyond models of national culture in information systems research. *Journal of Global Information Management,* 10(1): pp. 24-32.

Niederman, F., Boggs, D. J., & Kundu, S. (2002). International business and global information management research: Toward a cumulative tradition. *Journal of Global Information Management,* 10(1): pp. 33-47.

Straub, D., Loch, K., Evaristo, R., Karahanna, E., & Srite, M. (2002). Toward theory-based measures of culture. *Journal of Global Information Management,* 10(1): pp. 13-23.

Straub, D. W., Loch, K. D., & Hill, C. E. (2001). Transfer of information technology to the Arab World: A test of cultural influence modelling. *Journal of Global Information Management,* 9(4): pp. 6-28.

Felix B. Tan
Editor-in-Chief, Journal of Global Information Management
The University Of Auckland, New Zealand
Email: f.tan@auckland.ac.nz

# Preface

The influence of information technology can be felt around the world. From Ireland to Australia, governments, businesses and individuals are using technology to improve their lives. However, the use and development of technology is largely influenced by national and organizational culture, and, as businesses are becoming more global in nature, it is essential for academics, business leaders and technology researchers to understand the nature of information technology use and development in specific countries and in a general global sense. This book provides the most current research on global information technology management and will be beneficial to all those involved in the research or practice of information technology management across the globe. The chapters provide theoretical discussions of research in global information management and the field of human resource management on a global scale. Furthermore, the chapters provide practical examples of successful information technology management and outsourcing in the global marketplace.

Chapter 1 entitled, "The Relationship Between User Participation and the Management of Change Surrounding the Development of Information Systems: A European Perspective" by Tom Butler and Brian Fitzgerald of the University College Cork (Ireland) addresses the deficiency of process-based approaches to the study of user participation in information systems development and its impact on matters of organizational change. The chapter presents the results of a process-based study of these issues and then proposes a theoretical model which captures the institutional and development-related contexts that shape and influence the processes of user participation.

Chapter 2 entitled, "Global Information Systems and Human Resource Management: A Research Agenda" by Fred Niederman of the University of Baltimore (USA) explores five areas in which information technology and human computer designers or users interact within the context of global organizations. The five areas explored are: using information technology to support human resources strategy; using information technology to support the generation and distribution of organizational learning; using human resources management techniques and programs to support the work of IS professionals; using human resource management techniques and programs to support the work of global end-users; and national and regional policies to support technical and human resource infrastructures.

Chapter 3 entitled, "Applying Erlang Distribution for Software Size Estimation" by Derek Cheung and Ho-Leung Tsoi of the City University of Hong Kong (Hong Kong) discusses the weaknesses of the Program Evaluation and Review Techniques (PERT) model and proposes applying Erlang distribution to overcome the weaknesses. The chapter includes a practical description and example to illustrate the effectiveness of the alternative model presented.

Chapter 4 entitled, "Trajectories and Strategies in Global Software Outsourcing Relationships" by Richard Heeks and Brian Nicholson of the University of Manchester, S. Krishna of IIM-Bangalore and Sundeep Sahay of the University of Oslo investigates the means by which software clients and developers can achieve the benefits that higher value global software outsourcing offers. The chapter reports a longitudinal case study research with eight global software outsourcing relationships where the developer-half was based in India.

Chapter 5 entitled, "Emerging Research Themes in Global Information Management" by R. Brent Gallupe of Queen's University (Canada) and Felix Tan of the University of Auckland (New Zealand) reports one aspect of a wider study in global information management—emerging global information management research themes. The purpose of this chapter is to identify themes distinct to GIM research that are emerging from the literature with the aim of highlighting the potential areas of future research.

Chapter 6 entitled, "Undergraduate Global IT Education – An Experiential Approach Using the Concept of Fit" by Joan Mann of Old Dominion University (USA) demonstrates how to formulate a global information systems course for undergraduates that will be appropriate to the way undergraduates learn, include the skills they need at this stage in their career and create an affective change in their perspective. The chapter illustrates the importance of alignment between global strategies, organizational structures, the configuration of IS departmental areas, organizational information infrastructures and national/organizational culture.

Chapter 7 entitled, "European Electronic Service Infrastructure Building – Drifting Into the Future" by Åke Grönlund of Umeå University (Sweden) reports on a 3-year study of 9 cities in 8 European countries in developing a service infrastructure including organizational solutions for logistics and client-organization interactions. During the course of the study, the authors discovered twelve distinct crisis situations where goals environment and actors changed, and the process was found in a state of improvisation. The chapter discusses the keys to successful implementation and utilization of service infrastructures.

Chapter 8 entitled, "Comparative Study of the Usefulness of Online Technologies in a Global Virtual Business Project Team Environment" by Simpson

Poon and Shri Rai of Murdoch University (Australia) discusses online technology as a medium of educational service delivery. The chapter examines a case study where online technology has been used to coordinate virtual project teams around the world. The case study also shows that synchronous communications is not necessarily a better means to coordinate than its asynchronous counterpart.

Chapter 9 entitled, "Big-Bang ERP Implementation at a Global Company" by Nava Pliskin and Marta Zarotski of Ben-Gurion University of the Negev (Israel) describes the case study of Dead Sea Works, an international multi-firm producer of Potash and other chemicals. The case study describes the purchase and implementation of a big-bang ERP implementation of SAP R/3 and the organizations attempt to minimize risk through strategic purchasing and a steering committee. The R/3 system was implemented on time and on budget. The case examines the keys to successful implementation.

Chapter 10 entitled, "A Three-Tiered Approach to Global E-Commerce: Experiences of Nu Skin International" by David Paper of Utah State University and Ken Tingey of Opennet Corporation (USA) discusses a case study to research the strategic issues involved in Internet-based globalization. The chapter explores how the Internet and its related technologies can serve to help organizations better deal with the challenges of conducting global business, and identifies a set of heuristic "rules of thumb" that might be used to support Internet-based globalization efforts.

Chapter 11 entitled, "Growth and Consolidation in the Spanish-Speaking E-Commerce Market" by Roberto Vinaja of the University of Texas-Pan American (USA) analyzes the major challenges in the widespread adoption of electronic commerce in the Spanish-speaking population. The case study provides a general overview of related issues in global e-commerce and includes a description of the strategies followed by companies entering the Latin American market in order to illustrate some of the major cross-border issues.

Chapter 12 entitled, "ICT Diffusion and Strategic Role within Italian SMEs" by Giacomo Buonanno, Stefano Gramignoli, Aurelio Ravarini and Marco Tagliavini of the University of Cattaneo, Castellanza (Italy) and Donatella Sciuto of the Politecnico di Milano, Castellanza (Italy) focuses on the specific use of information and communication technologies (ICT) in small to medium sized enterprises (SME). The chapter discusses the current state of ICT employed within Italian SMEs. It further provides the results of a study of the relationship between ICT and SME strategy.

Chapter 13 entitled, "Critical Strategies for Information Systems Development Projects: Perceptions of Developers in Korea" by Chung Kim, Dan Peterson, Jerry Chin and Tonya Barrier of Southwest Missouri State University (USA) discusses the results of a study examining the views of IS developers from

Korea regarding factors perceived to be critical in IS development. The information of this study is of great importance as the software industry expands into Asia.

Chapter 14 entitled, "A Web-Based Data Warehouse Planning — A Large-Scale Project Management Information System in Taiwan" by Jia-Lang Seng of National Chengchi University (Taiwan) describes a data warehouse planning project to develop a large-scale management information system for a Taiwan government agency in managing a nationwide multi-billion dollar science and technology project management and control process.

Chapter 15 entitled, "Behind the Celtic Tiger: Key Issues in the Management of Information Technology in Ireland in the Late 1990s" by Joseph McDonagh and Andrew Harbison of the University of Dublin (Ireland) presents the results of a survey discussing the key issues in the management of information technology in Ireland. The results of the survey indicate that the important issues can be classified into four main areas of concern: technical and operational issues, strategic and planning issues, relationship issues with the organization and recruitment of IS staff.

Chapter 16 entitled, "Turning E-Commerce Theory into Action in Ireland: Taming the Celtic Tiger" by Ira Yermish and Dale A. Bondanza of St. Joseph's University (USA) examines a case study which becomes a possible strategy for entrepreneurial consultants seeking to provide expertise to rising e-businesses. The chapter discusses the demographics and technology issues facing Ireland, addresses the issues of the entrepreneur seeking to start an e-commerce consulting practice in Ireland and suggests additional steps to fill in missing details and develop a full-fledged business plan for action for young entrepreneurs.

Chapter 17 entitled, "Anxiety and Involvement: Cultural Dimensions of Attitudes Toward Computers in Developing Societies" by Roger Harris of the Universiti Malaysia Sarawak (Malaysia) and Robert Davison of City University of Hong Kong (Hong Kong) focuses on the adoption of information technology in developing societies. A case study examines computer anxiety and involvement with personal computers among six groups of students in different countries. Demographic and cultural factors weigh heavily in the results of the different anxieties and involvement of computers among the students.

Chapter 18 entitled, "The Context of Introducing IT/IS-Based Innovation into Local Government in Colombia" by Mahmoud M. Watad of Fairleigh Dickinson University (USA) examines the dynamics of introducing IT/IS into local municipalities in Columbia and addresses the factors that drive the introduction of geographical information systems into public organizations. The results are based upon data generated from documents and direct interviews in two municipalities. The findings indicate that middle managers are often the driving forces behind technological advances.

Chapter 19 entitled, "The Impact of Information Sharing on Order Fulfillment in Divergent Differentiation Supply Chains" by Troy J. Strader of Iowa State University (USA), Fu-Ren Lin of National Sun Yat-sen University (Taiwan) and Michael J. Shaw of the University of Illinois at Urbana-Champaign (USA) focuses on divergent differentiation supply chains which are commonly associated with industries such as fashion apparel. The chapter identifies the impact of information sharing by simulating order fulfillment in this supply chain type and finds inventory costs can be reduced while maintaining acceptable order fulfillment cycle times.

Chapter 20 entitled, "Leapfrogging an IT Labor Force: Multinational and Indigenous Perspectives" by Eileen Trauth of Northeastern University (USA) explores two key human resource issues for the global IT sector, ensuring a supply of qualified IT professionals in the host country and managing IT workers in a cross-cultural setting. The chapter is based upon an ethnographic study of the information sector in Ireland and the issues raised represent the experiences of American and Irish managers, Irish policy makers and Irish IT workers.

Understanding how to manage as a business or individual in the global technological world is becoming ever more important. Understanding differences in cultural paradigms, laws and governmental support, variations in infrastructure development is key to being successful in global information technology management. The chapters in this timely new book, all written by experts in their relative fields representing a variety of countries, present the most up-to-date research and theory in the field of global information management. This book is a must read for those interested in the theoretical dimensions of global information technology management or the practical challenges presented by operating across borders. Researchers, practitioners, managers and students will all benefit from the insights contained in this important new book.

IRM Press
January 2002

## Chapter 1

# The Relationship Between User Participation and the Management of Change Surrounding the Development of Information Systems: A European Perspective

Tom Butler & Brian Fitzgerald
University College Cork, Ireland

*While much is known about the general process of user participation in information systems development, its impact on matters of organisational change has not been the subject of systematic, in-depth investigation. In addition, researchers have typically adopted variance, rather than process-based approaches to the study of these related phenomena. This paper addresses these deficiencies and makes several important contributions to the literature. First, it presents the results of a comprehensive, process-based study of the relationship between user participation and organisational change in the development and implementation of information systems in a large organisation. Second, it presents a theoretical model which captures the institutional and development-related contexts that shape and influence the processes of user participation and management of change. Third, using the model as a framework to guide the research effort, this study illustrates that an organisation's institutional context plays a dominant role in shaping and influencing the content and process of user participation and management of*

Previously Published in the *Journal of End User Computing, vol.13, no.1,* Copyright © 2001, Idea Group Publishing.

*change in systems development. This particular finding has important implications for both research and practice. Finally, the model and its associated framework has been validated by the findings of this stud; it may, therefore, be used in future explorations of these important phenomena.*

## INTRODUCTION

Information systems development is a multi-dimensional change process that presents itself simultaneously within several related social environments—as a reality, it is socially constructed (Visala, 1991). The conventional wisdom within the information systems community argues that user participation is a core ingredient in this change process and is vital for successful outcomes in terms of both process and product (see Ives and Olson, 1984). However, two comprehensive reviews of research on the phenomenon of user participation revealed that the relationship between user participation and successful systems development is neither grounded in theory nor substantiated by research data (see Ives and Olson, 1984 and Cavaye, 1995). In addition, this paper argues that insufficient attention has been paid to the relationship that exists between user participation in systems development and the issue of organisational change surrounding the development and implementation of information systems: as a consequence, that relationship remains ill-defined and thus little understood. In order to address this deficiency, the study maintains that it is only by conceptualizing information systems development as a *change process* (Boland, 1978; Lyytinen, 1987), and by adopting a perspective that incorporates both user participation and management of change as being instrumental in determining the ultimate success of developed systems, can the relationship between these two concepts and their consequences be evaluated, explained, and understood. A conceptual model that incorporates the institutional context or framework within which the IS development sub-processes of user participation and management of change are effected, and which attempts to capture the interrelationships between factors that are posited to constitute these sub-processes is presented. The model's core components are drawn from institutional theory (North, 1990; Rowlinson, 1997), Cavaye's (1995) analytic framework, which was extended and elaborated by Butler and Fitzgerald (1997), and from seminal contributions of previous research on the phenomena of user participation and systems implementation (see Boland, 1978; Ives and Olson, 1984; and Orlikowski, 1993). The model helped formulate appropriate research questions to guide and direct case description, report its findings, and make appropriate conclusions.

As Orlikowski's (1993; p. 310) seminal investigation of the relationship between CASE and organisational change revealed, a process-based approach incorporating grounded theory *"allows a focus on contextual and processual elements as well as the action of key players associated with organisational change."* The constructivist paradigm not only incorporates a grounded theory perspective, it also offers researchers added rigor by providing an ontological, epistemological and methodological framework from which to conduct qualitative research (Erlandson *et al.*, 1993; Guba and Lincoln, 1994). Mature disciplines within the social sciences have accepted the need for a philosophical as well as a methodological rationale to underpin research; hence, a constructivist approach to research was adopted in order to apprehend the socially constructed reality that is information systems development in organisations (Visala, 1991; cf. Berger and Luckmann, 1966; Guba and Lincoln, 1994). The primary objective of this study, then, is to identify the critical elements that shape and influence the relationship between user participation in the development and implementation of organisational information systems and the process of change management surrounding the successful implementation of such systems. The research data is drawn from a case study of the development and implementation of two operational support systems in a large telecommunications company. Prior to presenting an analysis of the case data, the model and its analytical perspectives are first discussed.

## A MODEL OF USER PARTICIPATION AND MANAGEMENT OF CHANGE IN THE INFORMATION SYSTEMS DEVELOPMENT

While IS researchers such as King *et al.* (1994) and Kling and Iacono (1989) both incorporate institutional theories into their work, Orlikowski (1993) provides a graphic example of the importance of institutional contexts—environmental, organisational, and IS—and their impact on organisational change associated with the introduction of CASE. The role played by institutional frameworks in shaping organisational behaviour has been the subject of study and debate for some time now by economists and organisational theorists (see Rowlinson, 1997). Drawing on insights provided by North (1990), it can be argued that an organization's policies on systems development, particularly those related to issues of change management and employee participation, constitute a set of formal and informal criteria, the so called 'rules of the game', that help to mold human interaction in the pursuance of organizational goals and objectives. North argues that such policies shape and influence the institutional framework or context within which an organization's business routines are formulated and executed. Aaen (1986) has underlined the importance of managerial policy making in shaping the trajectory of the development process and its outcomes (cf. Ives *et al.* 1980). Hence, this paper argues that an organisation's policies on user participation in systems development and development-related change shape its institutional context through their formative influence on: (a) organisational culture and climate (Robey and Azevedo, 1994); (b) the type of participation (Mumford, 1979); (c) the degree of participation (Ives and Olson, 1984); (d) content and extent of participation (Hirschheim, 1983); and, finally, (e) the formality and influence of participation (Mumford, 1979) in systems development. With these institutional contexts (or 'rules of the game') in place, the manner in which development proceeds is argued to be directed and effected by:

1. Project-related factors
2. Process-related factors
3. User-related factors.

The content and manner in which the project-, process, and user-related activities are carried out is argued to be shaped and influenced by their institutional context (see, for example, Orlikowski, 1993). The proposed dimensioned sets of factors are based on Cavaye's (1995) analysis and have been extended and elaborated by Butler and Fitzgerald (1997). The interaction of institutional contexts and these factors are argued to determine system success in terms of user perceptions of system quality and user acceptance of the system. User acceptance of the implemented system is argued to be particularly dependent on the manner in which change is managed. The model illustrated in Figure 1 will, therefore, help illuminate the relationship between an organisation's institutional framework, project-, process-, and user-related factors, and development outcomes such as product quality and acceptance.

Boland's (1978) seminal work illustrates that a *change approach* (a bottom-up, participative strategy as opposed to the top-down *traditional approach*) to systems development attempts to have developers and users participate in joint problem solving with the objective of arriving at a systems solution through consensus. Resistance to change surrounding the implementation of the new system is thereby negated (Zmud and Cox, 1979). However, with few exceptions (see, for example, Ginzberg, 1981; Tait and Vessey, 1988; Krovi, 1993), the relationship between user participation in systems development and issues of organisational change does not appear to have been the subject of explicit, systematic investigation. Hence, the relationship between, and impact of user participation and management of change as separate constructs is often puzzling and difficult to interpret. This paucity in extant research begs to be addressed. The primary objective of this study, therefore, is to identify the critical elements that shape and influence the relationship between user participation in the development and implementation of organisational information systems and the process of change management surrounding the successful introduction and use of such systems.

## Research Questions

Based on the foregoing description of the above model, the first two of four research questions that help achieve the stated research objective are now articulated:

**RQ1:** What impact, if any, does an organisation's institutional framework have on the type, degree, content, extent, formality and influence of user participation in systems development?

**RQ2:** What role does the organisation's institutional framework play in the resolution of change management difficulties?

Successful systems development is a nebulous term; hence, it eludes direct evaluation. Accordingly, IS researchers employ surrogate measures to measure the success of development outcomes: Ives and Olson (1984), for example, propose system quality and system acceptance as appropriate '*outcome variables'*. Nonetheless, user satisfaction with developed systems has been widely employed by researchers as a surrogate for system success (Gatian, 1994; Cavaye, 1995). The dominant focus on user satisfaction and system success seems to capture implicitly, rather than explicitly, the fact that significant change has often taken place once a system has been implemented. Unfortunately, change management problems that often arise due to user resistance are generally ignored. Nevertheless, it is difficult to ignore the obvious, and several studies have underpinned Ives and Olson's contention that user participation may lead to increased *user acceptance*, a conceptual analogue for users' attitudes to the degree of change wrought by the introduction of new systems by:

- Allowing users to develop realistic expectations about system capabilities (see Lawrence and Low, 1993)
- Providing an arena for bargaining and conflict resolution about design issues (see Euchner *et al.,* 1993)
- Leading to system ownership by users (see Kozar and Mahlum, 1987)
- Decreasing user resistance to change (see Kozar and Mahlum, 1987; Hirschheim and Newman, 1988; Krovi, 1993)
- Committing users to the system (see McKeen *et al.,* 1994; Barki and Hartwick, 1994).

The forgoing points underline Regan and O'Connor's (1994) assertion that user resistance (or acceptance) centres more social change surrounding the introduction of systems, rather than on technical (quality-based) factors. The third and fourth research question posed here are:

**RQ3:** What critical project, process and user-related factors act to shape and influence (a) product quality and (b) product acceptance?

**RQ4:** Which of these factors related specifically to user participation, management of change, or both?

These then are the research questions which help guide this study.

## Research Philosophy and Strategy

A constructivist research approach, incorporating the hermeneutic method, was adopted for the qualitative, interpretive, case-based research strategy employed in this study (see Lincoln and Guba, 1985; Guba and Lincoln, 1994; and Butler 1998). This strategy involved an exploratory, single instrumental case study (Stake, 1994) with two embedded units of analysis (Yin, 1989)—that is, two systems development projects. Purposeful sampling was employed (Patton, 1990; Marshall and Rossman, 1989). The initial study was conducted in 1996, with a follow up study in 1998 to evaluate the outcomes of this organisation's approach to the development and implementation of corporate IS. The case design utilized has been described by Yin (1989) as both 'post-hoc longitudinal'—in respect of the original site-visits—and longitudinal—in respect of the overall study. Table 1 describes the study's research strategy in detail.

## Case Description and Research Results

As the Republic of Ireland's major telecommunications utility, Telecom Éireann provides universal telecommunication service and enjoys a monopoly in many areas of its business. Being a state-owned company, Telecom Éireann's majority shareholder is the Irish Government, which retains a 65% stake in the organization. The remaining shareholders include Telecom's employees, who obtained a 15% stake in mid-1998 as part of Telecom's overall change management strategy, and two European telecommunication's operators—KPN (PTT Telecom BV) of Holland and Telia (AB) of Sweden—who jointly own 20% of the company. Telecom Éireann entered into a strategic alliance with these companies in January of 1997; Telecom's labor unions favored this alliance because both KPN and Telia possess institutional frameworks based on the tenets of industrial democracy, as indeed does Telecom Éireann. Telecom's institutional context underwent radical change in 1998 when employees acquired a share in the ownership of the company in exchange for increased levels of participation and agreement on all change-related issues dealing with the introduction of new information systems.

There are ten companies in the Telecom Group, the majority of whom are wholly-owned subsidiaries. As of 1998, it employed in excess of 11,000 staff. Being a large company, operating in highly competitive national and international environments, it has dynamic information systems needs; these needs are fulfilled by its in-house information systems function—the Information Technology Directorate

(ITD): the within-case unit of analysis in this case study. The ITD is a centralized functional unit whose chief responsibility is the development, integration, maintenance, and support of all corporate information systems. Based in Dublin, the ITD has a staff of over 280 spread among its eight divisions. In 1994, however, the IS function was an obscure department within the Finance Directorate. The appointment of a new CEO saw the IS function elevated to directorate status in 1995 in order to enable it to play a pivotal role in the planned IT-enabled transformation of the company. As a consequence of this change, the power asymmetry that existed between the then IS Department and its business clients within the organization was effectively mitigated. The advent of this change in status, coupled with other related events, allowed the IS function to effectively manage its relations with business units and associated user communities within the organization. Senior IS management were, for the first time, able to take full advantage of the company's participative policies and set about building new particpative processes and structures that employed a mixture of participatory design (PD) and joint application design (JAD) in order to maximize the benefits associated with user participation for systems development. The two development projects described herein were two of the first to be developed by the IS function using this novel approach. Certainly, other systems had or were currently being developed using a weak form of PD coupled with JAD. However, these systems had little impact on the underlying business processes or on the role-related responsibilities and remuneration of staff; in these projects, the process of user participation, although important, was not deemed to be critical.

Subsequently, the company drew on its experiences with IT development projects, such as those described in the case, to refine its participative strategies in the development of its present portfolio of corporate IS. Of particular note here is that these experiences highlighted the need to ensure that change associated with the introduction of IS was managed effectively. As of 1998, the organization has modified significantly the participative dimension of its institutional contexts in light of these experiences. While user participation as described in the case remains central to Telecom's approach, the company's employees, through the auspices of the labor unions, have agreed to accept changes to fundamental business processes facilitated by new IT systems in exchange for a 15% stake in the company's equity. The remainder of this paper describes the context and process of user participation and its relationship to issues of change management with reference to two systems development projects—the Generic Appointment System and the Geographic Information System—and the using the proposed model (see Figure 1) as a reporting mechanism and analytic framework.

## TELECOM ÉIREANN'S INSTITUTIONAL CONTEXT

Table 1 summarizes the salient dimensions of this organisations institutional framework as it applies to systems development. Each of the following sub-sections describes the components in greater detail.

### Organisational Policy on Change

Since its inception as a state-sponsored organization, Telecom has adopted a participative approach to the implementation of organizational policies and decisions. This was recently underlined when the company reiterated its position viz. *"The process of consultation with unions in regard to all the implications for staff of technological change, is one to which the company remains fully committed."*[1] To give effect to this policy, the company has instituted several joint bodies; for example, the Computer Liaison Committee (CLC), whose members are drawn from both company management as well the labor unions, deals exclusively with issues surrounding the introduction of information systems within the organization. Prior to the development of information systems in Telecom, the business owners/initiators submit and present their proposals to the CLC. At this joint forum a broad framework and terms of reference for the development and implementation of the new system are established. Here the labor unions arrange for user participation and involvement at various levels. Issues related to management of change may be highlighted at this stage. In the past, the level of importance attributed to such issues depended on union representatives' knowledge of the impact which the new system would have on their members' work and remuneration etc. and, also, on the willingness of business managers to provide an accurate impact assessment. There was also the matter of 'unintended consequences' surrounding the outcomes of development plans; these matters would be addressed at later meetings of the CLC at the behest of either the IS function, business owners, or labor unions.

### Organisational Policies on the Type, Degree, Content, Extent, Formality and Influence of User Participation

In adherence to the company's chosen participative approach, each systems development project within Telecom has a designated business owner or project sponsor. For large projects, a development steering group (DSG) is formed from the constituencies of interest within the organization; managers from the relevant business areas and IT Directorate (ITD) normally comprise these groups. Two project managers jointly manage each project: a user project manager drawn from the business constituency, and a development project manager drawn from the ITD. The latter manages the physical development of the system; the former manages business user input into the project in areas such as the provision

*Table 1: Research Strategy*

| Activity | Description |
|---|---|
| Site Selection | Telecom Éireann was purposefully selected for study because it provided a typical example of a European company that had institutionalized the practice of employee participation and involvement in decision making; particularly in the area of the introduction of new technology and in information systems development. Hyman and Mason (1995) provide a thorough analysis of current thought and practice on this topic, and their work is supportive of this paper's contention that Telecom fits the profile of a typical European company operating in the private as well as the public sector. (US and non-European readers may not be familiar with the level of employee participation and involvement in European companies. The high level of employee involvement results from the degree of influence exercised by European institutions such as the European Community and the governments of individual countries through legislation in matters of employee relationships.) However, Telecom is unique in that it is one of several major European telecommunications utilities that are presently undergoing a significant degree of IT-enabled organisational change centered on meeting the demands of market deregulation and increasing competitive pressures. As a practitioner, one of the authors is intimate with the telecommunications industry in both Europe and the US, and recognized that Telecom offered a uniquely accessible site where the phenomena of interest could be observed and investigated fully. The two systems development projects studied (the embedded units) were also purposely selected because of the type and degree of user participation and involvement in evidence, and because both encountered significant change management problems. It was hoped that because of this they would provide fertile examples of the phenomena being investigated. |
| Data sources | Research into the selected case and its embedded units was conducted through the use of individual interview and documentary sources. Because the development teams on both projects were relatively small, it was possible to select for interview all team members—developers and user representatives. The model presented previously indicated the contexts and processes of interest to this study; hence, in order to fully investigated these dimensions a total of twenty-one interviews took place with key social actors from: (a) both development projects (e.g. development project managers, systems analysts and programmers); (b) the development environment (e.g. as described by the IS executive and his senior management team); and (c) the organisational environment (e.g. user representatives and user project managers who were considered to be representative of 'world views' in the relevant user constituencies). This focus on multiple environments helped capture the institutional contexts and attributes indicated in the model. Each interview was tape-recorded and lasted up to 2 hours in length. The initial study was conducted over a period of one month in the company's head offices. The subsequent follow-up study was more general in nature and involved two formal interviews with senior business managers, multiple informal conversations with a cross-section of managers, union officers and users, and comprehensive documentary analysis. |
| Data Analysis | The qualitative data analysis techniques of content and constant comparative analysis provided the necessary mechanisms for the required structural analysis (Miles and Huberman, 1994; Patton, 1990). So as to provide structure for the comparative analysis a set of initial (general) seed categories were first drawn from the model's component dimensions and the preceding content analysis. These seed categories enabled a relatively holistic description of the phenomenon to emerge. However, as the analysis progressed these were refined into tighter categories as the thoughts, ideas, and statements of individual actors were compared, and conflicts and difficulties that arose in the categorization exercise were addressed . Triangulation techniques were also extensively employed to provide insights into events, relationships etc. between primary data sets (Erlandson *et al.*, 1993; Patton, 1990). Several of within-case analysis strategies described by Miles and Huberman (1994)—e.g. checklist matrices and network analyses etc.—were used to identify saturated categories and, hence, complete the structural analysis. Descriptive matrices adapted from the model were used also to present and analyse categories in a condensed format; extended narratives were employed to provide additional detail and context. Being a member of the organisation chosen for study, one of the authors was what Bødker and Pedersen (1991) have termed a *"cultural insider"*. Hence, as a member of the general business/user constituency, the company's largest labour union, and presiding officer of one of the company's participative forums, he was intimate with several of the sub-universes of reality that comprised the overall institutional reality (Berger and Luckmann, 1966). This provided the researchers with valuable insights into the organisation's culture and climate and greatly aided in the interpretation of the case. |

and management of user-representatives, user groups, user test teams, and infrastructural resources etc. The development team normally consists of one or more user representatives from interested constituencies within the business and a team of developers from the IT Directorate. User representatives actively participate in most development activities, apart from programming and the technical aspects of systems development. Although key users are interviewed to elicit system requirements, user groups are also formed to provide the development team with a core group of users for further requirements analysis and to verify and ensure that the system, as developed, will meet these requirements. Because of the difficulty in involving all interested parties directly in the systems development, the company utilizes both PD and JAD approaches to the development process. The participatory mechanisms employed within the organization provided users with opportunities to express their 'world views', have political conflicts resolved, and helped negate potential power asymmetries between developer and user.

As Table 1 indicates, users participate directly and are consulted about design issues through project-based mechanisms such as JAD, the objective being to arrive at a consensus on such matters (see Mumford, 1979). User participation in the GAS and GIS development processes, for example, ranged from 'participation by advice' to 'participation by strong control' (Ives and Olson, 1984), depending on the organisational status of the end users participating. User representatives on the development teams participated as support for analysts in the requirements elicitation exercises with individual users and user groups. In both projects, user representatives were trained in the CASE tools and techniques, and participated in the use of these tools; Kozar and Mahlum (1987) comment on the importance of this aspect of user participation. User representatives also took an active role in the implementation of these systems. Users who did not participate on development teams did so at individual interviews and at group sessions with the systems analysts and user representatives during the requirements analysis phase: in the GAS project these users also participated in prototyping activities. In both projects users participated in testing the systems once developed. Users from the management constituency also played an active role either as project managers or as project sponsors (cf. Land and Hirschheim, 1983).

## PROJECT-RELATED FACTORS

As previously indicated, two systems development projects formed the embedded units of analysis in the study: the Generic Appointment System and the Geographic Information System development projects. Table 2 describes the project related-factors that the model in Figure 1 suggests as being relevant to the processes of user participation and change management. The following narrative

*Table 2: Telecom Éireann's Institutional Context for Systems Development*

| Institutional Context | Case Findings |
|---|---|
| Organisational policy on change | Since its inception in 1984, the organisation has maintained a participative approach to decision making and change. In terms of both structure and process, various participative fora exist to give them effect. The Joint Technology Committee (JTC) and the Computer Liaison Committee (CLC) are joined and oversee the introduction of new IT systems. In regard to systems development, user representatives are seconded to project teams, while user groups and individual users participate at JAD sessions. |
| Organisational policy on user participation | Telecom Éireann's policy on user participation reflects is commitment to participative decision making. Telecom's implementation of this policy lies in the manner in which it structures its development teams into user and IT project managers, developers, user representatives, and user groups for JAD sessions. |
| *Type of user participation* | All aspects of Mumford's (1979) typification of user participation e.g. consultative, representative, and consensual were in evidence in both projects. Individual users were by and large consulted on development issues; user representatives and user groups played a more active role that was both representative and consensual; overall consensus on development outcomes was obtained at the level of the CLC. |
| *Degree of user participation* | Ives and Olson (1984) argue that there are several degrees of participation. These range from no participation at all, to symbolic participation, participation by weak control, participation by doing, and participation by strong control. Both business managers (as users) and employees (through the CLC) exercised strong control, user representatives actively participated in both GAS and GIS, weak control was exercised by user groups, while the remaining users did not participate but, according to Barki and Hartwick (1989) were involved. |
| *Content and extent of participation* | User representatives were present throughout the development process. Individual users and user groups participated at key points in the development process: e.g. analysis and design, testing and implementation. |
| *Formality & influence of participation* | User representatives were co-opted into the development project team. Development steering groups were formed for management users, and user groups were formed to participate in requirements analysis, design verification and testing. Significant user influence was exerted, especially through labour unions and joint management/union forum. |
| *Organisational culture and climate* | The organisation's culture emanates from its status as a state-sponsored organisation. This culture is changing as competitive pressures, market deregulation, and the move to privatization as a quoted company, all act to change process and structure. The climate is reflected in the partnership approach, with workers being well remunerated and with excellent terms and conditions of employment. There were no incidents of industrial unrest in Telecom since the 1970s. The shared organisational culture of developers and users ensured that the team subculture was receptive to user participation in systems development. |

*Figure 1: A Model of User Participation and Management of Change in the Systems Development Process*

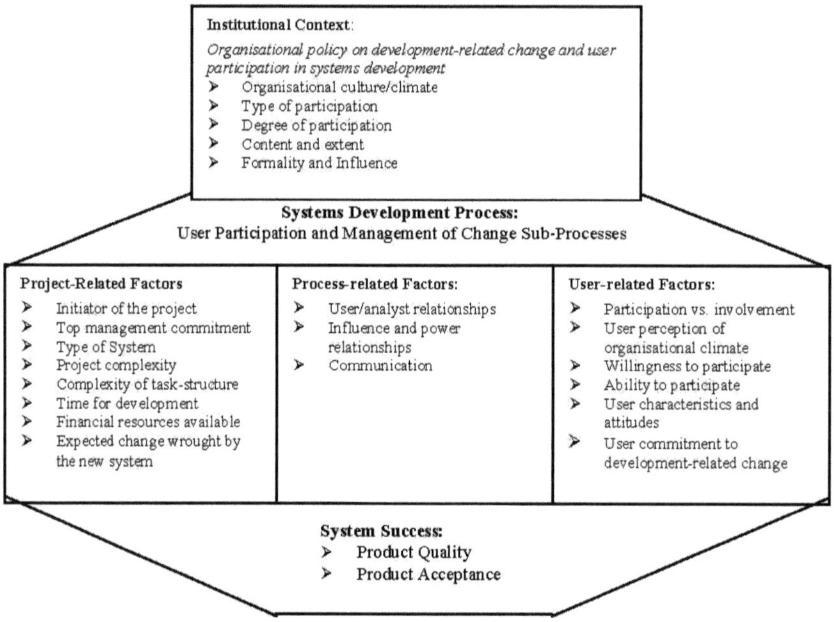


discusses these issues and provides a description of important processual features in both projects.

## General Project Characteristics: From Project Initiation to Implementation-Related Change

The Generic Appointment System (GAS) grew out of a business need in one key area of the company's operations—its telephone repair service. Business managers across the organization recognized the need to make efficient the manner in which repair service workloads were managed, and associated service appointments made with customers—more importantly, they recognised that there *was "a desperate need to radically transform [Telecom's] fundamental business processes, introducing the GAS was another step in that direction"* (senior middle-manager). A senior, non-executive business manager based in the company's Dublin HQ was this project's sponsor and initiator. However, according to a senior IS manager, the request for the system only received a response from and the support of the company's IS function when IS managers needed to choose *"a small, well bounded system, so that [they] could introduce and pilot-test [their] new application development environment (ADE), called IEF"*. Commitment from senior business and IS managers to this project was therefore quite high. One of the goals to be achieved by introducing this new system was the elimination of unproductive visits by operational staff to customer premises when customers were absent. The GAS would also assist supervisors in their task of allocating workloads to their repair teams, which consisted of telecommunications technicians. The GAS also supports the operation of the company's ten fault-handling and repair centers and the telecommunications technicians employed there. These internal and external groups therefore had a keen interest in the development and implementation of this system as it impacted on a range of their basic functions.

A development team that consisted of a user project manager, a development project manager, two analysts, the CASE vendor consultant, one programmer and a user representative carried out the development of the GAS; three user groups and several individual users formed the bulk of participating users from the constituencies of interest. A CASE-enabled rapid application development (RAD) approach saw development take place within a three month time period; that said, the implementation of the first phase of the GAS took a further six months. The GAS project operated within fixed budgetary and time constraints; however, neither of these materially affected the process or content of user participation. As a distributed IS, the GAS is comprised of eight relational databases that serve up to 180 Windows-based PC terminals in fault-handling centers around Ireland, and 400 additional terminals in operational depots nationwide. The GAS project came in on time and budget.

The Geographic Information System (GIS) was developed to provide a graphical database of the telephone network in the general Dublin area. Heretofore, the planning and drawing office functions manually recorded network-related details using paper-based records and maps. The business manager responsible for this project recognized that there would be significant improvements, in terms of economic and operational efficiencies, to be gained in using a GIS in this area of the company's operations. However, the implementation of the GIS meant that a radical change had to take place in one of Telecom Éireann's operational business processes. Accordingly, the development of the GIS posed significant challenges to the business sponsor, users and developers alike. On the one hand, there was the issue of change management associated with the radical change in work practices/roles of the users in operational units who performed telephone network mapping, planning, and record handling duties. On the other, there was the challenge of developing a highly complex and sophisticated information system within a proprietary application development environment. With the GIS a high degree of top management support existed in the first phase, but this waned in subsequent phases. As one business manager put it: *"This project became a political 'hot potato' because of the amount of hassle coming from the unions. I guess we should have foreseen it, you can't just do away with the draughtsmen, and expect them to go quietly... eventually [so and so] just wanted the whole thing to go away."* The complex technical, processual, and political factors were recognised from the outset by IS managers, and in an effort to avoid tying up scarce developer resources to a project that had all the potential to fail, they offered what the GIS project manager described as token developer resources to staff the project.

The GIS development team consisted of a user project manager, a development project manager, two analysts, three programmers, two user representatives, and a team of ten end-users whose primary role was to input graphical data and carry out test functions. User groups were also drawn from the two constituencies of interest—the drawing and planning functions. Consultants from the software vendor also participated in the development process. The GIS was built around a proprietary graphical database engine that served up to 40 high-end workstations. The first phase of the GIS development took almost two years to complete. The implementation and rollout of the first phase took a further year. The project failed to meet the scheduled completion date, and also exceeded budget. As with the GAS project, there appeared to be little or any constraint on user participation placed by time or financial considerations.

**Dealing with Project and Task Complexity**

While the previous sub-section has described several of the project-related factors outlined in Table 2, this sub-section focuses on the impact of two in particular—project complexity and task structure/complexity—which have been posited as having particular influence on the process of user participation. Recent research questions the need for comprehensive levels user participation across the SDLC (Guimaraes and McKeen, 1995); instead, Guimaraes and McKeen argue that there is little need for user participation when task (business process) and project complexity is low (see also, Cavaye, 1995). In many respects, the GAS was characterised by a low to medium degree of complexity in relation to task structure and a moderate level of project complexity: in respect of the GAS system, task structure refers to the operational task of making appointments with customers for equipment repairs or installation and associated work scheduling. GAS 'system complexity' was low to medium from a user perspective, however developers found certain aspects of the physical detailed design–project-related problems with network interface protocols that the GAS was using to communicate with the existing Fault Handling System (FHS)—to be problematic. The GAS crossed several functional boundaries, and it therefore led to a high degree of social, rather than technical, complexity in relation to task and project factors. IS and business managers were of the opinion that this necessitated a high degree of user participation. Also, due to a scarcity of in-house developers, users were encouraged to become more actively involved in systems development, particularly at the design and implementation stages. The GIS is a highly complex system in terms of system functionality and the nature of the business tasks it supports. The IT project manager commented on this: *"Without user participation on this project, well we just couldn't have done it, the requirements were so complex; mapping the network to the level of detail required, and then representing it graphically, and then updating and using the maps to plan the network and provide customer service, grappling with this level of detail and trying to computerize it was a horrendous task."* A GIS developer also commented on the technical aspects of the project: *"Working with the GIS vendor's proprietary development approach and programming language was bad enough, but add the complex requirements and the technical headache of representing them, and you get some idea of the challenge this system posed for us... look, there's a telco in [the US] doing the same thing for the past couple of years with a project team ten times as big, and we are trying to do the same with five people, crazy."* The practice of comprehensive user participation across the SDLC in the GIS and GAS projects was of obvious help to developers in coping with both project and task complexity, and it was greatly facilitated through the policy of on-site development at the business users' offices. Prior to the development of the GIS and GAS, most systems development took place off-site, that is, within the IS function's own business accommodation. Senior IS function management and development project managers recognized that there were significant benefits to be gained from on-site development at the users' place of business. For example, it was thought that this policy would provide additional opportunities for informal and indirect user participation, thereby improving user/developer communication and fostering good relations at all levels.

In each project, coordination and control of both developer and user activities was highlighted as being of particular importance in addressing issues of project complexity. Regular project meetings were considered by both developers and users in both GAS and GIS to be an important mechanism in the achievement of this goal. As expected, such fora helped developers to keep abreast of each others' progress and activities; however, the joint nature of such meetings provided user and development project managers with an opportunity to keep both user representatives and developers abreast of external issues such as industrial relations problems. Nevertheless, developers commented that the informal grapevine was far more informative with respect to keeping up to date with user-related problems on both projects. User representatives felt that taking part in these project meeting made them feel part of the development team, as the user representative on the GAS protect commented: *"For me, the project meetings were the 'icing on the cake' when it came to being on his project. While I got on with the developers, it gave me an opportunity to air my views and discuss issues with all of the IT people at one sitting; I also took this as an opportunity to sort out problems with [the user project manager]."*

## PROCESS-RELATED FACTORS

Table 3 presents the factors within the participation process that impact on the degree and effectiveness of user participation in systems development.

### User/Analyst Relationships

The high level and quality of participation in the GIS project was commented on by one developer: *"The team greatly benefited from the presence of user representatives. I was up to speed with user needs all the time."* These sentiments were strongly endorsed by developers in the GAS project also. Participating users were fully aware of the favorable attitude that developers had towards their contribution and responded accordingly: as the user representative on the GAS put it... *"the lads here really made me feel welcome and part of the team from the outset... I worked with John on the requirements elicitation in both the individual interviews and the user work groups... and was trained up on IEF, just like the rest of the development team."* Developers in both projects also articulated a need for more active participation by certain users, such as the draughtsmen on the GIS, and fault-handling centre technicians on the GAS, as it was felt that an increased level of participation by such users could have helped mitigate some of the contentious change management issues surrounding the implementation of the systems in the organisation.

### Influence and Power Relationships

Developers on both the GAS and GIS occupied positions of relative seniority to the operational staff who acted as participating users in both projects. There was no evidence of the 'not invented here' syndrome among developers, certainly, if it did exist, users did not mention it. The climate and culture of the IS function appeared to be egalitarian in the main, and users' opinions were treated with respect by developers, the GAS user representative captured this succinctly: *"To be honest, I was surprised at being treated like an equal, people at that level on my side of the house tend to be standoffish at best... the lads outside [in the user groups etc.] feel the same and get on well[with developers]. I guess the reason is that John and Don came from the technical side [originally], and know the score."* One of the GIS user representative also commented on this: *"There is a lot of ill-feeling out there, but none of it is aimed at the IT guys, the staff on the ground know that they are just doing their jobs, and although it is a bit like 'turkeys voting for Christmas' with the draughtsmen,*

*Table 3 Project-Related Factors*

| Project-Related Factors | Case Findings |
|---|---|
| Initiator of the project | In both the GAS and GIS, the initiators of the projects were senior, but not executive-level, business managers in the Dublin operational area. |
| Top management commitment | In respect of the GAS, a high degree of support existed from organisation and IS function management. With the GIS, on the other hand, a high degree of top management support existed in the first phase, but this waned in subsequent phases. There was also a lack of support from senior IS function management for the GIS. |
| Time for development | Although there was a very tight schedule set in both projects, it did not impact negatively on the degree of user participation. |
| Financial resources available | Budgetary resources did not affect the degree or quality of user participation in either project. |
| Type of system under development | Both the GAS and GIS are operational support sub-systems. |
| Project complexity | The GAS was a complex project, several functional groups were involved. The GIS, on the other hand, was a highly complex project and several functional boundaries were crossed. |
| Complexity of task-structure | Both systems supported operational activities that exhibited medium-level task complexity, and which were part of moderately defined business processes. |
| Expected change brought about by the system | The implementation of both systems meant a high degree of change for particular user constituencies in the operational areas concerned as new business processes were supported. |

*most of them will give all the help required... it's the unions job to sort out the problems with the company and get the best deal out of this [for all concerned]."*

**Communication**

Because of the on-site development approach in both projects, the presence of full-time user project managers and representatives, including the data capture team of 10 planners and draughtsmen on the GIS project, and the regular project meetings, communication between developers and users was not an issue here. There was on-site access to both business managers and users at all levels, formal and informal communication mechanisms abounded as one developer on the GIS project attested: *"I can step outside there and call on any one of the user representatives, or get the data capture people in here any time during the day and have them clarify things for me, or I can waltz off down the corridor and speak to a planner or a draughtsman, no problem. If I feel lazy, I can always use e-mail, and if people are out or at a meeting, I know I can always catch them in the canteen upstairs."* Both projects differed in the level of support for user/developer communication that their respective development approaches and CASE tools offered. In the GAS, the rapid application development (RAD) approach with its prototyping tools significantly enhanced user/developer communication at user representative level and at the JAD sessions. The GIS did not have the same RAD capability, and it was only during the testing and data capture that this type of helpful feedback occurred.

It is clear that the process-related factors mentioned in the model operated to produce systems that matched business needs, user requirements of a technical nature, and system preferences that did not impinge on the business objectives set for the GAS and GIS.

## USER-RELATED FACTORS

Beynon-Davies *et al.* (1997) argue that choosing the 'right type of user' to participate in systems development is critical to both the development process and its product; Leonard-Barton (1995) has also commented on this aspect of participation within wider organisational contexts. Choosing the 'right type of user' to participate in systems development was a problem that exercised the minds of business and IS managers and developers alike prior to the commencement of systems development on the GIS and GAS projects. IS managers and developers were eager to secure the most knowledgeable and proficient user project managers and representatives in order to make their *"lives that much easier"* in arriving at a full set of user requirements and in converting these requirements into a system that would be accepted by the business constituency. At a time when developer

*Table 4 Process-Related Factors*

| **Process-Related Factors** | **Case Findings** |
|---|---|
| User/analyst relationships | Very good across both projects. Relationships were enhanced by the existence of a common organisational culture and favourable development climate in project teams. |
| Influence and power relationships | Several institutionalised checks and balances existed which countered any power asymmetries or political opportunism that may have arisen. This was due to the implementation of organisational policy by all the constituencies involved in systems development. Positive management attitude toward and acceptance of user input was also of help here. |
| Communication | High degree of user/analyst communication was in evidence across both projects. In the GAS project this was greatly enhanced by on-site development training the user representative in IS development method and tools, and the prototyping approach adopted; in the GIS project, however, only some improvement in communication brought about by user training in SSADM. |

resources were scarce, issues like developer productivity and project life span were uppermost on the minds of IS function managers. This led one senior manager in the ITD to argue that *"if the ITD were going to commit scarce and valuable resources to a project, then business managers should do likewise."* In the GAS and GIS projects the selection of user representatives was perceived as a key issue due to the active role that they were expected to play throughout the development process, and in the subsequent testing and implementation of the developed systems. The other side to this coin was that given the potential for change management problems surrounding both systems, users' perception of the prevailing organisational climate, their ability and willingness to participate, their characteristics, attitudes, and commitment all proved pivotal to the eventual success of these development projects. As the unions had to be consulted on every issue, business managers could not appoint the individual user of their choice, particularly those favorably disposed, for whatever reason, to siding with management or who were not going to be impartial during development, as user representatives or members of user groups and test teams. Surprisingly enough the IT project teams ended up with user representatives and groups who collaborated wholeheartedly in the 'technical' development of the systems concerned as Table 4 illustrates.

In both projects, the development-related workshops normally consisted of developers from the relevant project team and end-users from only one of the user constituencies participating in the development process: the manner in which this participative mechanism was implemented possessed certain flaws, however. For example, group workshops on the GAS project tended to be used as a platform for political infighting between different user constituencies—internally-based fault handling centre technicians and externally-based repair team supervisors and team members—as certain users introduced arguments to oppose or alter system features favored by users from other operational areas who did not attend the sessions. User groups also tended to play on the stated objections of absent groups in order to influence the trajectory and outcome of the development process in their favor. For example, the user representative on the GAS project reported that: *"Staff at the fault handling centre felt that their jobs/roles were being whittled away and that the control of the fault handling system was being shifted to the repair teams."* This situation engendered a negative attitude towards the new system within one user constituency, and strongly influenced the deliberations of the Computer Liaison Committee (CLC). Because of the high degree of political conflict between the various groups, the user representative on the GAS project observed there was a need in future *"to have all the user groups affected by the systems development present at each of the workshops; this avoided the emergence of a 'them and us' situation between users... both the unions and management are to blame here."*

## CHANGE MANGEMENT IN THE GAS AND GIS

The issue of change management associated with the implementation of both the GAS and GIS was found to exert a critical influence on the trajectory and outcomes of the development process. Even though the development project teams were embedded within the user community, and user groups were employed in the elicitation and verification of requirements, in what could be described as a fully participative development exercise, change management problems arose in both projects during development, and, in particular, during implementation. Previous findings indicate that the level of participation observed should have resulted in positive user attitudes to system quality and full acceptance of the developed system (see Ives and Olson, 1984 for example). Nevertheless, this paper reveals that other development-related aspects of an organisation's institutional framework come into play and exert considerable influence on the development process and its product.

Even before systems are developed in Telecom Éireann, formal negotiations are entered into with the unions at the level of Computer Liaison Committee (CLC) regarding the implications of the systems for users and the input required from user constituencies. Business managers who initiate and have ownership of the project approach the CLC to have the issue put on its agenda; the matter is discussed and union members of the CLC then report back to their executive committees, who then inform the local branch organisation affected. The union executive usually agrees in principle and delegates responsibility for the provision of particular users to participate in the project to the local branches. Once agreement is reached and user resources allocated, development can begin. This process is, however, dependent on two things: (a) business managers presenting a full description of the system and its implications on the business process it supports to the CLC, and (b) the absence of subsequent inter- and intra-union conflict and political infighting regarding changes to the user constituencies affected. Both of these factors came into play in the case described.

Although the GAS had been accepted as developed by all the constituencies of interest, the CLC over-rode decisions taken and agreed by the user group. This situation arose despite the fact that one individual on the CLC had been involved throughout the development process as a member of the user group. A developer provided an explanation for this and reported that influential users who did not participate in the development process—i.e. technicians at the fault handling centres—had voiced their *"unhappiness with system features (and that this) prompted the CLC to say no to the implementation of the system."* Hence,

*Table 5 User-related Factors*

| User-Related Factors | Case Findings |
|---|---|
| Participation vs. involvement | In each project, only a sample of users from affected constituencies actively participated in systems development, i.e. user representatives and user group members. These users had strong favorable attitudes toward the processual and technical features of the system they helped create. Following the distinction made by Barki and Hartwick (1989), the subjective psychological state that reflects the level of importance and personal relevance of the information system to users reflects their involvement in the development process. By this definition all users felt involved because of the high level of representation on the project team and user groups, but also through the auspices of the CLC. |
| User perceptions of organisational climate | In relation to the GAS project, users felt that a favorable development climate existed. However, users involved in the GIS project were generally of the opinion that the organisational climate was negative; however, they felt that a favorable development climate existed. |
| Willingness to participate | In both projects users were eager to participate for several reasons, viz. self-advancement, political power interests, and technical curiosity. |
| Ability to participate | The use of dual project development teams (user and developer), and the use of user groups in JAD workshops greatly facilitated users ability to participate. |
| User characteristics and attitudes | Users generally held positive attitudes toward the social and technical aspects of systems development. However, user computer literacy posed problems in relation to their ability to fully participate. It was clear that the shared organisational culture was of benefit in accommodating different 'world views'; however, users of both systems were clearly from different constituencies within the organisation, and therefore possessed different characteristics and attitudes, e.g. internal and external technicians in the GAS and planners and draughtsmen in the GIS. |
| User commitment to development-related change | Depending on the user constituency concerned, the level of commitment to change differed among users. In the GAS project the internal and external technicians exhibited varying levels of commitment and enthusiasm as both groups endeavored to steer system features in particular directions which favored one group over the other. In the GIS, the draughtsmen and planning technicians were polarized in their commitment, the latter were highly committed, due to the increase in status and power bestowed upon them by the change, the former were less than happy because their roles and remuneration were going to be significantly affected. |

prior to its implementation at a trial site, several modifications had to be made to the GAS in order to address these objections. The problem here was that while the CLC agreed up front to the need for such a system, it is an unfortunate aspect of systems development that detailed requirements only emerge as a project unfolds and changes to previously-agreed functionality may be is required. If, as in the GAS project, the user representative and user groups settle on system functionality that is not agreeable to other users, and if the aggrieved parties possess or can apply enough political muscle to overturn such a decision, then perfectly legitimate user requirements can be discarded for what are essentially political reasons.

A very similar scenario existed in relation to change management issues that arose during the life of the GIS project: here, business management was aware of the potential for significant change management problems to develop when the system was implemented. These problems related to the radical nature of the change in work-related roles, responsibilities and remuneration of one of the user constituencies involved; and although these users participated in and were satisfied with the system as developed, they were unhappy with the consequences of its introduction. The issues here were quite complex as two different unions were involved, members of one, the draughtsmen, were losing out to the planning technicians, represented by the larger of the two unions. The business manager involved had left it to the CLC to sort things out instead of dealing with the issue prior to the commencement of development. The absence of adequate managerial attention to these issues meant that although both systems were developed with the cooperation of users, both projects encountered user resistance at the implementation stage.

## EPILOGUE

The problems described in the previous sub-section were not unique in Telecom Éireann. Such problems beset other instances of information systems development and implementation within the company. The weakness in Telecom's approach to change management at this time (1995/96) was commented on by the IT director who stated that: *"What we have been doing here at IT is just reacting to the business needs which up to now have been ad hoc and poorly articulated at best; this has been disastrous for us here. [The new CEO] has changed all that. One of my CSFs is to ensure that business process engineering gets off the ground in Telecom. I'm answerable to him on that. We all recognize the need to change the way we do business, if we don't we won't survive. It's as simple as that. Getting the business sponsors to sign up via a project charter is only part of the solution. We need then to take ownership... The unions are quite powerful, nothing gets done without their say so. That's the other challenge for us. I don't mean beating them over the head or anything... we need to get them on-side as well."* The results of a strategic business review conducted in the early part of 1995 helped chart future business strategy. As a direct result of this strategy, the CEO established a BPR unit in 1997, and since then the IT Directorate has been developing future methods of operation (FMO) based on reengineering operational business processes in concert with Bellcore (US) and the Business Process Design Directorate. In the first quarter of 1999, the fruits of this labour saw the introduction of several new information systems in the company, the component subsystems of which were associated with radical business process reengineering and change around IT. However, back in 1995, the CEO and the IT director admitted that this would be problematic. Hence the CEO set up a Joint Strategic Consultative Group (JSCG) with union involvement to give effect to a partnership approach to organisational change, particularly in relation to the implementation of information technologies, as it was intended to have IS enable Telecom's transformation. A framework agreement for the transformation of the company was drawn up in consultation with the unions at this forum. In this agreement, the finer details of the company's 'Organizing to Compete (OTC)' strategy were fleshed out. In order to achieve commitment to change, reduction in operating costs and increased quality of service to the customer, the company introduced an Employee Share Ownership Agreement (ESOP) that would give employees a 15% stake in the company. In return for this, staff and its unions would commit totally to the company's transformation strategies in relation to all issues of organisational change around IT.

It is perhaps ironic that the implementation GAS system, which caused of so much controversy among different constituencies of users, has now been integrated with the FHS on a company intranet platform that would allow, by the end of 2000, all operational and repair staff full access, via a GSM-enabled laptop computer, to both the fault handling and appointment systems. The upshot of this piece of business process reengineering was that all but two of the company's regional fault handling centers were closed by April 1999, and the staff redeployed to other duties. This would not have been possible without the ESOP deal and the accompanying agreement to all IT-related change by the union. The political role that the CLC played in the past has thereby been eliminated, along with many of the other obstacles to change described herein.

## DISCUSSION AND CONCLUSIONS

The previous section illustrated that Telecom Éireann's institutional context plays a pivotal role in shaping and influencing all aspects of the content and process of user participation in systems development (RQ1); it also provides ample evidence that it is institutional mechanisms that help resolve change management problems (RQ2). Clearly, as the section on change management and the Epilogue have illustrated, comprehensive organisational policies and structures are required to accommodate matters of organisational change surrounding the implementation of organisational IS. However, management in Telecom realized that even if organisational policies on user participation and change management are included as 'rules of the game' (in terms of institutional mechanisms and arrangements that structure organisational actors' roles and responsibilities in system development), they are no guarantee of actors' abilities or intentions to either 'play' competently or fairly. We have seen in this case that even within an institutional framework that possesses a mechanism to manage and agree change prior to development, in addition to an infrastructural mechanism that addressed issues of user participation, the interplay of project-, process- and user-related factors coalesced to produce systems of high quality, and which would have been acceptable to users, if business managers and the unions had been competent and fair in their dealings from the outset, and properly addressed issues related to change in business processes and their effects on organisational actors. Additional institutional arrangements, described in the Epilogue to this case study, have been introduced by Telecom's management to address these deficiencies. What can be deduced from this is that if organisations are to ensure that management of change proceeds smoothly, and the benefits of user participation are to be maximized, then a combination of process redesign—to have business managers focus properly on and provide a detailed plan of their requirements prior to development—and stakeholder agreement—to have organisational actors to buy into change upfront—is required.

The third research question (RQ3) inquired as to which of the model's factors were critical in shaping and influencing user perceptions of product quality and user acceptance of the end product. As we have seen, an appropriate institutional foundation provides the bedrock on which the process of user participation is based. This paper's findings indicate that issues of: (a) project complexity and degree of task structure (project-based factors); (b) user/analyst relationships and communication (process-related factors); and (c) users willingness and ability to participate (user-related factors), are critical for system quality from a user's perspective. While these issues have a general effect on *user acceptance* of the implemented system, institutional arrangements that exert a primary influence and greatly affect the level of *user acceptance* of systems are: (a) the expected change wrought by the new system (a project-based factor); (b) user influence and power relationships (a process-related factor); and (c) user commitment to development-related change (a user-related factor). In respect of RQ4 it would, as this study's findings suggest, be difficult, and perhaps dangerous, to attempt to disentangle and isolate factors that relate specifically to user participation and management of change. For ultimate system success both are required. Nevertheless, an answer to RQ4 has been provided *en passant* when describing the answers to the other research questions, particularly RQ3.

The model presented herein and empirically tested in a constructivist study of the related processes of user participation and management of change presents a significant contribution to understanding what are complex phenomena. It provides holistic confirmation of what were fragmented conceptualizations and findings in previous studies. This paper's findings have laid bare one of the more common myths of the IS field, namely that user participation may lead to increased *user acceptance* of systems by reducing *user resistance* to change (Ives and Olson, 1984). As the findings of this study have illustrated, an organization's institutional context is of primary importance in this regard. Finally, while this paper provides a much-needed, empirically-based understanding of issues surrounding user participation and management of change in systems development and implementation, the conceptual model presented herein can be employed by future researchers as a framework for investigating what are complex phenomena in order to establish a body of cumulative research on these important topics.

## REFERENCES


Aaen, I. (1986). Systems Development and Theory—In Search of Identification. *Quality of Work Versus Quality of Information Systems*, Report of the Ninth Scandinavian Research Seminar on Systemeering, Lunds Universitet, 203-223.

Barki, H. & Hartwick, J. (1989). Rethinking the Concept of User Involvement. *MIS Quarterly, 13*(1), 53-63.

Barki, H. & Hartwick, J. (1994). Measuring user participation, user involvement and user attitude. *MIS Quarterly, 18*(1), 59-82.

Berger, P. L. & Luckmann, T. (1966). *The Social Construction of Reality*. Doubleday and Company Inc, Garden City, NY.

Beynon-Davies, P., MacKay, H. & Slack, R. (1997). User Involvement in Information Systems Development: The Problem of Finding the Right User. In R. Galliers, C. Murphy, H. R. Hansen, R. O' Callaghan, S. Carlsson and C. Loebbecke (Eds.) *Proceedings of the 5th European Conference on Information Systems*, Volume II, University College Cork, Ireland. 659-675.

Bødker, K. & Pedersen, J. (1991). Workplace cultures: looking at artifacts, symbols and practices. In J. Greenbaum and M. King, (Eds.), *Design at Work: Collaborative Design of Computer Systems*, Lawrence Erlbaum Associates, Hillsdale, NJ, 121-136.

Boland, R.J. (1978). The Process and Product of Systems Design. *Management Science, 24*(9), 887-898.

Butler, T. (1998) Towards a hermeneutic method for interpretive research in information systems. *Journal of Information Technology, 13*(4), 285-300..

Butler,T. & Fitzgerald, B. (1997). A Case Study of User Participation in the Information Systems Process. In E. R. McClean and R. J. Welke (Eds.), *Proceedings of the 18th International Conference on Information Systems,* Atlanta, Georgia, 411-426.

Cavaye, A. L. M. (1995). User participation in system development revisited. *Information and Management, 28,* 311-323.

Erlandson D. A., Harris E.L., Skipper, B. L. & Allen, S.D. (1993). *Doing Naturalistic Inquiry: A Guide to Methods.* Sage Publications Inc, London.

Euchner, J. Sachs, P. & The NYNEX Panel (1993), The Benefits of Internal Tension. Communications of the ACM, June, *36*(4), 53.

Gatian A.W. (1994). Is user satisfaction a valid measure of system effectiveness? *Information and Management, 26,* 119-131.

Ginzberg, M.J. (1981). Key Recurrent Issues in the MIS Implementation Process. MIS Quarterly, *5*(2), 47-59.

Guba, E. G. & Lincoln, Y. S. (1994) Competing Paradigms in Qualitative Research. In N. K. Denzin and Y. S. Lincoln (Eds.), *Handbook of Qualitative Research,* Sage Publications, CA, 105-117.

Guimaraes, T. & Mc Keen, J. D. (1995). Successful Strategies for User Participation in Systems Development. In G. Doukidis, B. Galliers, T. Jelassi, H. Kremar and F. Land (Eds.), *Proceedings of the 3rd European Conference on Information Systems,* Athens, Greece, 879-899.

Hirschheim, R. A., (1983). Assessing Participative Systems Design: Some Conclusions from an Exploratory Study. *Information and Management, 6,* 317-328.

Hirschheim, R. & Newman, M. (1988). Information Systems and User *Resistance: Theory and Practice.* The Computer Journal, *31*(5), 1-11.

Hyman, J. and Mason, B. (1995). *Managing Employee Involvement and Participation.* Sage Publications, London.

Ives, B., Hamilton, S. & Davis, G.B. (1980). A Framework for Research in Computer-Based Management Information Systems. *Management Science,* September, *26*(9), 910-934.

Ives, B. & Olson, M.H. (1984). User Involvement and MIS Success: A Review of Research. *Management Science, 30*(5), 586-603.

King, J.L., Gurbaxani, V., Kraemer, K.L., McFarlan, F.W., Raman, K.S. and Yap, C.S. (1994). Institutional Factors in Information Technology Innovation. *Information Systems Research,* 5(2), 139-169.

Kling, R. & Iacono, S. (1989). The institutional character of computerized information systems. *Office: Technology and People, 5*(1), 7-28.

Kozar, K.A. & Mahlum, J.M., (1987). A User Generated Information System: An Innovative Development Approach. *MIS Quarterly, 11*(2), 163-173.

Krovi, R. (1993). Identifying the causes of resistance to IS implementation ; A change theory perspective. *Information & Management, 25*, 327-335.

Land, F. & Hirschheim, R. (1983). Participative Systems Design: Rationale, Tools, and Techniques. *Journal of Applied Systems Analysis, 10*, 91-107.

Lawrence, M. & Low, G. (1993). Exploring Individual User Satisfaction Within User- Led Development. *MIS Quarterly, 17*(4), 195-208.

Leonard-Barton, D. (1995). *Well-Springs of Knowledge: Building and Sustaining the Sources of Innovation*. Harvard Business School Press, Boston, MA.

Lincoln, Y. & Guba, E. (1985). *Naturalistic Inquiry*. Sage Publications, CA.

Lyytinen, K. (1987). Different Perspectives on Information Systems: Problems and Solutions. *ACM Computing Surveys, 19*, 5-46.

Marshall, C. & Rossman, G. (1989). *Designing Qualitative Research*, Sage Publications, California.

McKeen, J. D., (1990). Successful Development Strategies for Business Application Systems. *Management Science*, 36(1), 76-91.

McKeen, J. D., Guimaraes, T. & Wetherbe, J.C. (1994). The Relationship Between User Participation and User Satisfaction: An Investigation of Four Contingency Factors. *MIS Quarterly, 18*(4), 427-451.

Miles, M. B. & Huberman, A. M. (1994) *Qualitative Data Analysis: An Expanded Sourcebook,* Second Edition, Sage Publications.

Mumford, E. (1979). Consensus Systems Design: An Evaluation of this Approach. In N. Szyperski & E. Grochla (Eds.), *Design and Implementation of Computer Based Information Systems*, Sijthoff and Noordhoff, Groningen, Holland.

North, D.C. (1990). *Institutions, Institutional Change and Economic Performance.* Cambridge University Press.

Orlikowski, W.J. (1993). CASE Tools as Organizational Change: Investigating Incremental and Radical Changes in Systems Development. *MIS Quarterly, 17*(3), 309-340.

Patton, M. (1990). *Qualitative Evaluation and Research Methods* (2nd Edition), Sage Publications, London.

Regan, E.A. & O'Connor, B.N. (1994). *End-User Information Systems: Perspectives for Managers and Information Systems Professionals*, Macmillan Publishing Company, NY.

Robey, D. & Azevedo, A. (1994). The Organizational and Cultural Context of Systems Implementation: Case Experience from Latin America. *Accounting, Management and Information Technologies, 4*(1), 23-37.

Rowlinson, M. (1997). *Organisations and Institutions.* Macmillan Press Ltd., London, UK.

Stake, R. E. (1994). Case studies. In N. K. Denzin and Lincoln, Y.S. *The Handbook of Qualitative Research,* Sage Publications, CA, 236-247.

Tait, P. & Vessey, I. (1988). The Effect of User Involvement on System Success: A Contingency Approach. *MIS Quarterly, 12*(1), 91-107.

Visala, S. (1991). Broadening the Empirical Framework of Information Systems Research. Information Systems Research. In H. Nissen, H. K. Klein, and R. Hirschheim (Eds.), Contemporary *Approaches and Emergent Traditions, Proceedings of the IFIP TC8/WG 8.2 Working Conference,* Elsevier Science Publishers B. V. (North-Holland), 347-364.

Yin, R. K. (1989). Research design issues in using the case study method to study management information systems. In J. Cash and I. Lawrence (Eds.), *The IS Research Challenge: Qualitative Methods, 1,* Harvard Business School, Boston, MA, 1-6.

Zmud, R.W. & Cox, J.F. (1979). The Implementation Process: A Change Approach. *MIS Quarterly, 3*(2), 35-43

## ENDNOTE

[1] Statement of Company Position on Current Industrial Relations Issues, October 1995.

## Chapter 2

# Global Information Systems and Human Resource Management: A Research Agenda

Fred Niederman
University of Baltimore, USA

*A socio-technical approach to information systems requires recognition of the inextricable link between information technologies and humans as designers and users. This essay explores five areas in which information technology and human computer designers/users interact within the context of global organizations. These five areas are: using information technology to support the human resource strategy of global organizations, using information technology to support the generation and distribution of organizational learning, using human resource management techniques and programs to support the work of information systems professionals, using human resource management techniques and programs to support the work of global "end-users" or knowledge workers, and, finally, national and regional policies to support technical and human resource infrastructures.*

In recent years, there has been a flurry of research in the area of intellectual capital, knowledge management, and organizational learning. Much of this research has focused on three areas. First, papers argue that the successful acquisition and application of knowledge should contribute to the profitability of the firm. Secondly, articles asserting that strategies for accumulating and implementing knowledge must account for differences between explicit and tacit (or implicit) forms of knowledge. These types of knowledge are distinguished in terms of the degree to which they are codified

Previously Published in the *Journal of Global Information Management, vol.7, no.2,* Copyright © 1999, Idea Group Publishing.

or formalized for storage and dissemination. Where explicit knowledge has been or could easily be codified or formalized, tacit knowledge is embedded in human capabilities to perform an action as well as (or sometimes instead of) understanding how the action should be performed. Tacit knowledge has not been or would be very difficult to codify. Finally, here are studies that are concerned with an operational level of collecting, storing, and distributing information to implement organizational knowledge acquisition strategies effectively.

The thesis of this essay is that the development of intellectual capital, knowledge management, and organizational learning, particularly concerning transnational, multinational, and global organizations, should include consideration of both information systems and human resource management. It is important to clarify that intellectual capital refers to viewing the value of knowledge, skills, and abilities within an institution much as one would value a physical inventory. Knowledge management pertains to the organization of information and "know-how" in order for the firm to add more value or to add a given amount of value at lower cost or with more efficiency. Finally, organizational learning pertains to processes by which knowledge and "know-how" are gathered, institutionalized, and reinforced. In this paper, these concepts are viewed as distinct yet interrelated, much as income statements and balance sheets represent distinct facets of a firm's financial position, yet refer to the same underlying set of realities.

Consideration of intellectual capital, knowledge management, and organizational learning as related to information systems and human resource management of global organizations can be viewed through different lenses. First, information systems can be used to support a wide variety of human resource strategies, which in turn support the acquisition, development, and retention of intellectual capital resources embodied in organizational employees and other stakeholders such as customers, suppliers, and regulators. Secondly, information systems can be used to support a wide variety of mechanisms to enable the acquisition, storage, and distribution of information organized into explicit knowledge. Thirdly, human resource management programs can be designed to support the recruitment, training, and retention of information systems professionals at all levels of competence who carry with them varying (hopefully increasing) amounts of both explicit and more importantly tacit knowledge. Fourthly, human resource management programs, particularly in areas of training, can support knowledge workers throughout the organization in enriching their holdings of information technology related skills as well as the ability to translate these into domain specific programs that add value within their particular functional domain. Finally, nations can develop both information technology infrastructures and skilled personnel to support both high technology firms and traditionally low technology manufacturing and service firms to add value through information systems.

Although much of the discussion regarding intellectual capital, knowledge management, and organizational learning can apply to organizations existing within one domestic economy, a number of special issues are more pronounced in the global context. For example, transfer of knowledge, both explicit and tacit will almost certainly have significant interaction with multiple cultures in the global context. For explicit knowledge this may include issues such as language and electronic infrastructures to make such knowledge available to a variety of users. For tacit knowledge, this can include language, differences in cultural understanding and context, and difficulties in overcoming lack of physical proximity. It is not unusual for a central method of diffusion of new technologies to be based in transfer (permanent or temporary) of skilled personnel. In the international environment without a long-term planned global human resource management strategy, this can be very difficult (Roberts, Kossek, and Ozeki, 1998).

In the next sections of this essay, each of the five areas combining human resource management, information systems, and global organizations will be examined. It is hoped that this discussion will produce a deeper understanding of the issues involved, the managerial decisions that follow from knowledge in these areas, and of the potential for meaningful and productive research aimed at addressing questions that link human resource management and technical infrastructures within a global approach.

## FIVE-PART RESEARCH AGENDA

### IT Supporting Human Resource Strategy: Global HRIS

The traditional function of human resource management pertains to a variety of organizational activities ranging from labor relations and compensation to recruitment and performance evaluation. As Pfeffer (1998) points out, the quality of a firm's management of human resources can have significant impact on the firm's profitability. Roberts, Kossek, and Ozeki (1998) provide examples of how these issues take on a greater complexity and difficulty when applied to a global workforce. The key to achieving organizational benefits from human resource management derives largely from the skilled application of fairly traditional functions such as providing appropriate compensation and performance evaluation across a widely diverse and quickly changing inventory of employees. Computer-based human resource information systems largely have two purposes. First, they must streamline, automate, and (if possible) eliminate repetitive and routine functions with the intention of establishing lower costs and higher reliability for these functions. For example, they should transform payroll operations from a heavily labor dependent function to one that is automated to minimize errors and labor costs. Secondly, computer based HRIS should extract and store information from human resource transactions for improved management decision making. For example, storage of information regarding worldwide job applicants within a relational database that provides the capability of generating ad hoc queries can allow managers to locate high quality applicants from a large and diverse applicant pool.

Harrison and Deans (1994) identify and describe 12 essential components of a global HRIS. The identified components are: basic employee information, applicant tracking, recruiting, equal employment opportunity/ affirmative action (EEO/AA), position control, performance management, compensation, payroll, benefits, training, career development/skill inventory, and human resource planning. Each of these areas presents its own set of difficulties in expanding from a single country to multinational database. For example, a global benefits database may have to recognize that some of the potential benefits within one country may not be available within others. Even more significantly, benefits offered in one country may not be needed where the government of another country provides the benefit universally within its borders (e.g., some health care items). As Harrison and Deans point out, global organizations face a range of technical issues in order to gather and integrate these pieces of information across boundaries. As pointed out in the interview with Michael Lutz of Lucent Technologies (this issue), simply sending information to a single repository has significant difficulties. These are often overwhelmed by the difficulties in insuring that all data fields are used in the same way and that all data needed for each local organization is available without swamping the overall system. Global firms also face managerial, organizational, and cultural issues involved in using this information in new ways within the organization. Employees may find themselves facing increasing levels of competition for promotions or choice assignments. A plethora of highly trained employees in countries such as Ireland or India may effectively preempt assignments that would go to less highly trained employees in other countries. Additional managerial and organizational issues include: (1) providing the appropriate information for local use while still collecting data for aggregation and use at higher levels in the organization, (2) providing sufficient access while maintaining confidentiality and privacy, and (3) insuring that the non-technical procedures and reward structures are in place to use effectively the information generated by the system.

From a research perspective, the critical question for this area of global HRIS involves which combinations of technical solutions, organizational policies and managerial practices will result in superior human resource allocation within the firm while considering the variation from one firm to another in external context. This question presents a number of related issues. How are the technical resources of a firm, which are likely to be used by a range of functional areas for a large number of tasks, to be allocated or apportioned as the set used by an HRIS? What are the range of managerial policies and practices in the HRIS arena that can affect performance, how do these interact, and to what extent can these be represented by presence or absence of a polity or practice versus consideration of the spectrum of goals and degrees to which these are effectively implemented? What is the influence of the content of the HRIS, are there thresholds of size and complexity that bring contrasting costs or benefits, and when should an organization be satisfied with streamlining operations rather than generating significant managerial knowledge? What exactly is the dependent variable and how is it measured—particularly in terms of consideration of opportunity costs? If an IS professional is assigned to an overseas post and does a credible job, how is the value of that posting offset by the loss of potential value if that same person remained in his or her domestic position?

## IT Supporting Explicit Knowledge

Following the arguments of Nanaka and Takeuchi (1995), knowledge should not be thought of as permanently tacit or explicit. The challenge to organizations is first to aid employees in expanding their stock of tacit knowledge through actions and reflections in stimulating projects then to encourage the specification of that tacit knowledge in a format that makes it available for diffusion throughout the firm. Philosophers may argue whether knowledge that is genuinely tacit can ever be fully made explicit. However, from a pragmatic perspective, if even a portion of tacit knowledge can be made explicit in a way that is beneficial to the organization, this likely defines an objective worth pursuing. It is interesting that Nanaka and Takeuchi also stress the importance of employees applying explicit knowledge in real life situations as a basis for more quickly and effectively developing their tacit understandings and abilities.

It is debatable whether information technologies can, in principle, support tacit knowledge in an organization. Some would argue that once the knowledge is supported it is no longer tacit. It might be of greater benefit to investigate first how organizations can use information technology to support explicit knowledge in the organization and second how organizations can expand the store of explicit knowledge by drawing upon the tacit knowledge gained by employees through their varied experiences.

How does "explicit knowledge" differ from the realms of data and information that MIS departments have historically gathered, maintained, and distributed? From an organizational perspective the keys are first a tight link between the items and their applicability to current business issues, second the level of clustering or "chunking" of the items, and thirdly the formality and detail of the items. It is of great potential value for organizations to have collections of facts in database and data warehouse environments available for manipulation. It is also of value to generate standard reports to inform managers of specific events or state changes based on transaction data. However, an "explicit knowledge" program should aim higher than this. For example, Andersen Worldwide is reported to collect reports written for clients on a wide variety of subjects and making these available to other consultants for reference in currently active projects (Quinn, Anderson, and Finkelstein, 1998). The idea is to avoid consultants reinventing the wheel. Note that the consultant receiving the prior reports is not excused from the duties of scanning for relevant archived reports, understanding them, selecting portions that apply to the present client (and excluding those that do not), and adding his or her own submitted report to the collective body of explicit knowledge. If the system is productive, the time required for generating the new report should be reduced and the quality of the report should be improved. It remains to be determined whether the value of these benefits is sufficient to pay for the development, use, and maintenance of the archival system. For some firms, the costs may exceed the benefits until the archives grow to some threshold size (although, the opposite may occur if indexing and retrieval services do not keep up with the growth of materials) and employees master effectively using the archives and are sufficiently rewarded for adding to them.

Electronic performance support systems (EPSS) represent one approach to developing such systems for gathering, storing, and retrieving explicit knowledge. Gery (1991) describes in some detail the tools required to support such a system. These include advisory or expert systems, productivity software, applications software, help systems, interactive training sequences, and assessment and feedback systems. She also recommends including these seven knowledge types: procedures, processes, concepts, structures, principles, facts, and classifications. Mechanisms for implementing such systems are only expanded by the rapid growth and increasing utility of the Internet.

From the perspective of individual firms, research questions regarding EPSS

*Figure 1: Information Systems Support Global Personnel and Human Resource Management*

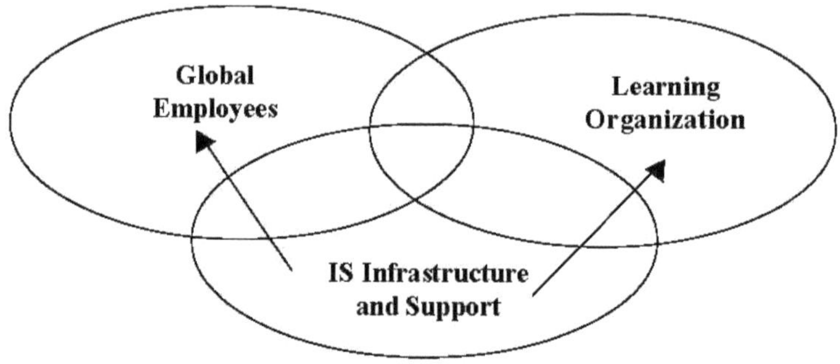


fall into three categories (see Figure 2). Under what conditions should individual organizations choose to begin an EPSS (or other explicit knowledge archival and retrieval program)? If a firm decides to embark on a program of this sort, how are the technical and organizational issues to be addressed to increase the probability of a successful implementation? For organizations with EPSS in place, what ongoing actions and policies increase its utility relative to its costs? From a broader market perspective, there are additional research questions. What factors, such as the ability to pilot test the new system or critical mass of users, influence the diffusion of this technology? Are there ways that multiple organizations can collaborate in the development and use of EPSS and the related question of whether firms that develop such knowledge can successfully market all or part of the "knowledge base" either to clients or commercially? Are there public policies that will encourage the generation, preservation, and use of knowledge created within firms?

Both of the first two areas, using IS to support global human resource management and using IS to provide performance support rely on information technology as a means of enhancing the productivity of personnel. These relationships are illustrated in Figure 1.

## HR Support For IS Professionals

Using human resource management to support information technology professionals begins with the realization that these positions require a particularly wide range of quickly evolving competencies. There has been much research work in this area of IS personnel, beginning with the landmark studies by Couger and Zawacki (1980). Their findings suggest that IS professionals in contrast to members of a broad range of other professions are more motivated by challenge and less motivated by interpersonal contact. These findings have been expanded upon by Myers (1991) who found variations between those IS professions primarily performing coding and computer operations in contrast to those performing more systems analysis and IS management tasks. Very recently Wynekoop and Walz (1998) have reinforced and provided additional detail to most of these findings. They state: "The differences found between traditional programmers and systems analysts and managers indicate the importance of studying, managing, and recruiting these groups differently, and also challenges the efficacy of the once-traditional career path of programmer to analyst to manager (p. 70)." They go on to note that analysts and managers with higher needs for achievement may need explicitly challenging career paths. In contrast, programmers, who were found to be less social and also fairly resistant to change, may respond well to being shielded from some of the more extreme shifts in new technology.

What are the implications of these findings for supporting global IS professionals? First, that IS professionals probably comprise more than one category and, therefore, need a range of human resource programs. The degree to which multinational organizations currently provide differentiated programs for technical, and in particular IS professionals is not clear. This represents a worthy area for future research investigation. Secondly, that the difficulties of working in varied nations and cultures should be presented to analysts and managers as stimulating challenges. Rewards and a sense of achievement should be cultivated not only for creating ideal technical solutions, but also for implementing these in a wide range of environments. This presents a range of research questions involving the best practices for introducing programs to shape these assignments as challenges and the nature of the outcomes when such a program is implemented. Thirdly, programmers accustomed to the norms, assumptions, methods, procedures, and requirement in one national and cultural environment may need training or orientation for dealing with the additional complexities of global cultures. Not only technical training, but cross cultural training and training in soft skills are important. Perhaps some will never rise to high levels of quality productivity in such an environment and would profit from programs that buffer them from such changes. Examination of best practices as well as measurement of the effects of programs of this type represent potentially fruitful lines of research investigation.

On the other hand, solid management practices well implemented by organizations may be sufficient for successful human resource management programs for both technical and non-technical staff. This was the essential finding of research by Ferratt and Short (1986, 1988) in examining IS staff in the insurance industry in the U.S. Midwest. This is similar to the findings of Finnegan and Murray (this issue) applying what is known to be good human resource management regarding times to IS workers. By the same logic, programs developed for motivating, retaining, and achieving high productivity with IS employees may be generalizable to a broader range of workers, particularly as employees around the world evolve to greater levels of knowledge work.

## HR Support For Global Knowledge Workers

Because the tasks of many knowledge workers who are predominantly involved with marketing, finance, accounting, or production involve high levels of interaction with computer information systems, the line between knowledge workers and information systems professionals grows increasingly fuzzier. In addition to HR providing support for IS professionals, there will need to be increasing levels of support for knowledge workers whose jobs are increasingly revolving around the creation, packaging, manipulation, and understanding of information.

Quinn, Anderson, and Finkelstein (1998) frame this issue in terms of managing professional intellect. They posit that four key ingredients in this area are: recruiting the best candidates, forcing intensive early development, constantly increasing professional challenges, and evaluating and weeding the workforce. From a global organizational perspective, this formulation can be expanded toward a worldwide consideration of best candidates with the implications for borderless recruiting systems. Such a system would at least aim to provide some sharing of applicants among varied organizational locations. A full version might recruit globally for all open positions — a tactic increasingly enabled by global Internet access to organizational job postings. Both early development and constantly increasing professional challenges require an ongoing concern for the career path and development of professionals. In this era of downsizing and flattening of management, this task in some organizations becomes a matter of self-preservation for individual employees. Human resource actions and policies involving senior managers, performance oriented decision making regarding promotion and transfer opportunities, and gathering, maintaining, and using information regarding employees are needed to provide coherent support to achieve objectives in the area of leveraging professional intellect. Evaluating and weeding the workforce also requires human resource actions and policies in the areas of developing realistic performance evaluations and organizational consequences resulting from that activity. It must be noted, though, that the content of performance evaluations can be perceived differently by people of different cultures and that the concept of performance evaluation may have different connotations and implications among different groups of people. From a global perspective, this raises interesting issues such as how to weight such results with the variations in business context among various operating locations. For example, in an environment where supplies received are of poor quality a production expert may require exceptional skills in creating low-maintenance final products although normal measures of productivity may show only average results.

The points raised by Quinn, Anderson, and Finkelstein (1998) are in addition to the more mundane human resource requirements of computer and information technology training for a global workforce. Where IS professionals operating globally may require training on a multitude of hardware and software products which are difficult to integrate, but, at best, provide local optimal results, knowledge workers are going to need training on the equipment and systems locally available. This means that the training and development component of human resources, if centralized within the global organization, will need to plan for and administer a wide range of training for essentially similar skills. Additionally, when knowledge workers shift location between operating venues, they need some retraining to use local versions of equipment and systems of varying similarity to those that are already familiar. These issues relate to "computer" training. Over and above these are issues related to the application of results from computer-based systems to their organizational activities. Davenport (1997) makes very clear the distinction between learning how to use computers and learning how to apply the results of using computers to creating productive outcomes. Clearly in organizations there is need for both sorts of knowledge, however, human resource training and development teams can be expected to excel when they treat these as two separate and important (it remains to be tested whether their importance is of equal magnitude) sets of skills.

As shown clearly in the work of Anakwe, Anandarajan, and Igbaria (this issue), the issues involved in supporting end-users or knowledge workers will display some commonalities as well as variation in attributes across countries. Some of the differences may be due to cultural issues such differences in power distance, individualism, masculinity-femininity, and comfort with ambiguity, following Hofstede's (1991) categories, but other differences may be due to straightforward issues such as per capita income, education levels, the distinctive economic competencies of a particular nation or area, or investment choices by world financiers. The implication is that developers of a global learning organization may face a larger than expected variety of design, implementation, training, and maintenance challenges.

Several research questions arise in this area. First, to what degree is the need for programs in this area significant? Although much talked about and apparently logical, perhaps the management of global knowledge workers is an area where organizations can muddle along with sufficiently productive results. To what degree do organizations actually participate in developing these programs? Even if organizations view these programs as potentially helpful to addressing a real problem, perhaps they are of low priority or perceived to be of such significant risk that firms just don't undertake them. To the extent that these programs are in fact

*Figure 2: Human Resource Management to Support Information Systems Personnel and Knowledge Workers*

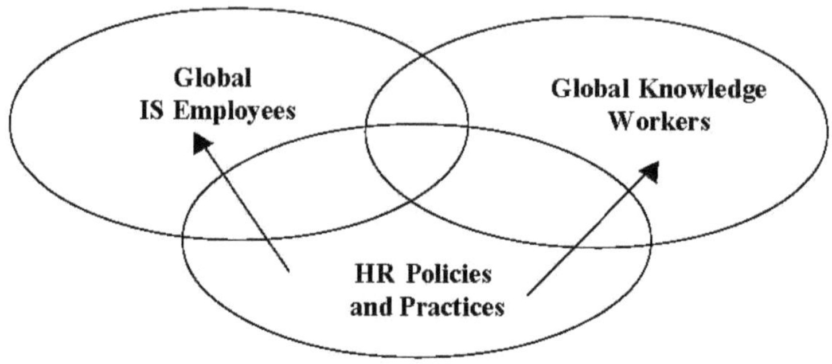


adopted by organizations, what determines the patterns of how such programs spread? What are the best practices that influence the effectiveness of these programs, and finally, to what degree do these programs provide benefit over cost to these organizations?

These next two areas, supporting global information systems and knowledge worker personnel rely on human resource management techniques and programs as a means to leverage investments in information technology and expand the opportunities for the organization. These relationships are illustrated in Figure 2.

## National And Regional Policies Supporting Technical And Human Infrastructure

It is difficult to generalize regarding the national and regional policies needed to participate in the information age. Conditions vary greatly between North America, Western Europe, Asia and the Pacific, the former Soviet bloc, the Middle East, the Indian subcontinent, Africa, and South America. It is probably safe, however, to postulate that all countries and regions that intend to fully participate in the knowledge-based economy will need to provide technical infrastructure and a pool of technically trained potential employees to organizations wishing to operate in their geographic area. As described by Trauth (this issue), there are levers that nations can pull to influence factors rendering their territory more positively disposed to the development of an information industry. In Ireland, the expansion of educational opportunities appears to have been influential in creating the context for a burgeoning information industry. However, such policies also create unintended effects, such as creating a surplus of trained personnel who leave for other areas.

Leapfrogging is the concept of investing in the newest technologies while bypassing expensive earlier versions. For example, many countries, including Singapore, South Korea, the Republic of China (Taiwan), and Hong Kong, have been leapfrogging information technologies as they build technical infrastructures. Singapore, for example, will soon have a fiber optic network throughout the entire country (Sussman and Lent, 1991, p. 165). Broadband integrated services digital network (B-ISDN) technology is being installed in several Asian nations including China. Since computerized trading is a requirement to be a leader in finance, China is planning broadband networks in Shanghai in hopes of making that city a major financial center (Engardio, 1994, p. 49).

Developing and maintaining both trained workforces and technical infrastructures are challenges that face all regions and countries. From a pragmatic perspective, organizations, both governmental and non-governmental have limited funds. It is not practical to invest in all technologies and it is never known with certainty beforehand which will eventually become obsolete and be discarded or never fully mature and which will become the basis for future global enterprise. From a research perspective, it will be of interest to investigate several issues. By what methods do organizations make these decisions? Is there a correlation between these decisions making methods and better results? Are there technical indicators that lead to better decisions? Are there decision making approaches or particular decisions that yield fewer negative unanticipated results along with those desired? From another perspective, how can collective needs and individual interests best be balanced as nations approach these decisions? To what degree should governments rather than the collection of firms and other organizations pay for these infrastructure and personnel development activities? Is there a correlation between government spending for science and technology programs and national wealth? Where there are strategies effective for one or two early cases, will they remain effective as commonly held policy? In other words, would Ireland's educational policies gain competitive advantage if it were joining dozens of other nations in such a policy rather than in breaking new ground? Finally, what kinds of interactions might there be between decisions regarding technical infrastructures and policies to encourage the development of technical personnel within organizations, nations and regional geographic entities?

This last area, pertaining to the national and regional contexts for providing support for both IS technology and trained personnel rely on the true integration of the social and technical aspects of information technology and organizations. These relationships are illustrated in Figure 3.

## CONCLUSION

This essay began with a discussion of intellectual capital, knowledge

*Figure 3: Political Regimes Creating the Context for Personnel to Populate IS and non-IS Industry Organizations*

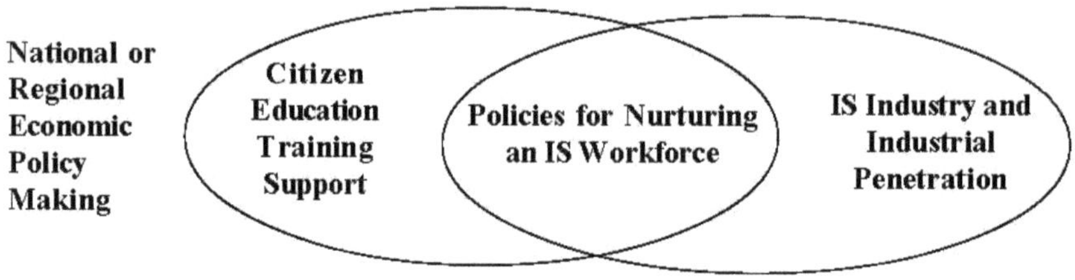


management, and organizational learning. It continued by examining five areas where there is interaction between human resource management and information technology in a global context. It also proposed for each of these areas a set of research questions that are focused on first the degree to which these issues are important; second the degree to which organizations actually have programs, in these areas; third the best practices where there are such programs; and finally the evaluation of the effectiveness of these programs. As a result, the essay proposes a large number of questions. It is the ultimate proposition of this paper that the interaction of information technology and human resource management in a global context is a largely unstudied area that will emerge as a critical one for the success of organizations and nations in the next century.

## REFERENCES

Couger, J.D. and Zawacki, R.A. (1980). *Motivating and Managing Computer Personnel*, New York, NY: Wiley.

Davenport, T.H. (1997). *Information Ecology,* Oxford University Press, New York, NY.

Engardio, P. "Third World Leapfrog." *Business Week* No. 3372 (18 May 1994): 47-49.

Ferratt, T.W. and Short, L.E. (1986). "Are Information Systems People Different? An Investigation of Motivational Differences," *MIS Quarterly,* 10(4), 377-387.

Ferratt, T.W. and Short, L.E. (1988). "Are Information Systems People Different? An Investigation of How They Are and Should Be Managed," *MIS Quarterly,* 12:3, 427-443.

Gery, G.J. (1991). *Electronic Performance Support Systems,* Gery Performance Press, Tolland, Massachusetts.

Harrison, J.K. and Deans, P.C. (1994). In (Eds. P.C. Deans and K.R. Karwan) *Global Information Systems and Technology: Focus on the Organization and Its Functional Areas,* Idea Group Publishing, Harrisburg, Pennsylvania, pp. 444-461.

Hofstede, G. (1991). *Cultures and Organizations: Software of the Mind.* London: McGraw-Hill Book Company.

Myers, M.E. (1991). "Motivation and Performance in the Information Systems Field: A Survey of Related Studies," *Computer Personnel,* 13(3), 44-49.

Nonaka, I. and Takeuchi, H. (1995). *The Knowledge-Creating Compnay: How Japanese Companies Create the Dynamics of Innovation,* New York, The Oxford University Press.

Pfeffer, J. (1998) *The Human Equation: Building Profits by Putting People First,* Harvard Business School Press, Boston, Massachusetts.

Quinn, J.B., Anderson, P, and Finkelstein, S. (1998). "Managing Professional Intellect: Making the Most of the Best," In (Ed. David A. Klein) *The Strategic Management of Intellectual Capital,* Butterworth-Heinemann, Boston, Massachusetts. Reprinted from "Managing Professional Intellect: Making the Most of the Best", March-April 1996, 71-80.

Roberts, K., Kossek, E.E., and Ozeki, C. (1998). "Managing the Global Workforce: Challenges and Strategies," *The Academy of Management Executive,* 12(4), 93-106.

Sussman, G. & Lent, J.A., (1991). (Eds) *Transnational Communications: Wiring the Third World.* Sage Publications, Newbury Park: California.

## Chapter 3

# Applying Erlang Distribution For Software Size Estimation

Derek F. W. Cheung
City University of Hong Kong

Ho-Leung Tsoi
Caritas Francis Hsu College
City University of Hong Kong

*The Program Evaluation and Review Technique (PERT) model is one of the most popular methods for estimating software size and, sometimes, development effort. It mainly relies on expert judgment to estimate the ultimate size of a project. However, the basic assumption of this technique may be incorrect. The issue addressed in this paper concerns the weakness of the PERT model and proposes applying Erlang distribution to tackle this problem. An example has been included to show the effectiveness of this new model.*

## INTRODUCTION

The major problem faced by software project developers is the prediction of the required resource level for development. It emphasizes estimation of the size of the system to be delivered in order that appropriate budgets and schedules can be agreed upon. Supposing the size of a software system and its development effort are underestimated, it means that the budgets and time are not sufficient to cover all processes that need to be developed. As a result, underestimated projects that do reach completion are often released prematurely in order to meet the budget, but these projects may omit some important features or system testing and result in

Previously Published in *Managing Information Technology in a Global Economy,* edited by Mehdi Khosrow-Pour, 

incomplete and unreliable systems. Jones (1991) claims that up to 15 % of new development projects are abandoned mid-stream, largely due to cost overruns. It is obvious that, without accurate estimates, the planning process is likely to be unreliable and result in budget and schedule overruns.

To tackle the problem, a number of software size estimation models have been developed in the past two decades (Albrecht, 1983; Boehm, 1981; Putman, 1978). Among them, the PERT model is a widely cited model used in software project management. The PERT model, based on the *Beta distribution*, was developed by the U.S. Navy in the 1950s to control the development of the Polaris submarine missile program (Pressman, 1992). Basically, it relies on expert judgment to estimate the ultimate size of a project. To offset the bias due to psychological and personal factors, it requires three estimates, namely Most Likely, Optimistic, and Pessimistic, to form a single expected value.

Although the PERT model is widely used, the assumption of this model may be incorrect. In the PERT model, user requirements and systems specifications are needed to estimate the software size. However, they may be very unstable in the early stage of the system life cycle. The estimation process is hindered by the imprecisions of assessment. Even small changes in these factors, such as resource availability, may heavily impact on the size of the final project. Since it is impossible to remove all types of imprecision and vagueness of system specifications in the sizing process, the most effective way is to indicate the probability of error in the estimates. Unfortunately, an extra error violates the basic assumption of the PERT model. In this chapter, several factors that cause imprecision will be discussed in detail. Moreover, it proposes to apply *Erlang distribution* to tackle the assumption problem in the PERT model. The last section of this chapter reports on the results of a real life application that applies Erlang distribution for doing the estimation. The result implies that the proposed model improves the way of estimation.

## Imprecision of Software Size Estimation

It is commonly recognized that size estimation needs to be taken as early as possible in the software development life cycle in order to cause a real impact on the development process. Unfortunately, there are a number of factors that cause imprecision in software size estimation, especially in the early stages of the software development life cycle. As a consequence, software size estimation has proved to be a difficult task in practice. Some typical examples are described as below:

### a) Uncertainty in Requirement Specifications

In the early stage of software development, user requirements may not be clearly known. Requirement specifications cannot be finalized until some iterations from the analysis phase to the design phase.

**b) Rapid Changes in Information Technology**

The development of software application is a dynamic environment that is characterized by many changing factors such as changes in information technology. It often leads to an improvement and will be an influence on the productivity rate. Thus the effort and time to develop software will be changed. Unfortunately, the level of impact is difficult to estimate before use.

**c) Difficult to Quantify Some Intangible Factors**

A software system is developed by many intellectual people whose activities, such as productivity, are hard to measure and determine.

## THE PERT ESTIMATION METHOD

In the PERT model, the estimator decomposes the development cycle into a number of phases and then statistically operates on the data to obtain estimates of the application. It enables the estimator to reduce the effect of uncertainty in estimated effort of the individual person and to obtain a better estimate of the overall value. In using this model, estimators are requested to give three estimates, namely optimistic ($O_i$), the most likely ($M_i$), and pessimistic ($P_i$). For instance, *Optimistic time* is the shortest possible time required for completing an activity or a subsystem. *Most likely time* is the time most frequently required for completing an activity or sub-system. *Pessimistic time* is the maximum possible time required for completing an activity or sub-system. The equation for estimating the expected value of total effort in the i<sup>th</sup> subsystem, $E_i$, is: $E_i = \frac{O_i + 4M_i + P_i}{6}$

where $O_i$ is the lowest possible number of $i^{th}$ subsystem,
$M_i$ is the most likely number of $i^{th}$ subsystem,
$P_i$ is the highest possible number of $i^{th}$ subsystem.

The equation for estimating the standard deviation of the number of in the $i^{th}$ subsystem, $s_i$, is:

$$\sigma_i = \frac{P_i - O_i}{6}$$

The total effort, $E_t$, and standard deviation for the overall system, $s_t$ (also called *sigma*), are:

$$E_T = \sum_{i=1}^{n} E_i \quad \text{and} \quad \sigma_T = \left(\sum_{i=1}^{n} \sigma_i^2\right)^{1/2}$$

In above equations, the mean tells you the estimated value and the standard deviation quantifies the risk for you. The standard deviation is an indication of the spread of the curve. To be 68% confident that a project will be completed at or below a given estimated value, you must take the mean and add to the standard deviation. Other confident levels are listed as below (Arifoglu 1993):

| | | | | |
|---|---|---|---|---|
| 99.8% | of being between | $E_T - 3s_T$ | and | $E_T + 3s_T$ |
| 95 % | of being between | $E_T - 2s_T$ | and | $E_T + 2s_T$ |
| 68% | of being between | $E_T - s_T$ | and | $E_T + s_T$ |

### Problem of Using the PERT Method

Despite the fact that the PERT model is widely used, it is not free from criticism. The assumption of Beta distribution is the major weakness to be criticized from a practical point of view. Basic property of Beta distribution is that it provides positive density only for the parameters for X-axis in an interval of finite length. If the value of X-axis is outside the interval, a negative density may appear. Negative density is impossible to be used in real life situations. For instance, there are no negative manpower or Source Lines of Code in software project development.

Experience and findings from numerous studies have shown that the development of the software project involves a lot of factors that are hard to control in this ever-changing world. Even small changes in one of these factors may cause a heavy impact on the expected value. Besides, human estimators tend to be too optimistic and the expected value will be biased toward the pessimistic estimated point (Luiz, 1990). As a consequence, a general point concerning the PERT method is that there is very small probability that the actual values do not fall beyond the optimistic estimates. On the contrary, there is very high probability that the actual values fall beyond the pessimistic estimates. Although the percentage for this overrun is very low (around 10%), we need to consider this situation in estimates, especially for risky projects. Considering that Beta distribution has a precise cut-off between "The optimistic" and "The pessimistic" points, it does not handle this practical problem.

## ERLANG DISTRIBUTION ESTIMATION METHOD

Gamma distributions (including exponential) provide a wide variety of probability models for continuous variables and are often appropriate in practice. Erlang is a special case of the gamma distribution. It was first studied by the Danish engineer Erlang in connection with the distribution of waiting times in telephone networks (Anderson, 1986) and is very widely used in engineering and science disciplines.

As already stated, most people tend to give estimates that are optimistic. As a result, most size estimates will not appear as symmetries. Most, in fact, will be skewed to one side of their mean. The project manager can assume that actual activity completion times are not normally shorter than the optimistic estimate, but they ought to allow for a certain percentage of activity times to take longer than the pessimistic estimate. The author admits that the percentage may be adjusted subject to the project managers' judgments or particular organizations. The way to determine the value is not fixed, but further discussion is beyond the scope of this paper.

For Erlang distribution in estimation, three estimates, namely optimistic time, most likely time and pessimistic time, and a percentage of error will be given by estimators. Once these parameters are given from users, the shape parameter of the Erlang distribution can be determined. Steps to derive Erlang distribution are shown as below:

Erlang Distribution: $f(x;\lambda,n)=\begin{cases}\frac{\lambda^{n}x^{n-1}e^{-\lambda x}}{(n-1)!} & X\geq 0\\ 0 & X<0\end{cases}$ mode = n(l - 1)

where n is called scale parameter and l is called shape parameter

**Step 1: Devise a comprehensive questionnaire that addresses the relevant project issues**

Similar to the PERT model, the practitioner relies on expert judgment to estimate the size of a project. Such estimation is based on the personal experience with past projects of similar characteristics. In order to minimize the subjectivity problem, a team of people with varying levels of expertise is formed to conduct the Erlang method. This step will consolidate the judgment of a group of professional software developers. Moreover, the project will be decomposed into three different phases; namely, design phase, coding phase, and testing phase. Estimation is made for each phase. According personal feelings, such as optimistic, most likely, and pessimistic, three estimated values will be given by the estimator. All information will be collected through questionnaires.

**Step 2: Solicit opinions from experts who understand the various relevant issues.**

**Step 3: Apply a simple statistical counting method to obtain and feedback the result to the participants.**

Various statistical data will be individually evaluated as shown below:

$$E_i=\frac{T_{oi}+4T_{mi}+T_{pi}}{6},\; E_T=\sum_{i=1}^{3}E_i\,,\; E_o=\sum_{i=1}^{3}T_{oi}\,,\; E_m=\sum_{i=1}^{3}T_{mi}\,,\; E_p=\sum_{i=1}^{3}T_{pi}\text{ and}$$

$$P_e=\frac{\sum_{i=1}^{3}P_i}{3}$$

where,

| | | |
|---|---|---|
| Optimistic man-power : | $T_{oi}$ | for $i^{th}$ phase, |
| Most likely man-power : | $T_{mi}$ | for $i^{th}$ phase, |
| Pessimistic man-power : | $T_{pi}$ | for $i^{th}$ phase, |
| Average man-power : | $E_i$ | for $i^{th}$ phase |
| Percentage of error : | $P_i$ | for $i^{th}$ phase, |
| Average percentage of error : | | $P_e$ for whole system |
| Total man-power : | $E_t$ | for whole system |
| Overall optimistic man-power : | $E_o$ | for whole system |
| Mode of man-power : | $E_m$ | for whole system |
| Overall pessimistic man-power : | $E_p$ | for whole system |

**Step 4: Ask the participants to reassess their judgment based on the feedback information and return with any appropriate comments.**

**Step 5: Repeat Step 3 and Step 4, until no major changes are found in the Participants' opinions**

The following information will be finalized after the end of the above processes:

Mode of man-power ($E_m$) = $n(\lambda - 1)$

$$\therefore \quad \lambda = (E_m + n) / n \tag{1}$$

$$\text{Expected percentage } (P_o) = 1 - P_e \tag{2}$$

$$P_o = \int_{min}^{max} f(x;\lambda,n)dx = \int_{min}^{max} \frac{\lambda^n x^{n-1} e^{-\lambda x}}{(n-1)!} dx \tag{3}$$

where n is called scale parameter and 1 is called shape parameter.

Using Incomplete Gamma Function table (Anderson 1986), shape and scale parameters can be determined by using computer iteration method.

## AN EXAMPLE

In this section, an example will be presented to show how to apply the Erlang estimation model for a real life software development. The example has been modified from a real-life project, which has been requested by a large jockey club and developed by an international software consultation organization. This project was designed to replace the existing telephone betting system on a proprietary hardware platform. The client request to develop this system on a Personal Digital Assistant (PDA) computer and several advance wireless transmission technologies will be included. After an intensive investigation, a decision has been reached applying Erlang distribution model to estimate the size of this application. The reasons are shown as below:

- One serious problem faced by the developer is that no sufficient historical data can be referred either from the company or public standard. It affects the accuracy in using many popular software estimation models.
- Object-Orient technology will be applied in this development and many current sizing models, such as Function Point, Feature Points, can be used in such platform.
- The application had been built on a new developed PDA platform working together with an advanced wireless digital modem. Unfortunately, the modem has not been released by the supplier while starting to develop the application.
- The development team consists of a mix of experienced software consultants and some newly recruited graduates. They do not have much experience, either in the PDA hardware platform or in wireless telecommunication technology.

In summary, there are many uncontrolled and unpredictable factors existing in the development cycle. According to the weakness of the PERT model mentioned in section I, the company derives an Erlang estimation model. It is expected to provide flexibility to the management for planning and controlling development and maintenance projects. Estimation process is shown as below:

Step 1: Devise Questionnaire to all members of the development team
Step 2: Solicit opinions
Step 3: Applying statistical method and feedback result to all participants
Step 4: Re-collect opinions from participants
Step 5: Repeat step 3 and step 4 until no major changes

The total expected value X (in KSLOC-thousand source line of codes) is aggregated by all individual stages:

Optimistic size : $E_o$ = 1 KSLOC
Most Likely size : $E_m$ = 3 KSLOC
Pessimistic size : $E_p$ = 7 KSLOC
Average Percentage of error : $P_e$ = 5%
Using (1) $\lambda = (\text{mode} + n) / n = (2 + n) / n$ (I)
Using (2) Expected percentage $(P_o) = 1 - P_e = 1 - 5\% = 95\%$ (II)

Thus, the probability that the project finish between optimistic and pessimistic time is 95%.

Using (3) Expected percentage -95%- $\int_{min}^{max} f(x;\lambda,n)dx = \int_{min}^{max} \frac{\lambda^n x^{n-1} e^{-\lambda x}}{(n-1)!} dx$

$$\int_0^{max} \frac{\lambda^n x^{n-1} e^{-\lambda x}}{(n-1)!} dx - \int_0^{min} \frac{\lambda^n x^{n-1} e^{-\lambda x}}{(n-1)!} dx = (\lambda^{n-1})[\int_0^{max} \frac{x^{n-1} e^{-\lambda x}}{(n-1)!} d(\lambda x) - \int_0^{min} \frac{x^{n-1} e^{-\lambda x}}{(n-1)!} d(\lambda x)]$$

Before using computer iteration method, let y = lx, then Incomplete Gamma function can be used in above equation.

$$95\% = (\lambda^{n-1})[\int_0^{max} \frac{x^{n-1}e^{-y}}{(n-1)!} d(y) - \int_0^{min} \frac{x^{n-1}e^{-y}}{(n-1)!} d(y)] = l^{n-1} [F(max,n) - F(min,n)]$$

Where F(x,n) is Incomplete Gamma function

By using computer iteration method, we find n = 5 and l = 1.22

The Erlang distribution is F(x:1.22,11) and the expected value (50%) is 3.5 KSLOC. It is very close to the result (3.3 KSLOC) yielded by the PERT method.

Finally, the actual size of the project is 7.3 KSLOC. The discrepancy between the estimated and real size values for the PERT model is +4 KSLOC (121%). The most interesting point is that the actual size of the project is greater than the pessimistic value in the PERT model. It is a good example to show that the assumption of the PERT model is incorrect in some cases, especially for some risky software projects.

## CONCLUSION

There is no doubt that project managers have to confront many different kinds of problems, such as technical, management, and personnel, in the software development cycle. They are responsible for ensuring that all development processes are controlled appropriately in order to meet the budget and schedule. Unfortunately, software projects regularly get out of hand in the development cycle.

As mentioned earlier, software project development has a dynamic nature. Therefore, changes are bound to occur. A good estimate is the first step to get a successful software project. In this paper, a technique, applying Erlang distribution for software size estimation, has been proposed. The Erlang estimation model is not a completely new method but it tackles the weakness of using the PERT model. The author admits that this model is still subjective, since it ultimately depends on expert judgment to estimate the ultimate size of a project. However, this model incorporates confidence concept to show the risk level in practice. Instead of a crisp value, the Erlang model provides an estimate with a different confidence level that allows project managers to consider the worst situation of the development effort estimation. Besides, project managers can select appropriate confidence levels for their projects.

From the example, it shows that the pessimistic estimated value in the PERT model may not be the upper boundary of a project. Overruns are common in the software industry. As a result, the assumption of the PERT model needs to be reconsidered. Besides, the Erlang approach has proved to be more informative than the PERT model. The project managers are free to select the appropriate confidence level for their project. With a higher probability level, the project management would be able to prepare a more realistic project plan to cope with the possible.

## ACKNOWLEDGMENT

The author is thankful to Dr. C. Yau for various insights concerning the ideas upon which this paper is built, and to Mr. Vincent Cheng and Miss Susan Addison for their helpful suggestions.

## REFERENCES

Albrecht, J. A., and Gaffney, E. J. (1983, November). Software Function, Source Lines Code, and Development Effort Prediction: A Software Science Validation. *IEEE Trans. of SE, 9*, (6), pp. 639-647.

Anderson, D. R. (1986). *Statistics: Concepts and Applications*. New York: West Publishing Company.

Arifoglu, A. (1993). *Methodology for Software Cost Estimation*. ACM SIGSOFT.

Boehm, W. B. (1981). *Software Engineering Economics*. New Jersey: Prentice Hall.

Jones, and Capers (1991). *Applied Software Measurements*. New York: McGraw-Hill.

Luiz, A. L. (1990). Software Size Estimation of Object-Oriented Systems. *IEEE Trans. of SE., 16*, (5), pp.510-522.

Pressman (1992). *Software Engineering*. New York: McGraw-Hill.

Putman, L. H. (1978, April). General Empirical Solution to the Macro Software Sizing and Estimating Problem. *IEEE Trans. on Software Engineering, 4*, (4), pp.345-361.

## Chapter 4

# Trajectories and Strategies in Global Software Outsourcing Relationships

Richard Heeks
University of Manchester

S. Krishna
IIM-Bangalore

Brian Nicholson
University of Manchester

Sundeep Sahay
University of Oslo

## GLOBAL SOFTWARE OUTSOURCING

Global software outsourcing (GSO) is the outsourcing of software development to sub-contractors outside the client organisation's home country. India remains the unquestioned leader, registering average annual growth of more than 40% over the last decade and developing some US$3.6bn-worth of software for foreign clients in 1999/2000 (Heeks, 2000).

Advice for potential GSO clients counsels starting small, starting at home, and starting with programmers (McFarlan, 1996). Many client organisations have followed this advice, putting a toe into the GSO waters through small-scale body shopping: for example, having Indian sub-contractor staff come over to the client site to complete a minor, non-critical piece of coding/conversion work.

Whilst minimising risk, however, this also minimises benefits. In the US, for example, onsite costs of India-related GSO undercut those for hiring US staff by only some 10-20%. In comparison, sending development work offshore to India will typically undercut by some 50%. Large projects offer a greater potential for savings than small ones. Likewise, the cost savings for hiring Indian analysts/project managers are typically US$1000-2000 per month greater than those for hiring programmers (Heeks, 1996).

Clients have therefore been keen to move up the value chain in order to increase the benefits of outsourcing. These benefits include not merely the greater savings just noted, but also improved access to local labour and to the local IT market.

However, moving up the value chain brings additional costs and risks. Clients and developers are therefore seeking routes through the cost/risk minefield to the benefits that higher-value GSO promises. The research reported here set out to investigate these routes, undertaking longitudinal case study research with eight GSO relationships for which the developer half was based in India.

## SYNCHING OR SINKING

In seeking to understand success strategies in GSO relationships, we were struck by contradictions arising from some prescriptive recommendations. Techniques that worked well for one relationship could be a cause for friction and failure in another.

The analysis of field data was therefore informed by a contingent perspective. Successful relationships were those in which a high degree of congruence was achieved between developer and client: we called this "synching." Unsuccessful relationships were those in which a low degree of congruence was achieved between developer and client: we called this "sinking."

We developed a dimensional framework for congruence. Using this, synching was more precisely defined as minimisation of gaps between client and sub-contractor along six 'COCPIT' dimensions: **C**oordination/control systems, **O**bjectives and values, **C**apabilities, **P**rocesses, **I**nformation, and **T**echnology.

Theoretical examples of complete congruence can be given for each dimension:

- *Coordination/control systems*: client and developer use the same management coordination and control systems; for example, having the same systems for staff monitoring and appraisal.
- *Objectives and values*: client and developer share the same objectives for their relationship and bring the same values to that relationship; for example, having the same organisational culture.

- *Capabilities*: this dimension is slightly different. One purpose of outsourcing is for the developer to provide human capabilities the client lacks. Congruence here therefore means that the developer's capabilities profile matches the requirements of the client; for example that, if the client needs ten Java programmers, the developer provides them.
- *Processes*: client and developer use the same work processes; for example, using the same software development methodology.
- *Information*: client and developer have access to the same information; for example, information relating to project requirements and timescales.
- *Technology*: client and developer use the same technology; for example, the same software and hardware platforms for development work.

## FINDINGS AND RECOMMENDATIONS

Table 1 summarises three of the cases studied.

Synching – creating congruence between client and developer along a number of dimensions – lays the foundation for higher project success rates in GSO and for higher-value GSO. Synching is a pre-requisite for moves up the 'trust curve' that lead to higher-value GSO. It is also a means of reducing the risks and costs associated with higher-value GSO.

Both client and developer managers must understand what synching means in their particular context, and must look for strategies that bring about synching. But they must also recognise the serious practical difficulties of synching. For example, whilst it proved relatively easy to synch technology and capabilities, it proved hard to synch some aspects of information and coordination/control systems and very hard to synch objectives and values. This created limits to synching.

### Limits to Synching

Western clients seem to fall too easily for the argument that, in a globalised world, distance, borders and place no longer matter. The experience of these cases suggests otherwise.

*Table 1: Client—Developer Relationships, Synching and Outcomes*

| Relationship | Degree of Synching | | | | | | Outcome |
|---|---|---|---|---|---|---|---|
| | C | O | C | P | I | T | |
| 'Pradsoft'—'Ameriten' | x | xx | √ | x | √ | √ | Insufficient synching leads to relationship failure |
| 'Global'—'Shiva' (pre-shock) | √ | — | √√ | √ | √ | √√ | Synching just sufficient for relationship development |
| 'Sierra UK'—'Sierra Bangalore' (pre-shock) | — | — | √ | √ | — | √√ | Synching sufficient for relationship survival but not development |

√√: very congruent; √: fairly congruent; —: partly congruent; x: slightly congruent; xx: not congruent

Distance does still matter, and it interferes with synching because of the difference between the formal/tangible and the informal/intangible aspects of GSO. Synching strategies in practice have been good at dealing with the former and poor at dealing with the latter. In particular, they have been poor at dealing with three overlapping issues: tacit knowledge, informal information, and culture.

In direct terms, the first two are part of the information dimension, and the latter is part of the objectives and values dimension. However, their importance derives from the way in which each indirectly affects all five COCPIT dimensions other than its own, as shown in Figure 1.

Technologies, specifications, processes, methodologies, skills, objectives and management systems can be transferred from client to developer. But all these have informational components consisting of two parts: explicit knowledge that can be laid out formally, and tacit knowledge that cannot. The former is readily transferred; the latter is not. Global, for example, went with GSO best practice and used clear, formalised requirements specifications. Whilst necessary, these were not sufficient because they incorporated a whole set of tacit assumptions and understandings that were not transferred: about the nature of the customer, about design and programming choices, about working practices.

Clients especially must therefore pay more attention to tacit knowledge. They must recognise its existence within all the COCPIT dimensions, identify its content, and look for ways to synch it between themselves and their developers.

Ready transfer of formal procedures also runs into problems because real software development requires constant divergence from formal guidelines and requires constant improvisation. Informality and improvisation require informal information which thus remains a critical resource for global software development.

Trying to focus on well-structured, stable projects enabled some case study clients to push a lot of information exchange into the formal realm that can be handled relatively well by ICT-mediated distance. But informal information could not be handled that way. Sierra found the cost, fragility and artificiality of video-

*Figure 1: The Influence of Tacit Knowledge, Informal Information and Culture*

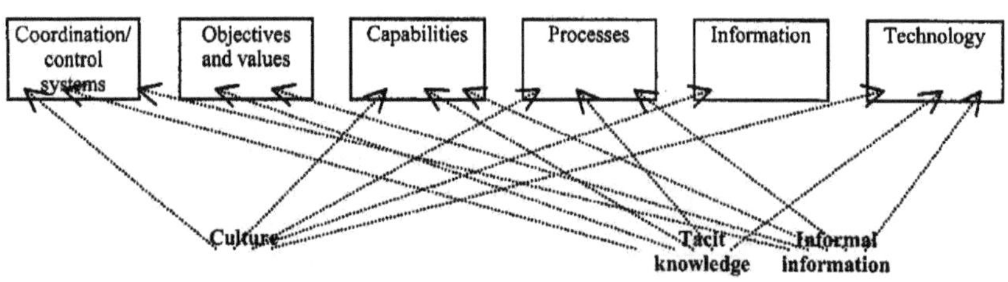

conferencing excluded informal conversations and restricted interaction to formal exchange of progress towards milestones. Email, too, only operated at an informal level once participants had physically met and built personal relationships. Travel and direct meetings will therefore remain a continuous and crucial element in GSO relationships to help more fully synch the information dimension.

Thirdly, players in this global game all still retain cultural values rooted in a particular locale. The overseas development centre is a powerful tool for synching and, thus, for raising project success rates and for moving up the GSO value chain. But it has its limits. Western systems, capabilities, processes, etc. can all be imposed. However, some cultural "stains" underpinning these dimensions are hard for this global tide to wash away.

## Living with Limits to Synching

How can Western clients retain the benefits of GSO and yet live with the limits to synching? Some pointers were noted above. These are specific examples of a more general need to create buffering mechanisms between the distant worlds of client and developer, and bridging mechanisms that allow "intangibles" to be exchanged between those worlds. Other examples include:

- *Management of expectations.* Sierra's overseas development centre foundered, in part, because UK expectations were out of synch with Indian realities on several COCPIT dimensions. Those expectations – of a Bangalore where the streets are paved with programmers, where technology can destroy distance, where Indian staff become British clones given the right environment – were built on too much media hype and too little cold, hard analysis. As the overly positive expectations crashed down to reality, they left in their wake a disillusionment with GSO.

  A key, then, to successful GSO is a realistic expectation of what can be achieved given the level of attrition, the limitations of the technological infrastructure, the cultural differences, and so forth. Global and Shiva built up a good in-house understanding of what can and cannot be achieved over a long period with a slow-growing expansion and trust curve. This was more successful than Sierra's "too much, too young" approach.
- *Using straddlers.* Straddlers are those with a foot in the world of the client and a foot in the world of the developer. Global and Shiva had many of these. Global, for example, used some of its India-born managers in GSO relationships. These staff have proven adept at understanding what can and cannot be brought into synch. They manage the dimensional gaps that remain, often acting as a buffer between the Indian developers and senior client managers. Shiva, too, despite attrition rates, has staff at all levels with significant experience of working in North America who are better able to see things from the client's

perspective. Those straddlers have also acted as conduits for transfer of tacit knowledge and informal information. Sierra, by contrast – perhaps because it is a smaller and more recent entrant – had no effective straddlers in either client or developer groups.

- *Building bridging relationships*. Straddlers bridge gaps within one individual. An alternative is to bridge gaps by building strong one-to-one relationships between individual clients and developers. Global and Shiva have been active in this, facilitating meetings and other informal fora for contact in which such relationships can develop. They have even proactively identified pairs of individuals to bring together. These relationships have been a valuable means by which to cope with continuing client/developer mismatch. They create a channel for translating between client and developer, and for passing informal information and tacit knowledge. They also create the 'synch' of mutual trust and understanding that can help overcome other dimensional differences.

In addition to their synching strategies, clients and developers must therefore also identify the buffering and bridging mechanisms that will help them deal with the limits to synching that still exist within global software outsourcing for some key COCPIT dimensions.

### Understanding the Synching Environment

A number of cases exposed the danger of external "shocks to synching," and they are a reminder that the client—developer relationship does not sit in vacuum. Instead, it sits within an environment of which some components are illustrated in Figure 2.

*Figure 2: The GSO Relationship Environment*

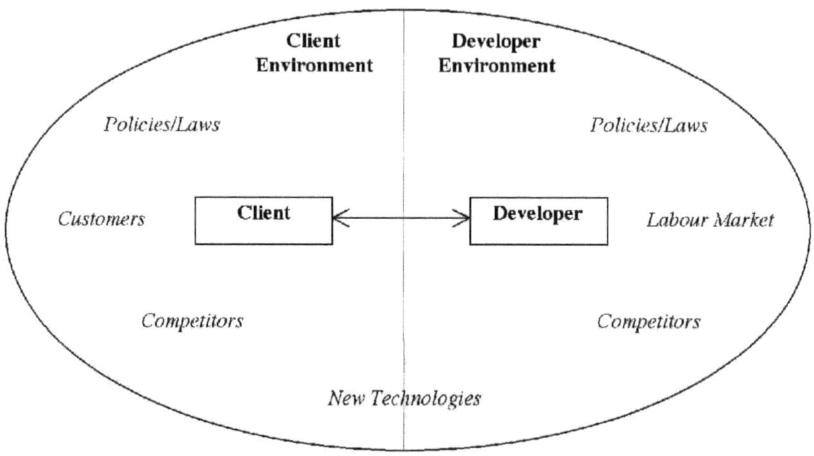

As well as synching with each other, clients and developers also seek to synch with other components of their environment. This creates pressures and tensions for the client—developer relationship.

These may be acute: the negative impact of e-business developments. Or they may be chronic: the problems of developers in meeting the demands of staff in the local labour market. These demands related to wages, development of state-of-the-art skills, travel, and length of project assignments. As Global and Shiva discovered, such demands may run counter to what the client provides, leading to attrition levels way above what the client expects.

There is no magic solution to meeting these environmental pressures, especially when they are unexpected. Explicit discussion of the pressures will be one step; diversification of both parties into other relationships will be another.

Managers must recognise that there is a seventh dimension to synching: time. Congruence must be actively sustained because, once created, it cannot ensure a permanently 'synched' relationship in a context of continuous environmental change. Clients and developers must therefore continuously re-examine their relationship and proactively move to address emerging mismatches.

## REFERENCES

Heeks, R.B. (1996). *India's Software Industry*, Sage Publications, New Delhi.

Heeks, R.B. (2000). *India's Software Industry*, IDPM, University of Manchester, UK. www.man.ac.uk/idpm/isi.htm.

McFarlan, F.W. (1996). Issues in Global Outsourcing. In P. C. Palvia et al. (Eds.), *Global Information Technology and Systems Management*. Nashua, NH: Ivy League Publishing.

## Chapter 5

# Emerging Research Themes in Global Information Management

R. Brent Gallupe
Queen's University, Canada

Felix B. Tan
University of Auckland, New Zealand

*Research in the field of global information management (GIM) is an area of information systems research that has grown tremendously in the 1990s. A large number of research studies have been published that have examined issues relating to the development, use and management of information systems in a global context. In an attempt to better understand the types of research that has been conducted, the research methodologies employed and the emerging research themes being addressed in the field, an in-depth study of the GIM literature was conducted. This chapter reports on one aspect of this wider study – emerging GIM research themes. The purpose of this chapter is to identify the themes distinct to GIM research that are emerging from the literature with the aim of highlighting potential areas of future research in the field.*

## INTRODUCTION

Global information management has received a great deal of research interest in the last ten years. As information technologies have improved and as organizations have adapted their operations to a global platform, research into how to use and manage these technologies on a global scale has increased in importance. Not only are established IS journals publishing an increasing amount in this area but there are now specific journals devoted to the major issues in the development, use and management of global information systems. Indeed, a ten-fold increase in

Previously Published in *Challenges of Information Technology Management in the 21st Century*, edited by Mehdi Khosrow-Pour, Copyright © 2000, Idea Group Publishing.

the number of papers published has occurred between 1990 and 1998 (Gallupe and Tan, 1999).

However, it was somewhat surprising to find that in a survey of the literature in this area that no systematic framework is guiding GIM research. It appears that current GIM research is being driven by temporal, "hot" issues in the field and not by a integrated and structured approach to knowledge accumulation. These "issues" studies are certainly important and help the field focus on the immediate challenges that practitioners face but in our view, this seems to be leading to research that may have little enduring value, that possibly duplicates itself, and that does not stretch the boundaries of what we know. What appears to be missing at this point is an overall research framework that will help guide future research into GIM and help organize and categorise research previously done. According to Palvia (1998), such a framework has yet to be developed.

In response to the call for a comprehensive GIM research framework, the authors conducted a detailed survey of published GIM research between from 1990 to1998. A publication outlining a 3-dimensional framework and a "call-to-action" resulted (Gallupe & Tan, 1999). This research framework builds on an earlier general IS research framework of Ives, Hamilton & Davis (1980). In their survey, Ives et al. (1980) classified general IS research into 5 categories (Type I to V), which employed research methodologies ranging from case and field studies to secondary (i.e., non-) data collection strategies. We argue that to better understand the research that has been done in the GIM field, a 3$^{rd}$ dimension

*Figure 1: Global Information Management Research Framework (Gallupe & Tan, 1999)*

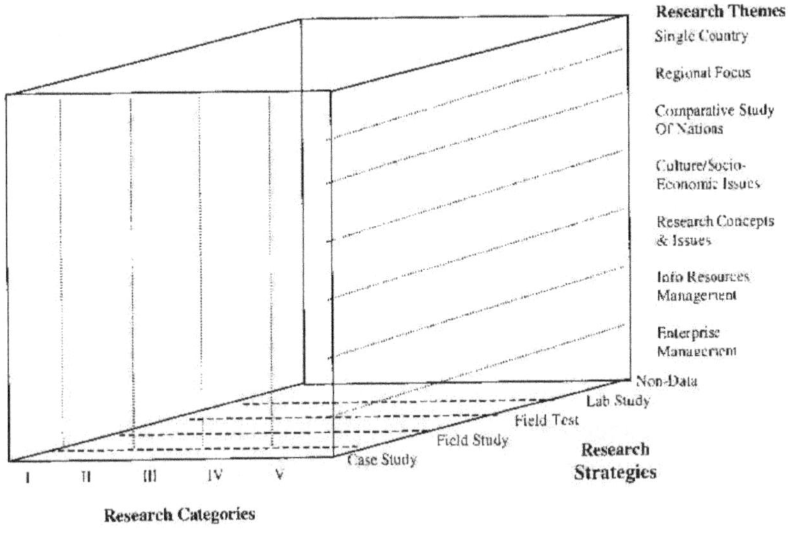

should be added to the framework allowing GIM researchers to pinpoint the research issues being addressed. This 3-dimensional framework is presented in Figure 1.

This chapter presents a deeper look at this 3rd dimension - emerging GIM research themes. The paper proceeds as follows. First, GIM is defined and research themes discussed. The research methodology is then outlined, results presented and discussed. Finally, directions for future GIM research are described based on emerging GIM research themes.

## A THEMATIC APPROACH TO GIM RESEARCH

A review of the IS literature could find no formal definition of global information management. Deans and Ricks (1991) refer to issues at the "interface of MIS and international business (p. 58)." Palvia (1997) refers to "global IT research" and describes a model to "assess the strategic impact of IT on a global organization engaged in international business (p. 230). In our study, we define global information management as the development, use and management of information systems in a global/international context. By "global" we mean those information systems that have impacts beyond a single country or country of origin. The term "global" is used in a general sense since no firm or information system is found in every country in the world. Global information management deals with technological, management and cultural issues such as IS development in different cultures, differing national communications infrastructures, differing IS quality standards, and many others. GIM research is the rigorous and systematic study of the development, use, and operations/management of a global information system(s) in a multi-country organizational environment. At the same time, traditional GIM research includes numerous single country studies focusing on the management of the information resource in a domestic context. According to Palvia (1998), these "first generation" studies have laid the foundation and helped define global IT. This paper has therefore included single country studies in the definition.

Much of the GIM literature that provides some kind of guide to research in the field has concentrated on identifying the "key issues" in the global management of information resources (Badri, 1992; Deans & Ricks, 1991; Ives & Jarvenpaa, 1991; Watson et al., 1997; Palvia, 1998). These publications survey various stakeholders involved in the research and practice of GIM and are useful in that they attempt to capture what these people think are the critical issues in the field at that time. For instance, in a survey of IS educators, Palvia (1998) identified and ranked in importance several GIM research topics such as country and environment, and regional issues. This paper builds on the work of Palvia (1998). It is different however, in that it takes a broader perspective of GIM research by analysing published GIM research and categorizing this research into emerging GIM research themes.

Although clear research topics have been identified in the literature, little has been said about the gaps in GIM research themes. So far, very little guidance is afforded to GIM researchers worldwide. This chapter therefore presents the findings of a survey of relevant published GIM research in journals that focus on the management of global information resources. The purpose of this chapter is to identify the emerging themes distinct to GIM research with the aim of highlighting potential areas of future research in the field.

## METHODOLOGY

The following approach was taken in this research.

a) A search for relevant GIM publications was the first step. It was felt that conceptual and empirical GIM research is more likely to be found in journals rather than other forms of publication such as books. Conference proceedings were

*Table 1: Emerging Research Themes Distinct to GIM Research*

| Emerging Research Themes In GIM Research | Description |
|---|---|
| Single Country Studies | The development, operations, management and use of IT in a domestic context. Does not include the management of information resources in a global context. For example, the adoption of IT in rural China. |
| Comparative Study of Nations | The comparison of IT development, operations, management and use between two or more countries. Culture is not a variable in these studies. For example, comparing the skills of systems analyst in Canada, New Zealand and Singapore. |
| Culture / Socio-economic Issues | The effect of national culture on IT development, operations, management and use. These are regarded as "pure" cross-cultural studies as opposed to those merely comparing nations. For example, exploring the effect of complex cultural dimensions on the level of IT transfer in Arab and East European nations. Also includes socio-economic issues relating to IT like government policy, legislation and economic factors. |
| Research Frameworks and Issues | Conceptual research offering frameworks, theory and research agendas on various aspects of global information management. Key issues studies in IS management from around the globe are included. |
| Global Information Resources Management | The development, operations, management and use of IT in a global context. Includes the management of information resources in a regional but not domestic context. For example, managing global IT outsourcing; motivating global IT development teams; and managing the introduction of telecommunications technology in Latin America. |
| Global Enterprise Management | Functional management of enterprises across national boundaries using IT. Includes the management of multinational and trans-national corporations. For example, the impact of IT on global supply chains, global distribution or global marketing. |

not examined because it was assumed that high quality GIM research papers would be published in journals. As such the following journals were referenced:

- Established IS journals where GIM related research has been consistently published (i.e., *MIS Quarterly*, *Information Systems Research*, *Journal of Management Information Systems*, and *Information & Management*);
- Niche journals which focus on publishing GIM research (i.e., the *Journal of Global Information Management*, *International Information Systems*, *Journal of Strategic Information Systems*, *Journal of Global Information Technology Management*, and *Journal of International Information Management*);
- Regional IS journals from around the globe (i.e., *European Journal of Information Systems*, *Scandinavian Journal of Information Systems*, *Australian Journal of Information Systems* and the *Asia Pacific Journal of Information Management*);
- Societal / Culture oriented IS journals which publish research dealing primarily with socio-economic issues and cross cultural dimensions of societies across the globe (i.e., *Information Technology & People* and *The Information Society*).

An examination of these journals is expected to provide a representative sample of the research done in the field. The period covered in the analysis is 1990-1998.

b) Based on the full text of the paper (and sometimes on the abstract if the study was clearly described), the authors independently identified distinct GIM research themes and categorised each article accordingly. If a paper could fit in more than one thematic area, the authors independently assigned the paper to one theme based on the key ideas in the paper.

c) Disagreements between the authors were discussed and a classification was agreed upon.

## RESULTS

Although the field of GIM has a number of similarities with the general study of IS, there are a number of distinct themes which differentiate emerging GIM research from the traditional scope of IS investigations. The search process identified 314 GIM related articles. A content analysis of the 314 publications sampled reveal six distinct GIM research themes. Table 1 presents these themes and their description.

As part of the study, the 314 GIM publications surveyed are categorised into these themes. Table 2 presents the results of this analysis.

## DISCUSSION

Single country "domestic" type studies dominates GIM research. Of the 314 publications analysed, 145 (46.2%) can be considered single country studies All of these studies focus on different aspects of managing information resources in a domestic context. Most of these studies use the case study or secondary research strategies. For instance, Kautz & McMaster (1994) presented a case study of an attempt to take a structured development method into use in the IT unit of a UK public sector organization. Wan & Lu (1997) provided an overview of computer crime and related legislation in the People's Republic of China using a secondary research strategy.

Studies comparing two or more nations prove to be another fruitful area of work. Forty-nine (15.6%) can be classified into this group. Several of these studies have compared the US to countries from Europe, Asia and the Middle East. For example, Abdul-Gader & Kozar (1995) examined the impact of computer alienation on IT investment decisions between the US and Saudi Arabia. Straub, Keil & Brenner (1997) tested the Technology Acceptance Model across three countries – Japan, Switzerland and the US. Others have compared two or more countries in the same region. For instance, Simon & Middleton (1998) analysed human resource management best practices in IS departments in Singapore, Hong Kong, Malaysia and the People's Republic of China. Tam (1998) examined the impact of IT investments on firm performance and evaluation in four newly industrialised economies in Asia. In all of these instances, field study (questionnaire survey and interviews) research strategy was used.

The "global information resources management" research theme accounts for 53 (16.9%) of the sample surveyed. Research in this category considers the development, operations, management and use of IT in a global context. It includes the management of information resources in a regional but not domestic context. Most of the research employ either case study, field study or secondary research strategies. Examples of these are: Ramanujan & Lou (1997) who conducted a field study on the issues involved in selective outsourcing of maintenance operations from an off-shore location; Gibson & McGuire (1996) who discussed quality control for global software development using secondary research strategy; and Trauth & Thomas (1993) who called for a global standards policy for EDI. The articles focusing on regions primarily employ secondary research strategy.

*Table 2: Emerging Research Themes Distinct to GIM Research*

| Emerging Research Themes Distinct to GIM | Research Themes Count | Research Themes Percent |
|---|---|---|
| Single Country | 145 | 46.2 |
| Comparative Study of Nations | 49 | 15.6 |
| Culture / Socio-economic Issues | 29 | 9.2 |
| Research Concepts & Issues | 23 | 7.3 |
| Global Information Resources Management | 53 | 16.9 |
| Global Enterprise Management | 15 | 4.8 |
| Total GIM Publications | 314 | 100 |

These articles describe IT related issues in a given region or prescribe guidelines for IT diffusion and transfer in developing countries. These publications are not studies comparing countries within the region. For example, a paper by Gibson (1998) identifies important developmental factors in the relationship between IT diffusion and economic development in Latin America. Loh, Marshall & Meadows (1998) discuss the ethical appropriateness of information and communication technologies for developing nations.

The "culture / socio-economic issues" research theme account for 9.2% of the 314 articles surveyed. What was surprising though is that a only 15 (3.7%) of the total sample surveyed are what we considered "pure" cross-cultural research as opposed to studies merely comparing two or more nations, but label themselves as multi-cultural studies. Field study is the predominant research strategy used in all of these publications. These 15 articles deal with and address the cultural dimensions and elements, which are inherent in the countries studied. For instance, Hill et al. (1998) focuses on the complex sociocultural constructs (beliefs and values) which can influence the level of IT transfer in Arab nations. Straub (1994) investigated the effect of culture on the use of email and fax technologies in Japan and the US. Harvey (1997) conducted an ethnographic study of national culture differences in IT theory and practice between Germany and the US using Hofstede's (1980) framework. Most of the publications dealing with the socio-economic issues primarily employ secondary research strategy.

The remaining two themes – "research frameworks & issues" and "global enterprise management" – together account for less than 13% of the research surveyed. Conceptual research offering frameworks, theory and research agendas on various aspects of GIM are categorised into the former group. For example, Nelson & Clark (1994) propose a research framework for cross-cultural issues in IS research. Martinsons & Westwood (1997) developed an explanatory theory of MIS in Chinese business culture. Deans & Ricks (1993) suggest an agenda for research linking IS and international business. Studies of key issues in GIM practice and research are also included in this grouping (Watson et al., 1997; Mata & Fuerst, 1997; Yang, 1996; Burn & Ma, 1993).

The "global enterprise management" grouping focuses on research dealing with the management of different aspects of the enterprise across national boundaries using IT. The published research surveyed used both field and case study methodologies. The bulk of the research, which falls in this grouping, is publications on multinational companies. For instance, Chidambaram & Chismar (1994) examined the use and investment patterns in U.S. multinational corporations. Cummings & Guynes (1994) compared the IS activities in U.S. and non-U.S. subsidiaries of transnational corporations. There is, however, a dearth of studies dealing with IT and global supply chain, human resource, marketing, manufacturing and distribution management. Exceptions are studies by Niederman (1993) and Sankar & Liu (1998).

A comparison between Palvia's (1998) research topics and this paper's research themes reveals two significant differences. First, one of the key thematic areas identified in this paper is "comparative study of nations." This is an important for on going GIM research as it offers the potential to highlight "best IT practices" in varying countries. A second difference is that in this paper "Technology" is not a major thematic area as it is in the Research Topics paper. The reason for this is that information technology is subsumed in all the major themes identified in this paper. It is the perspective of this paper that the technology is intended to support the management of global information, not the other way around.

## FUTURE RESEARCH DIRECTIONS

Single country "domestic" studies have been the predominant theme researched in the GIM field. This type of research tends to be descriptive and exploratory. According to Palvia (1998), these "first generation" studies are important especially during the early years of GIM research. "They have been useful in defining the global IT field ...(and have)...given us a reasonably good grasp of global IT" (p. 7). We believe that single country studies are as important today as they have been in the early years of GIM research, provided these studies include some insights for GIM practice and research. For instance, a study of the influence of government policy on the adoption of global neural networks in Latvia is only useful if it includes discussion on the implications of the study findings to GIM practice (i.e., to international companies planning to do business in the country) and research.

Opportunities also abound for research on themes relating to national culture and global enterprise management. Surprisingly, there is comparatively little published research in journals examining national culture. It is generally accepted that differences in national culture may explain variations in IS in different cultures (Deans & Ricks, 1991; Ein-Dor et al., 1993; Shore & Venkatachalam, 1995 & 1996). We therefore think that more research could explore how national culture can impact on global IS development, management and use as well as to better understand the characteristics of IS developed and used in various cultural settings. There is also very little published research into aspects of global enterprise management. Organisations, which span national boundaries, are required to operate rather differently from those in a domestic context. How can IT be used to best support and enhance the international competitiveness of these enterprises?

We urge our GIM research colleagues to embark on more investigations into the impact of IT on the global supply chain, human resource, marketing, manufacturing and distribution management. This aspect of GIM research may not be within the traditional scope of IS research, but we believe is equally compelling as organizations grapple with the management of their functional operations as they establish and compete globally. To this end, the *Journal of Global Information Management* recently published a special issue on global management of IT human resources (volume 7, number 2) and is working on another issue on IT support for global sourcing and procurement.

## REFERENCES

Abdul-Gader, A. H. & Kozar, K.A. (1995). The impact of computer alienation on information technology investment decisions: An exploratory cross-national analysis. *MIS Quarterly, 19*(4), 535-559.

Badri, M. A. (1992). Critical issues in information systems management: An international perspective. *International Journal of Information Management, 12,* 179-191.

Burn, J., & Ma, L. (1993). Critical issues of IS management in Hong Kong. *Journal of Global Information Management, 1*(4), 28-37.

Chidambaram, K., & Chismar, W. G. (1994). Telecommunication technologies: Use and investment patterns in U.S. multinational corporations. *Journal of Global Information Management, 2*(4), 5-18.

Cummings, M. L., & Guynes, J. G. (1994). Information system activities in transnational corporations: A comparison of U.S. and non-U.S. subsidiaries. *Journal of Global Information Management, 2*(1), 12-27.

Deans, P. C., & Ricks, D. A. (1991). MIS research: A model for incorporating the international dimension. *Journal of High Technology Management Research, 2*(1), 57-81.

Deans, P. C., & Ricks, D. A. (1993). An agenda for research linking information systems and international business: Theory, methodology and application. *Journal of Global Information Management, 1*(1), 6-19.

Ein-Dor, P., Segev, E., & Orgad, M. (1993). The effect of national culture on IS: Implications for international information systems. *Journal of Global Information Management, 1*(1), 33-44.

Gallupe, R.B. & Tan, F.B. (1999). A research manifesto for global information management. *Journal of Global Information Management,* 7(3), 5-18.

Gibson, R. (1998). Informatics diffusion in South American developing economies. *Journal of Global Information Management, 6*(2), 35-42.

Gibson, R. & McGuire, E.G. (1996). Quality control for global software development. *Journal of Global Information Management, 4*(4), 16-22.

Harvey, F. (1997). National cultural differences in theory and practice: Evaluating Hofstede's national cultural framework. *Information Technology & People, 10*(2), 132-146.

Hill, C. E., Loch, K. D., Straub, D. W., & El-Sheshai, K. (1998). A qualitative assessment of Arab culture and information technology transfer. *Journal of Global Information Management, 6*(3), 29-38.

Hofstede, G. (1980). *Cultural Consequences: International Differences in Work Related Values*. Beverly Hills: Sage.

Ives, B., Hamilton, S., & Davis, G. B. (1980). A framework for research in computer-based management information systems. *Management Science, 26*(9), 910-933.

Ives, B., & Jarvenpaa, S. L. (1991). Applications of global information technology: Key issues for management. *MIS Quarterly, 15*(1), 32-49.

Kautz, K., & McMaster, T. (1994). Introducing structure methods: An undelivered promise? - A case study. *Scandinavian Journal of Information Systems, 6*(2), 59-78.

Loh, P., Marshall, C., & Meadows, C. J. (1998). High-tech/low-tech: Appropriate technologies for developing nations. *Journal of Global Information Management, 6*(2), 5-13.

Martinsons, M. G., & Westwood, R. I. (1997). Management information systems in the Chinese business culture: An explanatory theory. *Information & Management, 32*(5), 215-228.

Mata, F. J., & Fuerst, W. L. (1997). Information systems management issues in Central America: A multinational and comparative study. *The Journal of Strategic Information Systems, 6*(3), 173-202.

Nelson, K. G., & Clark, T. D. J. (1994). Cross-cultural issues in information systems research: A research program. *The Journal of Global Information Management, 2*(4), 19-29.

Niederman, F. (1993). Information systems personnel as expatriates: A review of the literature and identification of key issues. *The Journal of Strategic Information Systems, 2*(2), 153-170.

Palvia, P. (1997). Developing a model of the global and strategic impact of information technology. *Information and Management, 32*(5), 229-244.

Palvia, P. (1998). Research Issues in Global Information Technology Management. *Information Resources Management Journal, 11*(2), 27-36.

Ramanujan, S., & Lou, H. (1997). Outsourcing maintenance operations to offshore vendors: Some lessons from the field. *Journal of Global Information Management, 5*(2), 5-15.

Sankar, C. S., & Liu, L. (1998). Study of job characteristics and organizational progressiveness posture in Singapore and Chinese companies. *Journal of Global Information Technology Management, 1*(3), 27-42.

Shore, B., & Venkatachalam, A. R. (1995). The role of national culture in systems analysis and design. *Journal of Global Information Management, 3*(3), 5-14.

Shore, B., & Venkatachalam, A. R. (1996). The role of national culture in the transfer of information technology. *The Journal of Strategic Information Systems, 5*(1), 19-36.

Simon, S. J., & Middleton, K. L. (1998). Asia's pending labor crunch: An analysis of human resource management best practices in IS departments. *Journal of Global Information Technology Management, 1*(3), 9-26.

Straub, D. (1994). The effect of culture on IT diffusion: E-mail and FAX in Japan and the U.S. *Information Systems Research, 5*(1), 23-47.

Straub, D., Keil, M., & Brenner, W. (1997). Testing the technology acceptance model across cultures: A three country study. *Information & Management, 33*(1), 1-11.

Tam, K. Y. (1998). The impact of information technology investments on firm performance and evaluation: Evidence from newly industrialized economies. *Information Systems Research, 9*(1), 85-98.

Trauth, E.M. & Thomas, R.S. (1993). Electronic data interchange: A new frontier for global standards policy. *Journal of Global Information Management,* 1(4), 6-16.

Wan, H. A., & Lu, M.-T. (1997). An analysis of Chinese laws against computer crimes. *Journal of Global Information Management, 5*(2), 16-21.

Watson, R. T., Kelly, G. G., Galliers, R. D., & Brancheau, J. C. (1997). Key issues in information systems management: An international perspective. *Journal of Management Information Systems, 13*(4), 91-115.

Yang, H. (1996). Key information management issues in Taiwan and the US. *Information & Management, 30*(5), 251-267.

## Chapter 6

# Undergraduate Global IT Education — An Experiential Approach Using the Concept of Fit

Joan Mann
Old Dominion University, USA

*There are special considerations which are important when developing and teaching a course on information systems for international business, especially when the course is designed for undergraduates. This chapter demonstrates how to position and organize a global information systems course for undergraduates which will 1) be appropriate to the way undergraduates learn, 2) include the skills they need at this stage in their career and 3) create an affective change in their perspective. The primary approach used to produce these results is to demonstrate the importance of alignment between global strategies, organizational structures, the configuration of IS departmental areas, organizational information infrastructures, and national/organizational culture, as well as, alignment between national concerns, national information infrastructures and national information policy.*

## INTRODUCTION

Business schools and their accrediting bodies have long recognized the need to internationalize the curriculum. For example, in the United States the primary business school accrediting body created a standard in 1974 that called for all business curriculums to include an international perspective. Moreover, the federal government has put into place funding initiatives to support the internationalization of business education. For instance, it has designated several universities to be Centers for International Business Education and Research (CIBERS). A

Previously Published in *Challenges of Information Technology Management in the 21st Century*, edited by Mehdi Khosrow-Pour, Copyright © 2000, Idea Group Publishing.

primary mission of any CIBER school is to assist other universities in internationalizing their curriculums.

Information systems (IS) departments within colleges of business have started to internationalize but few universities, however, have a course at the undergraduate level. The primary objective of this paper is to present a course plan for an undergraduate global IS course that emphasizes the importances of *IS-organizational context alignment*. The first step, however, is to make an argument that a global IS course is needed and that it must be designed with undergraduates in mind.

## IMPORTANCE OF GLOBAL IS EDUCATION FOR UNDERGRADUATES

Undergraduates from most IS programs expect to become programmers and/or systems analysts. Curriculums therefore, tend to emphasize useful technical knowledge and analysis skills. Coursework in other business fields are incorporated so students will be able to design systems for businesses. Using this same logic, a case can be made for the importance of undergraduate global education.

The globalization of business will have a large impact on how the IS function operates. Global organizational information infrastructures are built upon national information infrastructures and therefore, the number of vendors and standards increases substantially. IS staff will find the array of technological choices both more diverse and more constrained. At the development project level, undergraduate IS majors need to be aware of this complex technological environment and given tools for making sense of it.

Our students are more likely to be relocated outside their home country. Users, technicians and executives will come from a variety of national backgrounds. Undergraduate IS majors need to learn how to function well in this complex social environment, especially when teams will increasingly communicate electronically.

Not only that, the turbulent global environment means IS functional areas must adjust people and technology to constantly changing organizational strategies and structures. Even though undergraduate IS majors will be primarily working on individual projects, a long term perspective is needed. One that connects systems being built today to future organizational needs. For example, implementation is a traditional topic covered IS curriculums. In a multinational enterprise (MNE), diffusion and technology transfer issues become just as important because systems will rarely be used in one geographic location.

So, if undergraduates need a global IS course, then how should we adjust the curriculum accordingly and how should it be taught?

## POSITIONING THE GLOBAL IS COURSE

One way to prepare IS majors for working in MNEs is to make repeated connections between the operations of MNEs and the technology that is available. For students to comprehend these connections, they must have an understanding of technological and organizational concepts. Therefore, the global IS course should be positioned towards the end of the student's career. An added benefit to positioning the course later, is that the business strategy course will be also taken about that time. These courses will become mutually reinforcing.

## CUSTOMIZING THE COURSE FOR UNDERGRADUATES

There are two issues to consider when preparing a global IS course. The first, is to understand the way business undergraduates learn. Using the Myers-Briggs personality type instrument, researchers have found that business undergraduates tend to be extroverted and sensing.[1] Extroverted students learn best when they are able to "talk things out." Sensing means they learn concrete concepts better than abstract ones (see trees first, have trouble seeing the forest), they want to know the "why" before the "what," and prefer experiential learning.

The second issue to consider is that IS undergraduate curriculums are project-oriented rather than enterprise-oriented. This means curriculums focus more on effectively creating systems than managing information resources. This focus is appropriate because it matches the personalities of business undergraduates and because they often have very little business. They can visualize the organizational benefits of a specific system and understand the importance of IS planning (see the trees and their relationship to the forest) but would find it difficult to prepare an enterprise wide IT plan (start at the forest level and build the trees).

The course described below, therefore, is customized to take advantage of the learning styles and backgrounds of IS majors and uses their project level training as a springboard to understanding how projects relate to the enterprise as a whole.

## USING AN EXPERIENTIAL (EXTROVERT + SENSING) APPROACH

Given that our students are extroverted, sensing and project-oriented, how best to design the global IS course?

Suggestion 1: Regularly explain where you are going and why. This suggestion comes directly from the need for these Sensing students to know "why" before "what." Figure 1 shows a diagram that can be useful for this purpose:

Each number represents a course module. Students are told that one needs to understand all four major entities (the environment, firms, nations, IT) and their

*Figure 1: Visual Representation of How the Course is Structured*

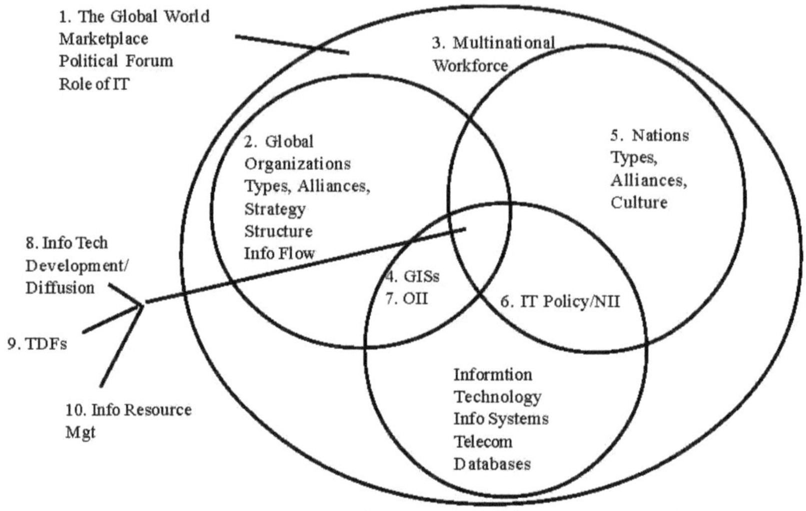


dynamics to be successful. At the beginning of each module the diagram is shown again and the new module is connected to the previous and future modules.

Suggestion 2: Use a seminar approach by including regular small group discussions. This both supports the extroverted learning style and allows students to use business knowledge gained from other courses which supports sensing — gives them the why, brings abstract from the concrete.

Suggestion 3: Give students articles to read before class and have them write-up questions before class. It is difficult for a lecture to capture the richness of the global environment, the global operations and national conditions. The readings let the student *experience* what it is like "out there" and ensures they are prepared for discussions.

Suggestion 4: Make each student an expert. Have them research an industry, a country and a type of global information system so that each knows something that the other students do not. Encourage students to share what they know of the world from international travel, friendships, work experience.

Suggestion 5: Have them experience how MNEs operate. Use listservs, chatrooms and e-mail to host discussions with students or even better have them communicate with people internationally.

Suggestion 6: Design the course around the concept of fit (see below)

## USING THE CONCEPT OF FIT

It is very important for students to understand how IT impacts the global environment and the actors within it. This may be difficult for students. When asked to analyze an industry, many undergraduates fail to understand that while an industry encompass individual firms (trees), an industry report must identify patterns across firms (the forest). To adjust for this problem, the fit approach to learning has two phases:

1. Coverage of individual concepts (the trees)
2. Exercises/discussions/exams that helps students make connections between concepts (build the forest).

For example, in the organizational context module, we cover various global strategies, structures and how structures help define information flows (trees). Students are then asked to match specific structures to the different strategies. For example: A multi-national strategy (focus on localization) is matched to an area structure (geographic divisions each with their own functional departments). When students grumble that this is not IS related, then we discuss whether the information systems built would be different for these different organizations. We also do an exercise that diagrams information flows in MNEs (remember, they are sensors) and how flows change over time. This helps them to visualize (sense) the complexity and dynamism of MNEs. It also solidifies the connection between strategy, structure, information flows and information systems.

When we cover different types of global managers (product, function, country) we match the roles of each manager to the structure and information needs of an MNE. We also have a short session about working in another country and how IS development is increasingly being performed remotely. These two topics reinforce that they may find themselves relocated and will have to adjust to fit this new environment.

In the module on nations, we cover the different levels of development and ways nations interact with each other and with MNEs. It is emphasized that nations have developmental objectives which involve IT. The fit concept is used when we try to determine whether a nation's information policy and infrastructure fits the its goals. For example, Austria's goal is to become the information hub for Europe. This policy fits their central location.

The use of the fit concept is also useful when discussing the IT components of the course. Each student is asked to read a paper on a different type of global information system (ex: customer support). We then have a discussion on whether global ISs are truly different from domestic systems. Students invariably mention that global systems are different because their organizational context is different (fit!). In the connectivity module, we examine how an organizational information infrastructure (OII) can help a firm implement its strategy and improve the functioning of its structure. Once again, the emphasis is on fit. Does the OII fit the firm's strategy/structure? It is also stressed that OIIs must be flexible because organizations are always adapting to a dynamic environment.

When examining IS development, we first cover the role of culture in any development project. Two cases are used to emphasize the importance of cultural fit and we examine issues of fit when developing graphical interfaces based on a text by Fernandez on Global Interface Design.

Finally, to bring all the threads together, we discuss global IT management. The students read an article that lists several IT changes made by firms and we trace them back to changes in strategy/structure. We also do a matching exercise that has the students determine which IT configuration and IT organizational structure fits each strategy/structure configuration. By the end of the course, the student can no longer envision the development process without thinking about the organizational context, including its international aspects.

## CONCLUSION

If one uses the concepts outline above, students will be exposed to important topics in a way which requires little memorization. This is important because given the technical nature of most IS curriculums, students tend to see this course as marginal. By keeping preparation time low but including assignments that ensure students are, in fact, prepared and by having them learn through discussion, the course should become more interesting and have a greater long term effect on the students.

The use of fit to encourage students to become more abstract can also be beneficial. Students learn they can take the individual bits of knowledge from other classes and put them together to examine their organizational context. The generalizations they create and the process used to create them can then be used throughout their lives.

## NOTE

[1] Brightman, H. and Hightower, R. (1989). Improving Teaching in the EMBA Classroom: Some Empirical Findings, *Organization Behavior Teaching Review*, 14(1).

## Chapter 7

# European Electronic Service Infrastructure Building – Drifting Into the Future?

Åke Grönlund
Umeå University, Sweden

*Delivery of electronic services requires a service infrastructure including organisational solutions for logistics and client-organisation interactions. We report a study covering three years of efforts by nine cities in eight European countries in developing such solutions.*
*We found twelve distinct "crises," situations where goals, environment, and actors changed, and the process was found in a stage of improvisation. The overall process was largely unstructured and improvised. Stabilising factors were central government policies (national, EU), the general technical development, and market demands.*
*The Web was in practice a "cuckoo in the nest," intruding in ever more activities and introducing unexpected new demands. Web projects were seen as technical projects; more important success factors pertaining to users and organisation were largely neglected. There was typically an infrastructure link missing, a body competent of managing the whole process of bundling services from different service providers.*

## INTRODUCTION

Many local and regional governments have followed companies in setting up Web information systems (WIS). The purposes are often boldly stated and cover economic, social, and democratic ambitions:

"The public administration shall use IT for making *more effective operations* and providing good service to companies and citizens. More *rational work routines*, more *effective organisation* and *co-operation in the public agencies*

Previously Published in *Challenges of Information Technology Management in the 21st Century*, edited by Mehdi Khosrow-Pour, Copyright © 2000, Idea Group Publishing.

shall *improve services* and at the same time *reduce costs.* [......] IT shall be employed to *develop contacts and interplay* among the public, businesses and the public administration. Citizens' and businesses' public sector *contacts* shall be made *simpler* and more *open.*" (Gov. Bill No 1995/96:125; Toppledarforum, 1998. My italicisation and translation)

Such goals are inscribed in central government IT strategies in many countries. At the local level, however, strategies are often lacking, or outdated (Håkansson, 1996; Hansson & Johansson, 1997). This is a problem, because much of the development is in the hands of local authorities.

What happens during the process of developing electronic services; what hurdles are there, and how to get past those?

Case stories abound (IDPM, 1998), but they usually only describe ongoing projects with little analysis or comparison. Research typically falls into one of two categories: (1) exploratory sites working under non-typical conditions (i.e., Virginia Tech, 1998; Tsagarousianou et al., 1998). In such cases, special conditions typically make results less general; extraordinary resources, subsidies to providers and/or users, special user populations, etc. (2) single special issues of a technical character are covered; fraud prevention, integrity of data in electronic document exchange, public access to official records, etc.

While important, such studies give only fragmentary insights. What about comprehensive strategies for using IT for achieving the bold purposes? At the end of the day, governments must find ways of relating to WIS as a strategic medium. How can development methods be found, that is, how can early experiences be shared, evaluated, related to societal goals, and debated so as to make the transition of services into the electronic medium smooth? It seems the development so far happens much ad hoc (Greeves, 1998). Strategy documents of the kind cited above abound, but knowledge about implementation beyond things like guidelines for home page design are largely lacking.

For a city to provide services electronically is a much more complex task than it is for a company, as it encompasses a wide variety of activities and includes elements of democracy and exertion of authority as well as economic goals.

We report an attempt to examine the hurdles to successful use of the electronic medium in a local government setting.

## INFRASTRUCTURE DEVELOPMENT: "IMPLEMENTATION," "DRIFTING," OR WHAT?

Many see technology deployment in an organisation as a rather straightforward matter of "implementing decisions" (Broadbent & Waill, 1997). From that point of departure, the problem concerns how to make things happen the way they were intended to; decisions must be "kept alive" Latour (1996, p 86). Ciborra

(1997), on the other hand, claims that technology "drifts;" during the "implementation," numerous decisions are made, by many actors, that make the original "grand plan" obsolete. New appropriations of the technology are made along the road, which makes the outcome considerably different from the intended, even "almost out of anybody's control" (ibid, p 76). Others claim that these reappropriations are in fact not free and unpredictable, but are structured by the "ongoing effort to keep the (gradually metamorphosed) infrastructure alive" (Monteiro & Hepsö, 1998, p 270).

Given this spectrum of views, we are here concerned with how cities made the central governments' grand plans happen.

## METHOD

This chapter summarises a three-year (1996-1999) empirical study of nine European cities in eight countries developing telematic services within the Infosond project (EU contract UR 1017). The empirical material is presented in more detail in Grönlund (1998), Grönlund & Forsgren (1999), and Grönlund (2000).

A large number of cases were collected. Project leaders in the nine cities were interviewed at their bimonthly meetings. We did a usability study of selected systems in 1997 (Grönlund, 1997) and studies of use, management, and economic evaluation in 1998 (Schijvenaars, 1998). Further materials were project reports, minutes from meetings, and documents from the cities, like infrastructure plans and regulations.

In spring of 1998 a group including researchers and people from the cities reviewed the development process and analyses led to the "dozen crises model," briefly presented below.

Over the following six months, the material was further analysed. On three occasions, the report was presented to a wider audience of project leaders in the cities: the model, the description of the crises, and the use of the cases. This was done to ensure people understood our model and agreed with our interpretations, and that we had used their cases appropriately.

## DUE PROCESS AND "THE DOZEN CRISES MODEL"

According to Actor-Network Theory (ANT), the creation of *facts* comes about during processes of negotiation among *actants* (human or non-human actors). Facts are not found, they are introduced into a social context as *claims*, which are translated by enrolment and inscriptions of networks of actants. A growing network of actants are involved as claims get strengthened and weakened (McMaster, Vidgen, & Wastell, 1999). Actants are heterogeneous; the strength of the network relies on the strength of the actors' aligned interests.

*Due process* is the process by which a claim gains support or gets excluded.

Introduction of a claim brings *perplexity* to the network (or organisation) where it is introduced. Follows a period of *consultation* where the claim is considered by the actors in view of other claims and existing facts. On the agenda is the claim's place in the *hierarchy* of things. If accepted, the claim becomes institutionalised; a *fact*. If not accepted, it gets rejected; in order to be reconsidered it will have to go through the due process once again (McMaster, et al, 1999).

Although networks evolve constantly, it is feasible for the purpose of making the process visible to focus on certain instances in the development. *Freeze frames* represent *inscriptions* of a technology in a particular moment.

The "crises" represent the "freeze frames" of the development we saw. Each of them represents a completed step in the development in terms of activities and goals in the cities. In that way they represent stable networks (shared views of the technology). Seen in a larger perspective, they were only steps towards the goal of proficient electronic management.

"Stability" means that people in the organisation felt a problem encountered was solved. It usually was, but because challenges ahead were not foreseen, the solutions often contained a seed to problems during the next stage of development.

## FINDINGS

We identified the following 12 "crises" (unforeseen situations requiring profound changes) on the road towards proficient electronic service management:

1: "Start-up….of what?" (Initial motivations differ from final)
2: "Thousands of pages….." (Institutionalisation of production)
3: Messy appearance (Graphical design, Web organisation)
4: Parallel systems (How to phase out old routines?)
5: Choice of future technical platform
6: Cross-departmental integration of data resources
7: Staff motivation
8: Poor usability
9: Where is the payoff? (How to measure costs and benefits?)
10: From monopoly to service provider (Organisation faces a role change)
11: Where are the users? ("There is not much use yet – when will it come"?)
12: "Administrative tribal struggles" (social groups feel their domains intruded upon)

In the following, we briefly describe the nature of the crises; the main actors and the main conflicts. We exclude crises 7, 8, and 9, which are not so interesting in terms of actors and networks.

### Crisis 1: Start-up….of what?

Cities enter the Internet largely because others are there. Goals are specified, but are usually short term and, at best, only remotely related to better services or more efficient logistics. Examples of early goals include providing jobs to unemployed (creating Web pages), establishing a Web presence, keeping up with others, etc.

Systems are typically initially set up by some enthusiast(s) and funded by project money, which means they do not automatically outlive the project. If a project is successful, there appears a need to integrate it in other IT and organisational projects going on. This typically does not happen easily.

As a consequence of the lack of goals, it is hard to focus and provide good contents. A typical Web project starts off with the Chief Information Officer (CIO) in the organisation publishing things like the company brochure on the Web. (S)he does that alone with technical support from someone; a consultant, a project group rallied up from a local university media education programme or from a programme for unemployed, or from the computer department.

Only after this is done, the debate on the purpose of the Web system. Why should we be on the Web? The CIO comes up with a number of reasons, typically requiring organisational change at business departments. At this point, it becomes clear to several people in the organisation that the Web system will cost, if they go along with the CIO's suggestions. So they begin asking why they should.

In this phase, there is just a small group of actors. Legitimacy of the project is found in places at best vaguely related to the overall goal (not yet formulated). People who are needed to formulate such an overall goal are typically not involved, or are involved in a way that in effect conceals that goal. Thus, success or failure of the project is at this stage measured against the "wrong" goals.

### Crisis 2: Thousands of pages.....institutionalisation of Web production

Once a Web site is established, there is typically no special budget allocated to further Web related work. The task of maintaining all the pages quickly overwhelms the small team of active staff. As a result, the task of overseeing the accuracy of the information is delegated to the business departments.

Additional problems are that Web technology changes, which means systems have to be redesigned and demands for information and services increase. Appears a need for automation and integration with existing systems.

Production may be a centralised or distributed. There is a need both for delegation of work and co-ordination, but the decision on which way to go can typically not be decided freely. There are traditions, and there may be changes of direction ensuing from political decisions that entail some actions and prohibit other.

At this stage, tradition typically conquers innovation. We see a number of actors entering the scene; local entrepreneurs promoting departmental interests, central bureaucrats and politicians striving for control, and technical consultants promoting their products, which have different focuses and overlapping scope. There is typically not yet a network of people in the organisation with the goal of "proficient electronic services management," but there are several networks with departmental interests. These now engage in discussions on how to "share the burden" of production of the new services. Thus, the organisation of production is negotiated on a "something for everyone"- basis, not in terms of efficient logistics or improved services.

### Crisis 3: Messy appearance/Web organisation

As systems grow and information providers multiply, responsibility for information provision and updating is delegated. This results in each department wanting to do things its own way. Arises a co-ordination problem; top management wants the organisation to appear in a coherent and stylish form on the Web, following a corporate profile. This interferes with work already done, as well with different ambitions at different departments.

Arises a conflict between departmental interests in service quality, customer contacts, and perceived business mission, and top management's strive for uniformity and rational production overall. Often, a "graphic profile committee" is set up to resolve the conflict. Sometimes - rarely - independent evaluation is involved. This often happens when some external project (for instance EU funding for some user-oriented service) so requires, only occasionally by own initiative.

### Crisis 4: Parallel systems

By now, the graphic profile committee has made the system look nice. Contents are produced efficiently, and at least some people use the services. But all manual operations are still in place and run like before. The Web system has not replaced anything. Arises the first cost crisis: "Why is there no process reengineering? The banks introduced automated teller machines, and this made less people go into the offices, allowing for rationalisations. Why does not this happen with our services?"

Re-engineering does not happen without determined measures being taken, since the preserving forces in an organisation are typically stronger than those striving for innovation. One problem here is the scope of reengineering; Process? Department? Organisation? Again, networks at department level are invoked to preserve status quo.

### Crisis 5: Choice of future technical platform

Actors: The IT department, different consultants, different business departments, EU projects.

Conflicts:

1) There are technology related choices to make in order to reduce the risk of having to make drastic changes in the future – what database, what Web platform, etc. Still vivid in this debate is the controversy between proponents of the Web and of proprietary systems, like kiosk systems.
2) EU projects often come with a more offensive agenda than that of the IT department. While the former is often directed towards use and service reinvention, the latter is typically defensively focusing on security, standardisation and maintenance.
3) System metaphors. Different business departments typically employ consultants from different trades; Web design/PR, databases, infrastructure, business solutions, etc. These all claim to be in the "IT" business, but they have very different working metaphors for what they build– a library, a database, a newspaper, a brochure, a traffic system, etc. When different departments who have walked the WWW road some time with different consultants at some point have to merge their IT activities, clashes in world views occur. By then, the metaphors have been implemented in software and operations to a considerable extent, making shifts expensive and complicated.

It is typically not until this phase that IT infrastructure strategies are discussed with Web services as an ingredient.

**Crisis 6: Cross-departmental integration of data resources**

Producers often find that it would be useful to have access to some data possessed by other city departments. For instance, tourist information could be better presented by using maps, which already exist as GIS systems owned by the city planning office or the like.

Appears a need for sharing resources, but how? The answer in practice depends a lot on what legacy systems there are, but also on how far the different departments have come towards telematic service publishing. Those who have come far are likely to want to pursue their way of doing things (often by old technology). Those who have not come so far are typically more likely to try new ideas. Neither is willing to become subordinate to the other.

Problems of this kind are often created by organisational innovations like departmental reforms and internal trade regulations that make cross-departmental use of resources expensive to the individual department.

At this stage, typically another committee has to be set up to negotiate use of "common but divided" resources. Substantial knowledge and visions in the technological field are required in this committee; pure mediation may well lead to old technology prevailing and/or the committee becoming just a forum for inter-departmental struggles.

### Crisis 10: From monopoly to service provider

City organisations, and staff, are used to work in a monopoly environment. They are not used to thinking about what users look for, how they look, and so on. The Internet culture, totally different in this respect, as well as deregulation leading to competition with other service providers has begun to demand a change of attitude.

This crisis is only resolved when the city finds its role in the new environment. Different cities may assume different roles; for instance, rural towns often play a more important role in supporting local small business than large cities in prosperous regions do. Only when the role of the city is found can the appropriate roles of the actors in the organisation be properly designed.

### Crisis 11: Where are the users?

Many ask themselves about their Web systems: "There is not much use yet – when will it come"?

In our study, we saw that the "media chain" in which the organisations under study worked was missing a crucial link. The development in Web business is towards concentration and exclusion; a few global companies are gaining ground, not least due to the policy of many Internet providers to grant exclusive contracts. Small companies working on a local or regional basis are largely excluded. A problem in the environments we studied - typical for Europe - is that cities are not concentrating on finding and packaging local content, one of the crucial catalysts for getting people on-line. So far, there is not the equivalent of a book publisher or television or film distributor for emerging on-line services, not with a regional focus and ambition. The small company or organisation that want to go on the Web will have to go alone, which is more often than not beyond its capacity.

### Crisis 12: Administrative tribal struggles

Man has only recently emerged from a tribal existence. Loyalty to your clan was the leading guarantee of human survival for thousands of years. This legacy sits deep within our subconscious, and is the source of much of the excitement, but also antagonism, at work.

One of the principal problems of the modern society is the difficulty of identifying a tribe to belong to. We feel many sympathies and loyalties, and thereby identify ourselves with many different "tribes" of like-minded people. However, it is not uncommon that our various tribes end up in conflict among themselves. In such a situation, our loyalties are torn, and we are forced to take sides, often against our will.

The stated goals and business practices of many professional organisations are widely disparate. Municipalities may openly wish to serve their members efficiently and well, always seeking for ways of improving their service, and cutting unnecessary costs. As a professional community, however, city officials wish not to rationalise away city jobs, because that would decrease state subsidy to the city, as well as its relative economic importance in a region. It would also aggravate the city's financial position, by pushing former employees into the ranks of the unemployed.

In such a situation, members of the corresponding "professional clans" are torn between loyalty to their explicit professional mission and that to their professional community.

Introducing telematic services amounts to changing established service patterns. The professional staff influenced by such a perturbation will be divided by their tribal loyalty. If the tribal chief or council concludes that such services will do more harm than benefit to the tribal community, the telematic service will prove cumbersome and awkward to use, and definitely too immature to be seriously considered for operational adoption.

## DISCUSSION AND CONCLUSIONS

We found twelve distinct "crises," situations where goals, environment, and actors change, and the process is found in a stage of improvisation until new stability is achieved. We saw that:

1) The Web was in practice a "cuckoo in the nest," intruding in ever more activities and introducing new demands on them.
2) This generally came as a surprise.
3) There was typically a missing link in the organisations' infrastructure to deal with electronic services. Web projects were considered technical projects; though in fact issues pertaining to users and organisation were most important, they were initially largely neglected. The missing link was a "electronic service manager" (ESM), a body competent of managing the whole process of bundling services from different service providers and publishing them in a coherent fashion, providing support to service providers during the process of inventing, refining, and evaluating services, improving logistics, and conducting the necessary but typically ignored activities of analysis of the quality of services, and policy making.

All along, there were a couple of stabilising factors. Most important, we think, was the general technical development and market demands; The Web has continued to grow and improve technically, and people today demand Web services also from government.

A second stabilising factor was central government policies. All the studied countries had national government policies of the type cited in the introduction. Also, in every city there were several EU funded projects where electronic services and the Web were understood as the way to go. These projects influenced cities by their goals and activities, but also because they over time have fostered a cadre of "Web champions," city staff with international contacts and visions of technology use for service improvements.

Of the crises identified, ten out of twelve (all except no 5 & 9) belong to the area of users and usability of services and the ensuing demands on organisations. Yet projects were pursued as "technical projects," not as service projects. Why was no business innovation entity (the "ESM") set up? We have discussed the development in terms of how different actors influenced the view of technology at different stages of the development. We saw that these views changed considerably over time, and that the changes were not planned or foreseen. They evolved as new actors entered the scene. There was never an overall tactical plan which outlined "the road towards proficient electronic service management," just plans for "implementing a Web system," "making our system usable," etc. The cities acted reactively, often successfully, but typically not proactively.

Was infrastructure "implemented" or did it "drift" due to events down the road; we saw that there certainly was a lot of drifting if we see the different crises in relation to the overall goal of professional telematic service publishing. But that goal appeared only late in the process. There was no grand plan. There were a number of activities, including plans for short-term developments, which together led to the development of an infrastructure.

Looking at each "crisis," the drifting was not appalling. Certainly there were conflicts, and certainly departmental interests, or even those of individual actors, made a great difference. But in most places solutions to the problems encountered at a particular crisis were found, that were digestible for all parties.

There is a great lack of strategy in the field of electronic services in local governments in Europe. This is a major problem in the light of the importance of that sector and the challenges it is currently facing.

The electronic services business in Europe's public sector is a cottage industry. It needs to be modernised. The solution is not more home pages but a service perspective professionally and visionary implemented.

## REFERENCES

Broadbent, M., & Waill, P. (1997). Management by Maxim: How Business and IT Managers Can Create IT Infrastructure. *Sloan Management Review*.

Ciborra, C. (1997) De Profundis? Deconstructing the Concept of Strategic Alignment. *Scandinavian Journal of Information Systems*, 9(1), pp 67-81.

Greeves, R. (1998). *The Penultimate Mile: Local and State Governments Collaborating to Serve Citizens Through Information Technology.* http://www.excelgov.org/techcon/sldoc/index.htm.

Grönlund, Å. (1997). *Report on usability and estimated usefulness.* Infosond Deliverable D8. Antwerp: InformatikCentrum. http://www.infosond.org.

Grönlund, Å (1998). *Make IT happen – a tale of a dozen crises.* Infosond project, Report D10. Antwerp: InformatikCentrum.

Grönlund, Å (2000). *Electronic Service Management - a practitioners' guide.* London: Springer (In press).

Grönlund, Å. & Forsgren, O. (1999). *Portalbyggarens Klokbok.* B1999:2. Stockholm: NUTEK.

Håkansson, S. (1996) *IT-strategies in Swedish Cities.* Stockholm: Näringsdepartementet, Struktursekretariatet.

Hansson, L., & Johansson, M. (1997). *Municipal IT-strategies in Västerbotten.* Umeå university, dept. of Informatics, report SPC 97.55.

IDPM (1998). *Government and public sector pages.* Institute for Development Policy and Management, University of Manchester. http://www.man.ac.uk/idpm/devtlinx.htm#itgov.

Jundin, P. (1997). *Personal communication,* October 10.

Latour, B. (1996). *Aramis or the Love of Technology.* Cambridge, MA: Harvard University Press.

McMaster, T., Vidgen, R.T., & Wastell, D.G. (1998). Networks of association and due process in IS development.. In T.J. Larsen, L. Levine, & J.I. DeGross, *Information Systems: Current Issues and Future Changes.* Proceedings of ICIS 1998, Helsinki, Finland. IFIP, Laxenburg, Austria.

Monteiro, E., & Hepsö, V. (1998). *Diffusion of Infrastructure: Mobilization and Improvisation.* In T. Larsen, L. Levine, J.I. DeGross (Eds.), *Information Systems: Current Issues and Future Changes,* pp 255-274. IFIP, Laxenburg, Austria.

Schijvenaars, T. (1998). *Report on economic evaluation and system use.* Infosond Deliverable D8. Antwerp: InformatikCentrum. http://www.infosond.org.

Toppledarforum (1998). *Government proposition 1995/96:125. Measures for broadening and developing use of information technology* http://toppled.nutek.se/itprop.html.

Tsagarousianou, R., Tambini, D., & Bryan, C. (1998). *Cyberdemocracy: technology, cities and civic networks.* London: Routledge.

Virginia Tech (1998). *Blacksburg Electronic Village Research.* Virginia Polytechnic Institute and State University, http://www.bev.net/research. Visited 1999-09-28.

## Chapter 8

# Comparative Study of the Usefulness of Online Technologies in a Global Virtual Business Project Team Environment

Simpson Poon and Shri Rai
Murdoch University, Australia

*Lifelong learning is quickly becoming an integrated part of today's working life because of the demand for keeping up to date with latest developments due to rapid change in technology and business. To maintain flexibility and quality, online technology is often used as a medium of educational service delivery. In this chapter, we examine a case where online technology has been used to coordinate virtual project teams (in an educational setting) around the world. The experience from the case study is that although online technology promises to offer an independent learning environment anytime and anywhere, only some aspects of the technology are useful depending on the nature of the task. In addition, there are behavioural and cultural issues, which can be exacerbated by underdeveloped personal relationships due to constraints imposed by online technology. The case study also shows that synchronous communication is not necessarily a better means to coordinate than its asynchronous counterpart. A challenge faced by online education providers is to find the right mix and how to come up with a framework that will provide optimal results.*

## BACKGROUND

The funding for universities in Australia has dropped from 89-90% fifteen years ago to about 50% by the end of 1999. At the same time, demand for education has increased simply because of the rapid pace of technological change and new business innovation. In the area of Electronic Commerce (E-commerce), the demand for courses has grown rapidly, from hardly any subject offered in 1997, to almost all universities offering some form of E-commerce education. This is not only an Australian phenomenon. Universities around the world have come to realise that e-commerce education is very much in demand. Most business schools now have students enrolled in e-commerce.

Murdoch University is one of the four universities in the city of Perth, Western Australia. Perth is one of the most remote capital cities in Australia and in the world, being more than 4,000 km from the country's other capital cities such as Sydney or Melbourne. Despite the tyranny of distance, Perth has a vibrant group of universities and is particularly active in E-commerce education.

Murdoch University is a medium-size university by Australian standards and is located in Murdoch— a suburb of Perth. The university is named after an academic and essayist Sir Walter Murdoch. "Planning for the university began in 1970. The second university to be established in Western Australia and the seventeenth in Australia, it was constituted on 25 July 1973 by an act of Parliament of Western Australia. In 1975, its inaugural year, Murdoch University enrolled 672 undergraduate students, had 6 full Professors, a dynamic young staff on a 254 hectares campus. Today, Murdoch offers more than 50 undergraduate programmes (majors) of study, has over 11,000 students enrolled, employs over 50 full professorial staff and includes a regional campus located at Rockingham" (*http://www.murdoch.edu.au/about/murdoch.html*).

According to the introduction on its Campus-Wide Information System (*http://www.murdoch.edu.au/about/murdoch.html*):

> Murdoch University combines the best in university teaching with outstanding graduate satisfaction and a flexible outlook that let students design the degree they need for their future. The only institution in the country to have scored 5-stars for graduate satisfaction from the Good Universities Guide, for five years in a row, Murdoch has also been rated the best teaching campus of all Australia's public universities by an independent national survey of university graduates. This high standing was enhanced by two awards in the Prime Minister's 1998 Australian Awards for University Teaching. These honours constitute a "user's bouquet" through which its students declare their belief that Murdoch provides one of the very best learning experiences in Australia. This progressive and dynamic university offers courses in: Biological and Environmental Science; Veterinary and Biomedical Science; the Law; Business and Information Technology; the Humanities; Mathematics; the Physical Sciences; Engineering; Education; and the Social Sciences.

*Figure 1: Student Enrollment Numbers in Distance Education at Murdoch University*

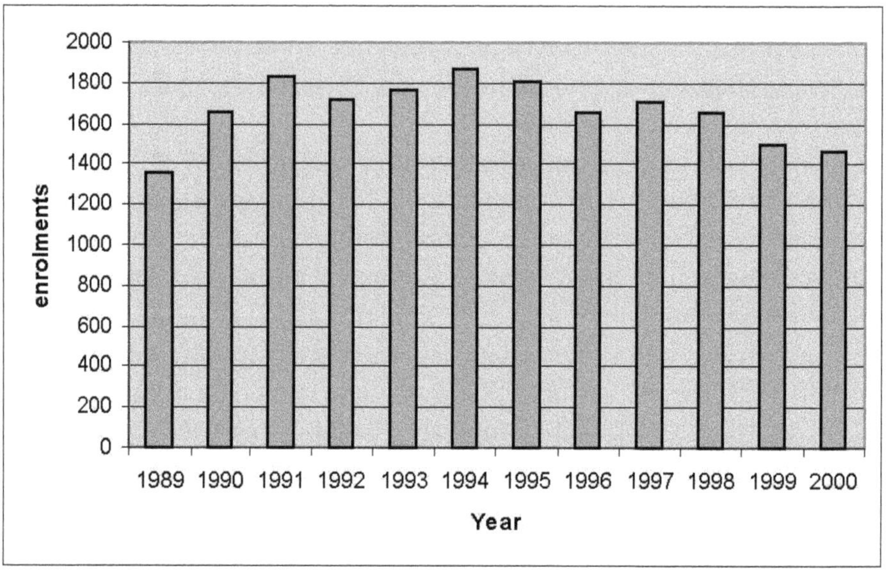


In addition, this community-orientated university offers a particularly flexible degree structure which encourages students to boost their job prospects by completing two programs of study, e.g., Commerce and Mass Communication, Chemistry and Mineral Science.

Distance education at Murdoch commenced in 1975 (inaugural year) with six subjects. In semester 1 2000, there were 214 units on offer covering many discipline areas. Figure 1 shows the enrollments for distance education since 1989.

E-commerce teaching is mainly done through the Business School. The school is offering a Master of Electronic Commerce and a specialisation in its undergraduate degrees. The Master of E-commerce is a cost-recovery program for executives and the fee as of year 2000 is approximately US$10,000 per year. Cost-recovery courses of this type are popular because they address emerging issues or areas with a high demand.

## SETTING THE STAGE

The use of online technology to support educational activities has received much attention due to the popularity of the Internet (Goldberg, 1997). The Web has opened up enormous opportunities to allow teaching and learning to be carried out in innovative ways where geographical distance and time constraints are no longer barriers. The traditional model of having everyone sitting in the same place to listen to a lecture is no longer necessary. With a computer and a modem, teachers and students can interact with each other anywhere anytime. Michael Zastrocky (research director/academic strategies) of the GartnerGroup concludes that "Students are 'going to school' less and getting more content electronically". His forecast is that this trend will increase. He also points out that there is a 70% chance that "Competition will push 80% of traditional higher education institutions in the US to deliver 60% undergraduate content through distance learning by 2003" (GartnerGroup internal presentation).

Breaking through geographical limitations and allowing learning to be done remotely is not new. The development of distance learning and education by sending paper-based teaching materials to students and/or with open learning programmes on TV has been an effective way to provide off-campus education. In fact, under certain circumstances, this is still the most effective way to diffuse education to remote communities. What makes the Internet and online technology so unique is the ability to use a low-cost medium to provide highly interactive teaching and learning. Such interactivity can be synchronous or asynchronous. Synchronous interaction normally happens in real time, such as using online chat systems, which allow teacher and learner to interact instantaneously. This kind of environment is important if the mode of study is of a highly interactive nature such as a class discussion. Asynchronous interaction happens when communication is not happening instantaneously. An example of such interaction is using electronic mail or bulletin boards, where both the teacher and learner send messages to each other to be read and responded to at leisure. Asynchronous interactivity provides a non-intrusive way of communication such that both parties can have time to reflect before responding.

Murdoch University puts a lot of emphasis on distance education. Distance education means students studying externally without physically coming on campus for lectures or tutorials. Traditionally this was done using paper-based courseware that was sent off via postal systems. For some subjects, such as those taught in the areas of IT and Media Studies, sending off floppy disks, CD-ROM, audio and videotapes as well as printed materials were the only means to conduct distance education. The External Studies Office (ESO) manages distance education at Murdoch University. The office acts as an intermediary between students and academic staff. Students enrolled in distance education are called "external students".

Over the latter half of the 1990s, the Internet has provided valuable opportunities for educational institutions to deliver off-campus education online. Murdoch University, given its strategic position, has seized the opportunity by initiating the Murdoch Online project. According to Murdoch's description:

> Murdoch Online provides a gateway into the University's use of the Internet for teaching and learning purposes…
>
> We seek to integrate the Internet's capabilities for rapid communications, multimedia delivery and access to worldwide online resources, with capabilities in quality of teaching, and in personalised, individualised communications and support. Diversity of users is one of our most valued characteristics, as online units provide for full time and part time students, for on campus and external students, for novices and experienced users, and for school leaver and mature age students.

Using a commercially available virtual learning environment, WebCT (www.Webct.com) as well as a range of other means (in-house), the university provides over 90 subjects in Web-based mode. The Murdoch online project involves the migration of Web-based courses to WebCT. Students can study a significant portion and in the future the whole degree through WebCT. For example, the Computer Science degree can be obtained by study in fully distance education/online mode. Such flexible and accessible approach to education delivery is gaining popularity. There is a strong demand—particularly among students who are currently employed—for shifting from traditional on-campus mode to online mode of education.

For subjects in the areas of e-commerce and Internet studies, using online technology as a delivery medium is a natural development. One advantage of online delivery is that students who are doing such subjects either already have experience in using the Internet or will be doing so as they progress through their course of study.

The management of the university (both at the chancellery and deans' level) sees online education as a strategic development and with significant competitive implications. The Teaching and Learning Centre (TLC) manages the online initiative and the group consists of technical support personnel, online course designers, educational specialists working together with academics to achieve the best outcomes for both staff and students.

The Teaching and Learning Centre and the External Studies Office are somewhat independent of each other. It would be interesting to observe their impact on each other as the online project managed by the TLC nears completion. For example, the ESO requires all assignment submissions from external students to go through them because they provide assignment tracking and reporting facilities (to ensure good turn-around time for submitted work) as well as providing administrative support for external students and academic staff dealing with external students. WebCT provides facilities for assignments to be submitted directly to the academic staff marking the assignment. With the use of WebCT, there is then no outside body ensuring the timely return of the marked assignments.

## CASE DESCRIPTION

In 1999, two new subjects in the area of electronic commerce were established. One was at the introductory level, providing students with initial understanding of the Internet and related technology such as the Web browser, HTML, TCP/IP protocols, multimedia applications over the Internet. Here we call it Subject A. The second subject (relates to this case) was about understanding the business models of E-commerce and the management, legal and governmental issues related to E-commerce and the Internet (here we call it Subject B). Given both subjects involved teaching about the Internet and E-commerce, the logical approach was to conduct much of the teaching and tutorial activities using the Internet itself. Subject A was run using WebCT. As Subject A did not involve team-based projects, not all the findings from this subject would be relevant here. However, we believe that some of the findings from Subject A would be generally relevant to all online coursework. Over 80% of the students preferred to attend classes as well as have all the course material online. Just over 1% percent wanted to attend classes with nothing available online and about 10% did not want to attend classes with everything available online. As most of the students were enrolled in "internal mode" (class attendance needed), the students seemed to be saying that they wanted (if given the choice) "face-to-face" contact as well as having course material available on demand. Therefore it appears that pure online mode of enrollment is attractive to students who for some reason (perhaps working) or other are unable to attend classes. This conclusion substantiates the authors' experiences when handling queries from students wanting to enroll.

In Subject B, a slightly different alternative was used. Instead of using a preestablished environment, standard Internet tools such as e-mail, FTP, subject Webpages, ICQ and mailing lists were used. WebCT was not used as there were many project participants who were external and it would have been difficult to insist on all the participants using WebCT because of the steep learning curve for the participants and the costs involved for the organisers.

The key assessment item in Subject B was a virtual team project. Students from 12 universities in Australia, the United States, Latin America and Europe (see Figure 2) worked together in small teams, each with four students (one per university), on development of a business plan. Each global virtual team (GVT) was free to choose a business venture, but the plan was supposed to be about a venture that used E-commerce in one or more aspects of the company (e.g., marketing, delivery or customer support). This case study examines the communication issues involved in working on a project in a GVT environment. The communication issues were not part of the subject or of the project itself. We were interested in finding out how effective faceless communication could be. The unique nature of the GVT project provided us with a good opportunity to carry out this study.

*Figure 2: The Set-up of the Global Virtual Team Project*

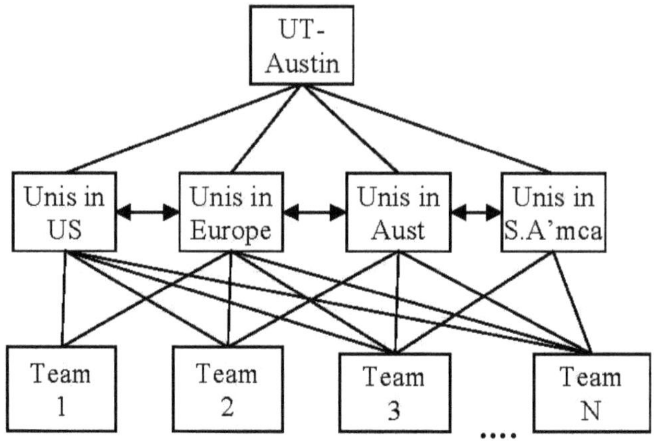


We studied the communication technologies used and looked at some related communication issues unique to working in virtual teams.

Six online technologies were used to support the working of each Global Virtual Team (GVT) and they were:

(1) Subject mailing list
(2) Team mailing list
(3) Person-to-Person (P-to-P) email
(4) Interactive chat
(5) Forum (a.k.a electronic messageboard)
(6) Group Webpage (display of completed work)

*Subject mailing list*

A subject mailing list was set up within each local university. The purpose of the subject mailing list was to enable announcements to be made to the class and mutual support to be built for resolving subject matters. All local students were asked to sign-up onto the subject mailing list so that announcements and updates would not be missed. The subject mailing list was set up using Majordomo listserver software and an archive of the mailing list was kept for historical reference. Students had permission to send email to the list but they were told that no personal attack and slandering was allowed and common netiquettes were to be observed at all times. The list owner (in this case the subject coordinator) had the right to remove anyone who repeatedly violated such rules.

*Team mailing list*

Team mailing lists were set up for all participating teams at the GVT command headquarters (University of Texas, Austin). UT-Austin was responsible for technical support and coordination of these lists. Each team mailing list was only accessible by team members within the team and the coordinator at each university - normally the professor of the subject. Student members, who were often from different universities, used the team mailing list to familiarise themselves and to coordinate their project work during the course of the GVT exercise.

*Person-to-person email*

Person-to-person email was used to directly communicate with an individual within a team or within the local class. This was necessary if two persons were sharing a task for the GVT or within the class. The functionality that was offered by person-to-person email was similar to that of mailing lists in the sense that it was just another email system.

*Interactive chat*

An interactive chat forum was set up by one of the North American universities to allow synchronous interactive communication. This was done because there were times when team members needed to discuss issues where interactivity was important (e.g. finalising issues when closer to due dates). The chat forum was Web-based and to send off a message, participants needed to click a 'submit' button every time. This was cumbersome and many eventually turned to more responsive tools such as Internet Relay Chat (IRC) or I Seek You (ICQ).

*Forum*

Forums are a kind of electronic message board. They were set up to allow team members to communicate by posting notes and replies. The key difference between forums and mailing lists is that forums require users to actively access the messages while the mailing lists normally deliver the messages to a user's email inbox. The forum used in GVT was set up by one of the North American universities using a Web-based interface.

*Group Webpage*

Group Webpages were hosted on a central Webserver by a North American university. The purpose of the group Webpages was to provide a team noticeboard to post work completed at different stages. Group Webpages were not used for normal communication but were used to demonstrate completed stages of the work. The Webpages would then serve as a basis for discussion and improvement of the work. All groups were required to post their final product onto their own team's Webpage.

In order to study the usefulness of individual technologies during the course of the GVT exercise, Murdoch students participating in the GVT project were asked to complete a survey of the usefulness of each online technology at the beginning, halfway and end of the project. The rating was from 1 (totally useless) to 5 (most useful). Because of the evolving nature of the team relationship, we suspected that different technologies might be useful during different phases of the project. We also collected survey data on student experiences on communication issues. Correlation analysis was carried out on the "usefulness of technology data". Student responses from the beginning of the project were correlated with their responses at the mid-stage of their project. The responses from the mid-stage of the project were correlated with the responses at the end of the project. This showed trend in students' responses to a particular technology. Figures 3-8 show the histograms of student responses for each stage of the project for each technology that we studied. The x-scale is the response (1 - 5) and the y-scale is the number of students who gave a particular response.

## Results and Analysis

*Subject mailing list*

The usefulness of subject mailing list was found to be decreasing during the course of the project (see histograms in Figure 3).

At the beginning of the GVT project, the subject mailing list was fairly useful (mean = 3.4) mainly due to the need to receive subject information and guidance from the local subject coordinator. Since most students were new to the experience

*Figure 3: Usefulness of subject mailing list over time*

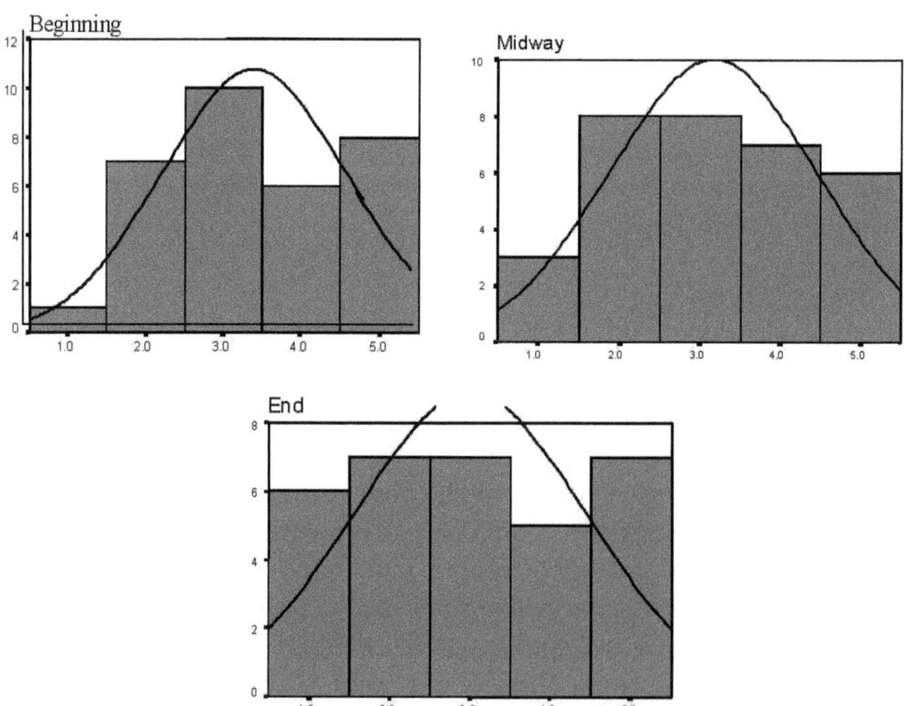

of working in a Global Virtual Team, the subject mailing list also served as a common source of shared information. As the subject progressed, students were involved more in GVT teamwork and the usefulness of the subject mailing list decreased. The usefulness decreased to 3.2 and then 3.0 at the end. Figure 3 shows a pattern of the responses (normal curve) shifting to the left of the scale as time went on. Correlation analysis of the responses between beginning and midway (coef. = .756, p = .000), then midway and the end (coef. = .900, p = .000) shows that the experience was consistent with the majority of the class.

*Team mailing list*

The usefulness of the team mailing list was always higher than the subject mailing list, presumably due to the need to work closely with overseas team members. The team mailing list was almost the only feasible way to do so. The mean values of usefulness in the beginning, midway and end are 4.2, 4.4 and 4.4 respectively, indicating its importance is increasing over time (Figure 4).

Correlation analysis indicates the change in experience from beginning to the end is strongly correlated (beginning to midway: coef. = .819, p = .000; midway to end: coef. = .875, p = .000) again suggesting a consistent pattern amongst the class.

*Figure 4: Usefulness of Team Mailing List Over Time.*

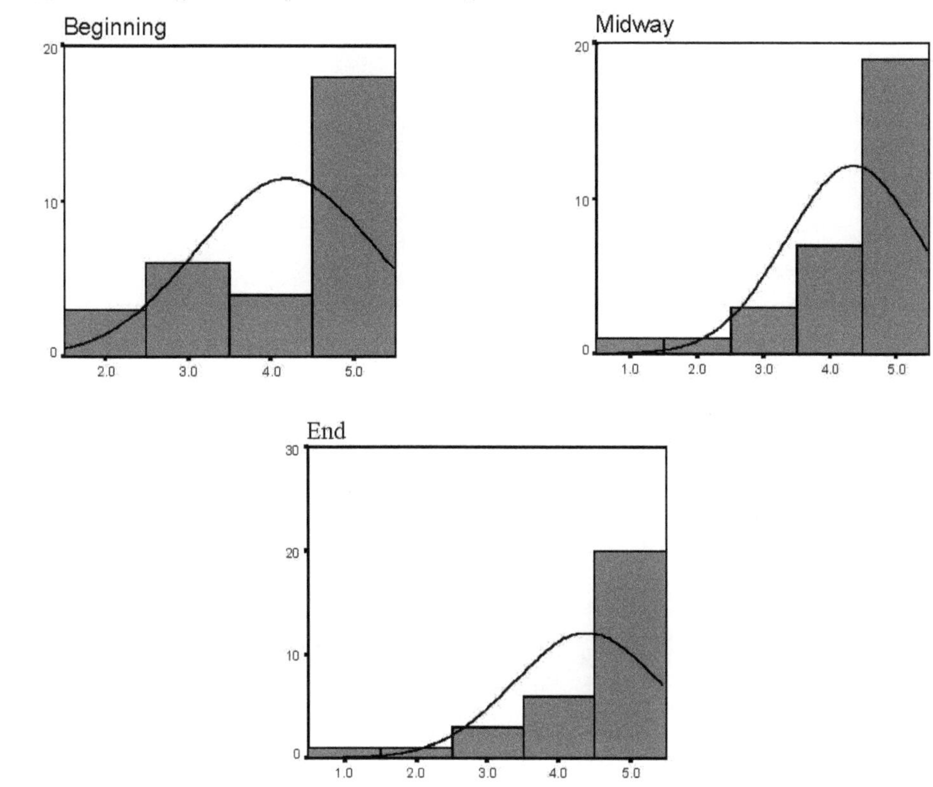

Histograms in Figure 4 show that the responses are skewed towards the high end, suggesting that the GVT mailing list was very important to the project and the trend was an increasing one as compared to the subject mailing list which had a decreasing trend. Together with the outcome from Figure 3, it shows that as the project progressed, the subject mailing list became less useful.

*Person-to-Person email*

The usefulness of person-to-person (P-to-P ) email was an interesting phenomenon to observe because it did not show a monotonic trend from beginning to end. P-to-P e-mails usefulness varied from being not particularly useful (midway) to more useful (end). At the beginning, usage of P-to-P email was consistent with a mean usefulness of 2.84 but dropped to 2.65 midway and increased to 2.97 (Figure 5) at the end of the GVT experience.

This variation can be due to the way project teams were set up and the allocation of tasks within a team. In the beginning of the project, there would have been more group discussion while towards the end, individuals were allocated

*Figure 5: The usefulness of person-to-person email over time*

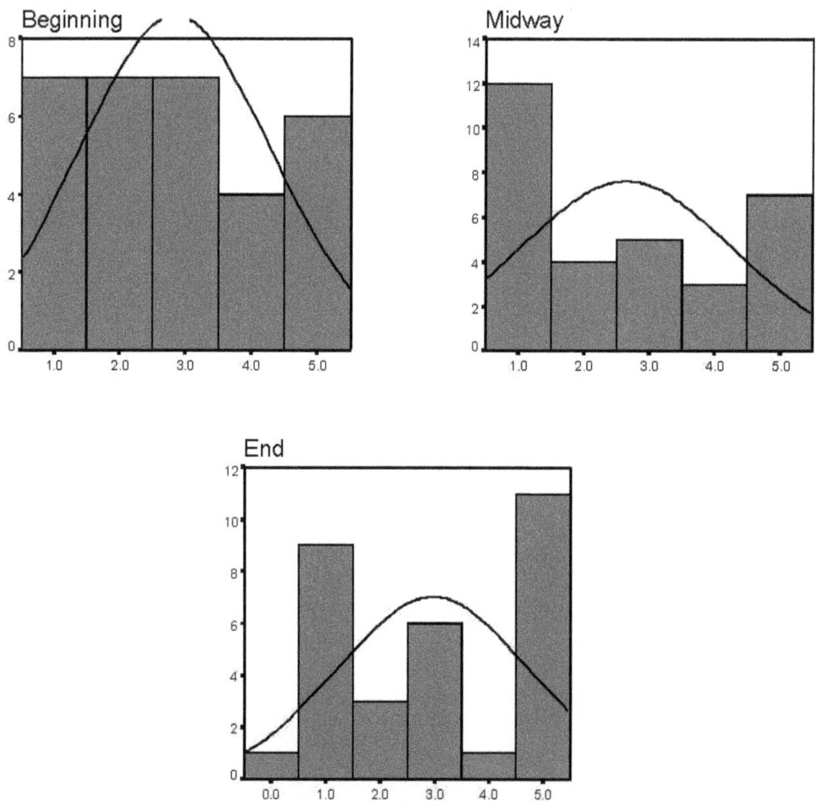

specific tasks. With specific responsibilities, P-to-P email was more useful because there was a specific person to contact to complete the project as is depicted in the last histogram in Figure 5.

*Interactive chat*

The usefulness of online chat was limited partially due to the slowness of the technology and it was cumbersome to use. In fact this led to the use of alternative chat facilities such as IRC and ICQ. Histograms in Figure 6 reflect the unpopularity of the official chat facility with the mean usefulness falling from 2.66 in the beginning to 2.58 midway but rising back to 2.68. The skew towards the low end of the scale (all three diagrams in Figure 6) reflects the average usefulness of the technology.

According to general feedback from the class, synchronous communication was not really essential for the business plan development process. This change of pattern was consistent within the group with a correlation coefficient of .847 (p = .000) when measuring the change in usefulness from the beginning to midway and .795 (p = .000) from midway to the end. Apart from the usability of the original chat tool, many from the class found that having to meet and chat online at different time-

*Figure 6: The Usefulness of Online Chat Over Time*

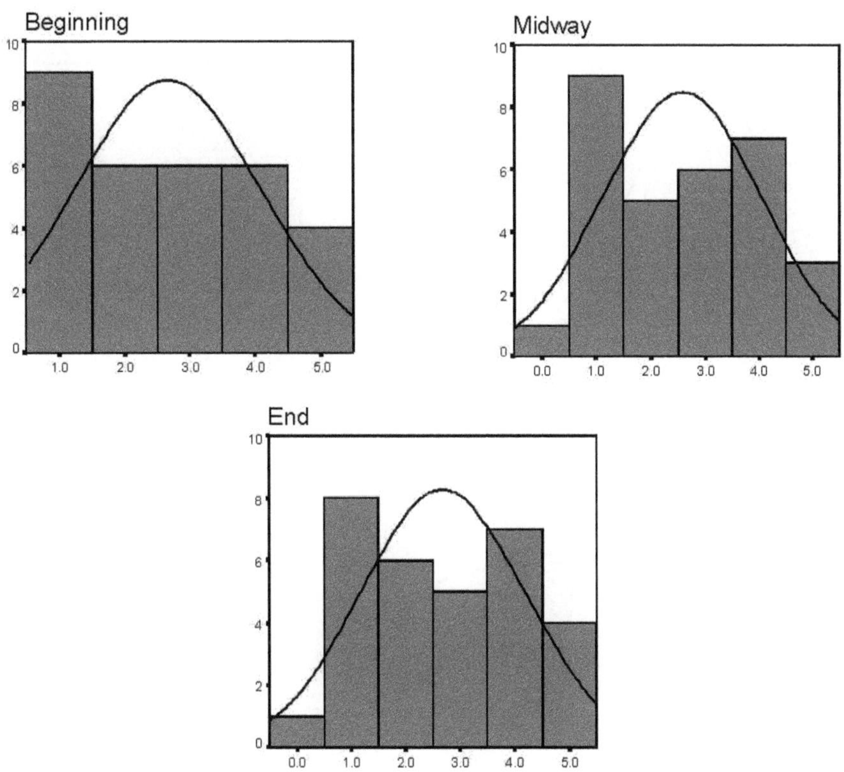

zones was inconvenient because the Australians need to stay up/wake up early in the morning while the Americans had to stay up late in the evening. Many only resorted to interactive chat during the last stage of the project when a final consensus was needed.

*Forum*

The usefulness of forum was also low. The forum was not originally part of the support system because it was thought that with the other technologies in place, there was no need to have the forum implemented. Eventually, it was set up because some teams had difficulties accessing email.

For the local students, forum was in general not used because the team mailing list was sufficient to handle most correspondence. The mean usefulness of the forum had dropped from 2.6 (beginning) to 2.3 (midway) then to 2.0 (end) with a strong skew towards the low end of usefulness (see Figure 7). This change in pattern was consistent among the respondents with correlation coefficient of .639 (p=.000), from beginning to midway and .748 (p=.000) from midway to end.

*Figure 7: The Usefulness of Forum Over Time*

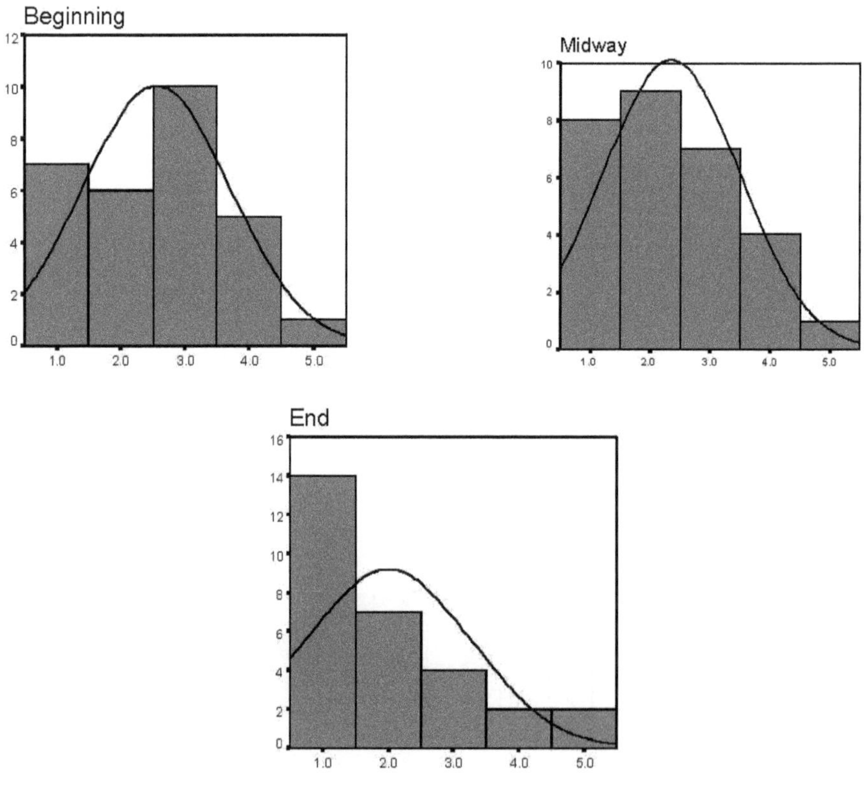

*Figure 8: The Usefulness of Group Webpage Over Time*

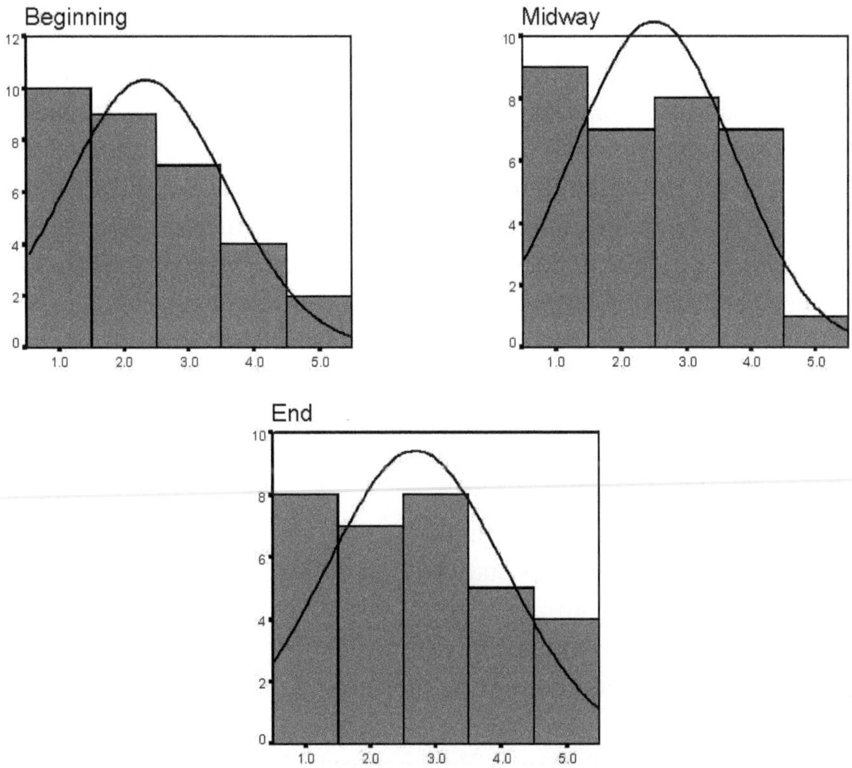


*Group Webpage*

Group Webpages were used continuously to display the completed work during the course of the GVT exercise. The mean usefulness of group Webpages had grown from 2.3 in the beginning to 2.5 midway and to 2.7 at the end (see Figure 8).

This is understandable because as the GVT exercise approached the end, more and more materials were posted on the Webpages and members could discuss how to finalise the submission. This increased use of the Webpages indicated the Webpages' usefulness.

Based on the Global Virtual Team exercise in which we participated, our results show that the usefulness of an online technology depends on the type of work being carried out and the stage of the project. It also depends on the requirements of the project as set out by academic staff. As a project whose final outcome was to produce a business plan, we found that the most useful technology was a team-based group mailing list while the least useful technology was a group forum. This could be because the mailing list was already playing the role of a virtual forum so there was no need to have an extra place to post information. Results from another study (Subject A) indicate that the forum was the second most important tool during the beginning and midway of an introductory course where group work was not required. It went to third place at the end.

Although the concept of virtual project teams is innovative, it does create new challenges from a course management perspective. These challenges are due to the lack of understanding amongst team members. Traditionally, a project team can meet face-to-face and communicate using synchronous and highly interactive medium such as teleconferencing. In the case of the standard Internet applications, communications are often confined to asynchronous modes such as email (including mailing lists) and bulletin boards (e.g.. newsgroups, forums). Even though synchronous communication is possible, it is often not feasible due to bandwidth limitations – particularly links to Latin America.

Other communication issues that arose during the project were:

- insufficient understanding of the cultural/personal background of the team
- overcoming time-zone differences
- the lack of standardised assessment weighting
- differences in work load expectation of the subject of which this project was part
- accessibility to Internet connectivity
- use of appropriate technology at the right time

## Cultural/Personal Background of the Team

Working in a multicultural team can be an enriching experience for a team because it can bring together different views which are often valuable to the project. However, without developing prior understanding among members, cultural diversification can become a barrier to collaboration. Some teams experienced cultural differences so severe that it became a debilitating factor to the team. The so-called cultural factor can be more than just social norms, mannerisms and attitudes toward individuals. Some teams found that even the different subjects they were doing created different expectations from their assessors. In the virtual team project, there were students doing different degrees (and/or majors) and the expectation of the subjects was possibly different. For example, a student enrolled in an IT degree would be putting more emphasis on the technical aspects of the project than those enrolled in a Master of Business Administration. Such differences can be a complementary asset but can also create tension.

## Time-Zone Differences

Another barrier to collaboration, particularly in a set up like this, was the overcoming of time-zone differences. When project team members had to work synchronously, it was often necessary for them to be online at the same time. With time differences up to 20 hours, some discussions were happening very late at night for some team members and in the early morning for others. This may explain why asynchronous communications such as email and forums were the most useful tools during the project. Only a small number of teams considered synchronous communication over the Internet useful, although this might have been different if voice and video communications were made available.

### Lack of Standardised Assessment Weighting and Workload Expectations

Due to the various subject structures, team members from different universities found it difficult to demand equal commitment from their teammates. For example, some subjects were allocated only a small percentage of their total assessment weighting to this project, say 10%, while others were allocated a significant portion (up to 35% or more). Such differences automatically created an 'expectation gap' where students who were doing subjects with a high assessment allocation saw the project as being more important than the others did. This often formed the spark of dispute amongst team members. Added to that was the importance of studies to the individual team members. Some team members were studying part-time with a heavy work schedule, while others were studying full-time and this project was part of the last subject to be passed before graduation.

### Internet Access

Accessibility to resources that allow team members to work online was another major barrier to effective collaboration. Although basic Internet access was available to members in project teams, regular and reliable access was not guaranteed. Some team members only had access to Internet facilities during laboratory time (two hours per week) and the line speed left a lot to be desired. After hours access was not always possible. When heavy demands were put onto Internet access as a key to effective collaboration, such as exchange of large document files, real-time chat and frequent email communication, not being able to carry out such tasks effectively was a major frustration. Even though in cases when accessibility was guaranteed, the differences in line speed, reliability and places where such access could be made were critical. The differences between laboratory access, often up to 100Mbits/sec, as compared to modem access, which can sometimes be as low as 14.4Kbits/sec, became in some cases the difference between working effectively and a major impediment.

### Use of Appropriate Technology at the Right Time

The usefulness of each online technology usually varies over time and therefore it may be helpful to provide different online technologies at different stages of the project by mapping the technology to the communication needs at the time. This is especially important when project participants are not fully knowledgeable about the appropriate use of each technology. Such participants become confused by the plethora of technologies available and frustrated when they use a technology at the wrong time and get little or no satisfaction from the usage. Knowledge of the use of the appropriate technology has implications for distance learning in general.

## CURRENT CHALLENGES/PROBLEMS FACING THE ORGANIZATION

The challenges facing virtual project teams both as an educational exercise and in a real working environment are how to coordinate teams which rely on the Internet to support team activities and making team members feel that they share the same team spirit. This may involve creating a new governing structure which enables teams to work together through a collaborative spirit instilled by team members themselves rather than external forces. Due to the differences in expectations, virtual teams are often more vulnerable to failure because of problems related to the coordination and alignment of team members' expectations. It is proposed that more thorough "ice-breaking" exercises are carried out by team members before the start of any significant cooperative venture.

In general, coordination among faculty members can also be a challenge. Faculty members usually teach in environments where coordination is within their university's structure. When cross-university projects such as this one take place, a 'meta-level' virtual team of faculty members needs to be created in order to ensure good coordination. Although a forum for such a meta virtual team was there, it was not used extensively to its full advantage. The development of meta-level coordination team is important when a tightly knitted structure is needed in teams with fluid organisational form where team members have competing needs.

Virtual teams are more than just using coordinating technology to enable disparate groups to communicate. It has more to do with how such technology can make up for the lack of face-to-face communication and understanding. In such cases, the development of a team spirit, a common objective for the team, self-governing policies and finally a membership sharing the same vision and goal would facilitate proper functioning of the team. Although interactive technology such as video and audio conferencing over the Internet makes this possible, it is the sharing of intellectual and emotional space that forms strong bonds within a team. How to really develop such sharing processes is still a challenge in virtual team environments.

## FURTHER READING

Berg, Z. & Collins, M. (1995). Computer-Mediated Communication and the Online Classroom: Overview and Perspectives. *Computer-Mediated Communications Magazine*, 2(2), also at URL: http://www.december.com/cmc/mag/1995/feb/berge.html (Feb. 2, 2000)

Cantor, J.A. (1995). *Co-operative Education and Experiential Learning;* Wall & Emerson, Inc., Canada.

Clement, D.H. & Nastasi, B.K. (1992). The role of social interaction in the development of higher-order thinking in Logo environments. in De Corte, E. et al (eds); *Computer Based Learning Environments and Problem Solving,* 229-248, Springer-Verleg, Berlin, Heidelberg.

Cook, L. (1991). Co-operative Learning: A Successful College Teaching Strategy. *Innovative Higher Education,* 16(1), 27-38.

Goldberg, M.W. (1996). CALOS: An Experiment with Computer-Aided Learning for Operating Systems. *Proceedings of the ACM's 27th SIGCSE Technical Symposium on Computer Science Education.*

Johnson, D.W. & Johnson, R.T. (1990). Co-operative learning and achievement; in Sharan, S (ed); Co-operative learning: Theory and Research, 23-37, New York: Praeger.

Kerr, N.L. (1983). Motivation Losses in Small Groups: A Social Dilemma Analysis. *Journal of Personality and Social Psychology,* 45(4), 819-828.

Kerr, N.L. & Bruun, S.E. (1983). Dispensability of member effort and group motivation losses: free rider effects. *Journal of Personality and Social Psychology,* 44(1), 78-94.

Salomon, G. & Globerson, T. (1989). When teams do not function the way they ought to. in Webb, NM (ed). Peer Interaction, Problem-Solving, and Cognition: Multidisciplinary Perspective; *International Journal of Educational Research,* 13(1), 1-119.

## REFERENCES

Goldberg, M.W. (1997). Communication and Collaboration Tools in WebCT. *Proceedings of the conference Enabling Network-Based Learning,* May 28-30, Espoo, Finland.

## APPENDIX

### Participating Universities and number of Participants

| Institutions | Participants |
|---|---|
| American University | 25 |
| Copenhagen Business School | 28 |
| EAESP - Fundação Getúlio Vargas | 54 |
| Kent State University | 12 |
| Louisiana State University | 25 |
| Murdoch University | 38 |
| University of Berne, Switzerland | 4 |
| University of São Paulo - Brazil | 7 |
| University of Tampere | 35 |
| University of Texas at Austin | 16 |
| University of Vaasa | 3 |
| University of Western Australia | 39 |

Chapter 9

# Big-Bang ERP Implementation at a Global Company

Nava Pliskin and Marta Zarotski
Ben-Gurion University of the Negev, Israel

## EXECUTIVE SUMMARY

Dead Sea Works is an international multi-firm producer of Potash and other chemicals whose sales for 1998 were about $500 million. In 1996, the Information Systems group convinced top management to pursue a big-bang ERP implementation of SAP R/3. To reduce project risk, risk management was practiced. First, only modules that matched the functionality of the then-existing systems were targeted, avoiding as much as possible software modifications and process reengineering. Second, a steering committee was set up to handle conflict resolution and set priorities throughout the project and top users were given responsibility with implementing modules within their respective functions. R/3 went into production on July 1, 1998, six months ahead of schedule and without exceeding the $4.95 million budget.

## BACKGROUND

Dead Sea Works Ltd., one of 15 member companies of Israel Chemicals LTD, is a producer of Potash and other chemical products from the mineral-rich Dead Sea, Israel's greatest natural resource. Situated at the lowest place on earth, it lies in a valley whose southern part is suitable for evaporation pans and enjoys ample sunlight for most of the year. This combination of chemical riches and topography that is amenable to practical use fired the imagination of Theodore Herzl, the father of modern Zionism. After hearing of a plan to extract minerals from the Dead Sea, during his 1896 visit to Palestine, Herzl described in his book, *Alt*

Previously published in the *Annals of Cases on Information Technology Applications and Management in Organizations, volume 2*, 

*Neuland,* a Jewish State whose economic strength would be derived from the treasures of the Dead Sea.

Moshe Novomeisky, chemical engineer, came from Siberia to Palestine at the beginning of the century inspired by *Alt Neuland* to turn this vision into a reality. In 1930, he obtained from the British Mandatory authorities a concession to extract minerals from the Dead Sea, established the Palestine Potash Company LTD, and constructed a plant in the northern part of the Dead Sea. In 1934 evaporation pans and a chemical plant were constructed in Sodom as well. This became the foundation for today's Dead Sea Works (DSW) which, since it was reestablished in the 1950s, has increased production steadily to its current level of close to three million tons of Potash per year. In addition to Potash, DSW produces Magnesium Chloride Flakes and Pellets, Salt, Bath Salts, Magnesium Metal, Chlorine and Bromine.

Instead of mining, as do most of its competitors, DSW extracts Potash from the Dead Sea. The production process begins with the pumping of Dead Sea water to 105 square-kilometer salt pans, where the solution is concentrated. An additional forty square-kilometer pans are then used to crystallize materials, which after settling on the pan floor, are pumped by harvesters directly into refineries. In this process, DSW takes advantage of the energy of the sun, another important natural resource in the region. Artificially, these drying processes would require 10 million tons of oil per year.

Wherever Potash is produced, transportation is a major expense, as was the case for DSW because of the 900-meter altitude difference between the factory at Sodom, the lowest point on earth, and the nearest railway terminal. Since this gradient rules out the possibility of a direct rail link and the remote location makes road transportation expensive, DSW chose to build from Sodom to the railway an 18-km conveyor belt, whose incline at some points reaches 18 degrees. Since its completion in April 1987, DSW's transportation costs have declined substantially.

Currently, the multi-firm DSW Group (see Exhibit 1) is distributed internationally (e.g., Europe, and China) and within Israel (e.g., Sodom, Beer-Sheva, and Eilat). In Europe, DSW has been involved in several joint ventures. In 1996, DSW established Dead Sea Magnesium LTD., investing with Volkswagen (65%, 35%) close to \$500 million. Yearly production capacity at the new plant has already reached 25,000 tons and, by the end of 1999, is expected to grow by 50%. In 1998, DSW has partnered with Eurobrom B.V. in Clearon Holding Corporation and acquired from the Spanish-Companies Authority, jointly with two Spanish partners, Grupo Potash, a producer of one million tons of Potash per year sold mainly to the Spanish and French markets. In late 1998, DSW joined the Chinese government in building a Potash production plant that will eventually produce 860

*Exhibit 1: DSW Multi-firm Structure*

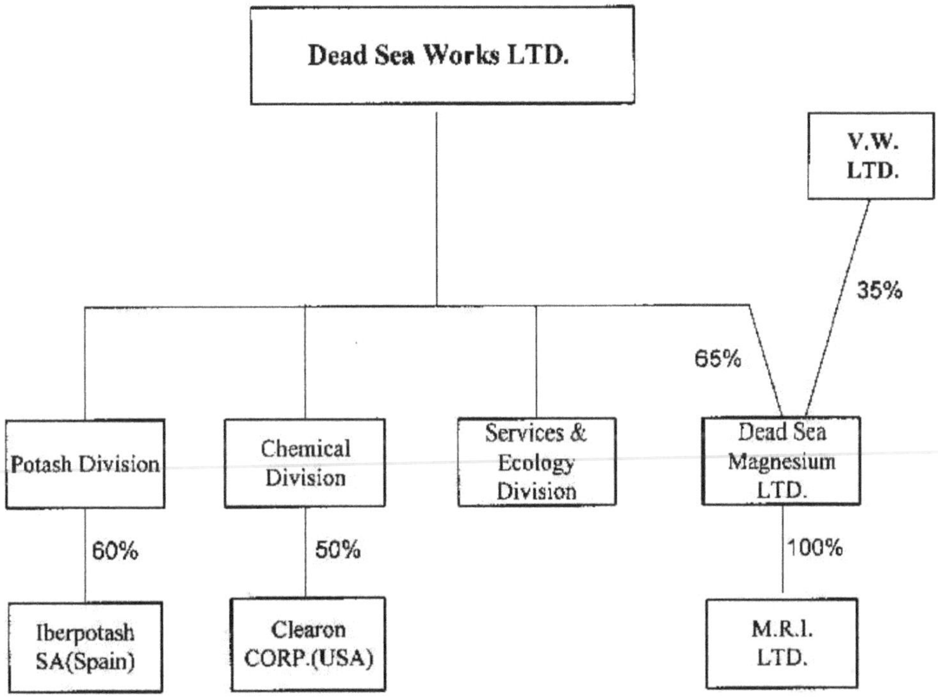


thousand tons every year. This joint venture is expected to increase DSW's sales to China, a market whose Potash consumption per year grows at a rate that exceeds the world's rate.

Due to these expansion activities, the number of DSW employees almost doubled over the last two years, reaching 2,458 at the end of 1998. Potash sales worldwide, which reached about 42 million tons in 1998, are growing at a rate of 3% each year (see Exhibit 2). The American and Asian markets are the biggest Potash consumers and importers, since their own production does not meet the demand. DSW sells to 50 countries on every continent, protecting itself from local market fluctuations, and is expected to increase its share in world Potash production from six to nine percent. Even though 10% of its sales are directed to one of its clients, DSW is not dependent on any one customer. It is also free from dependence on any supplier and any material that is not included in its license from the Israeli government.

Because DSW is a global company, its financial performance depends on trends in the world economy, including the economic conditions in South East Asia, Russia, and changing attitudes toward the environment. Despite periods when

*Exhibit 2: World-wide Potash Sales*

Total - 2,850 thousand tons

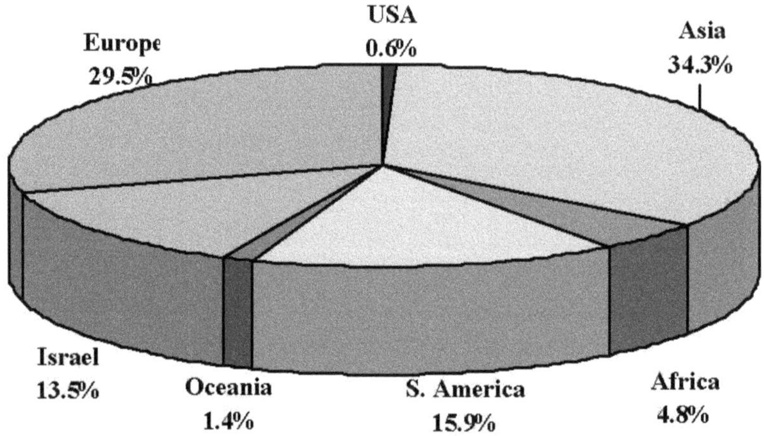


world Potash prices reached record low levels, DSW recorded a profit every year since 1970, while many of its competitors have faced difficulties (see Exhibit 3).

Strategically, DSW is focused on persistent growth by taking the following measures. First, DSW is constantly expanding Potash production in Sodom and elsewhere, while lowering costs. Second, DSW is accelerating business development around the world (e.g., in China and in Spain), including joint ventures in Salt and Chloride Aluminum. Third, DSW is investing in development of power and water resources. Finally, in the marketing arena, DSW is paying attention to widespread distribution of sales, sensing customer needs and responding to them.

## SETTING THE STAGE

Information Systems (IS) at DSW were custom-developed specifically for DSW since the early 1970s, when the IS unit consisted of an IS manager, six data-entry clerks, three operators, and four programmers. The first functions to be automated, in batch mode, were accounting, costing, and budgeting. The number of users in those early days was 20. When hardware was upgraded in 1978 to IBM's System 3 Model 15, each functional IS was operating in isolation, using its own removable disk.

A technological turnaround took place in the 1980s in terms of both hardware and software. The IBM 4331 mainframe computer running the VSE operating system was acquired, setting the stage for later upgrading to an IBM 4341 and a 4341-31 that operated in parallel respectively for IS production / operations and

*Exhibit 3: DSW Performance Highlights*

| FINANCIAL FIGURES (in thousands of U.S. dollars) | 1993 | 1994 | 1995 | 1996 | 1997 | **1998** |
|---|---|---|---|---|---|---|
| Revenues | 245,342 | 295,910 | 362,280 | 357,567 | 453,185 | **493,710** |
| Net Income | 17,852 | 28,013 | 35,171 | 39,840 | 39,645 | **49,901** |
| Total Assets | 526,462 | 628,092 | 742,182 | 934,450 | 983,598 | **1,245,460** |
| Shareholders' Equity | 305,703 | 320,109 | 333,005 | 351,372 | 368,415 | **418,468** |
| Dividend Paid | 21,792 | 13,907 | 13,917 | 21,220 | 36,808 | **9,680** |
| Capital Expenditures | 134,400 | 156,400 | 216,400 | 306,321 | 126,460 | **57,816** |
| | | | | | | |
| Earnings $ Per Share of NIS 1 par value (in $ U.S.) | 0.06 | 0.10 | 0.12 | 0.13 | 0.13 | **0.17** |

| QUANTITY FIGURES (in thousands of Metric Tons) | 1993 | 1994 | 1995 | 1996 | 1997 | **1998** |
|---|---|---|---|---|---|---|
| Potash Production | 2,182 | 2,099 | 2,207 | 2,492 | 2,481 | **2,860** |
| Potash Sales | 1,968 | 2,317 | 2,451 | 2,178 | 2,861 | **2,916** |
| Casted Magnesium Production | - | - | - | - | 7.2 | **24.5** |
| Casted Magnesium Sales | - | - | - | - | 5.4 | **15.1** |
| Table Salt Sales | 58 | 63 | 80 | 86 | 104 | **99** |
| Magnesium Chloride Flaked and Pellets Sales | 51 | 74 | 81 | 90 | 68 | **63** |
| Bath Salts Sales | 1.3 | 2.2 | 1.9 | 1.9 | 2.2 | **2.1** |
| Pan Salts Sales | 268 | 235 | 178 | 191 | 261 | **219** |
| Number of Employees | 1,629 | 1,654 | 1,670 | 2,030 | 2,245 | **2,103** |

for development/testing. In the software arena, the purchase from Cincom of Total, a hierarchical Data Base Management System (DBMS), and Mantis, the associated development tool, set the stage for upgrading to the Supra DBMS a few years later. With a DBMS in place, DSW recruited a number of systems analysts and expanded its IS group to include ten programmers, adopting a systems approach and building integrated systems that replaced fragmented ones.

Microcomputers also made their way into DSW during that time. Except for the first stand-alone system, all operated in terminal-emulation mode. By the nineties, microcomputers ran applications developed by the IS group, as well as user-developed applications.

Until the early 1990s, the following process was in place for IS renewal and maintenance: The IS group took the initiative by conceiving an idea, and then approached top management for a budget to turn the initiative to reality. This changed considerably when a new manager took charge of the IS group in 1992. Under his leadership, users were encouraged to take IS initiatives and to convince top management to allocate the needed resources. This approach allowed better budgeting, in advance, according to user requests. Consistent with this approach, the mission of the IS group was redefined as service provision and workshops for IS personnel were conducted to teach how to become high-quality service providers.

Additional changes, which had a major effect on IS management at DSW, occurred after a new CEO, who perceived the information resource as instrumental to business success, took office in 1993 and restructured DSW (see Exhibit 4). The IS group turned into a division, instead of a department, and its head was promoted to Vice President of Information Systems (VPIS), reporting to the Senior VP and CFO, instead of the VP of Accountancy and Control. A consulting company hired around that time helped DSW in its IS planning processes. For the long term, the consultants helped put together a five-year IS investment plan. For the shorter term, they recommended that users be allowed to seek IS services outside DSW and, at the same time, be charged for IS services that they opt to acquire internally from the IS group.

In 1993 the mainframe hardware was upgraded to ES/9000 with the VSE/VM operating system. In the communication arena, LAN and WAN infrastructures were installed, using Novell and TCP/IP. By 1995, the IS environment at DSW was serving 700 satisfied users (100 used dumb terminals and 600 used microcomputers). Despite the fact that the functional information systems at DSW were consistent with major business processes, it became clear that major IS changes were called for due to technological and organizational reasons. Technologically, the Supra DMBS was no longer being supported and the systems at DSW were not state-of-the-art. Organizationally, the functional systems, which were developed to support the functioning of a single firm, could no longer support a multi-firm enterprise.

By 1995 the CEO and VPIS were well aware of the seriousness of the Year 2000 (Y2K) problem and concluded that for DSW, as for many other firms, resolution of the Y2K problem was critical to survival (Kappleman, 1997; Violino, 1997). Acknowledging that Y2K compliance was more of an organizational than a technical concern, the CEO charged the VPIS and the VP of Strategic R&D with planning how best to invest DSW financial and other resources to meet this challenge.

Like many organizations, DSW sent letters to all its suppliers in order to find out what their response to the Y2K problem was. In parallel, outside consulting services were recruited to help deliver Y2K solutions for embedded systems at a cost of about $0.5M. Both these activities are beyond the scope of this case study, which is focused as a main issue on a big-bang implementation of an Enterprise

*Exhibit 4: DSW Structure*

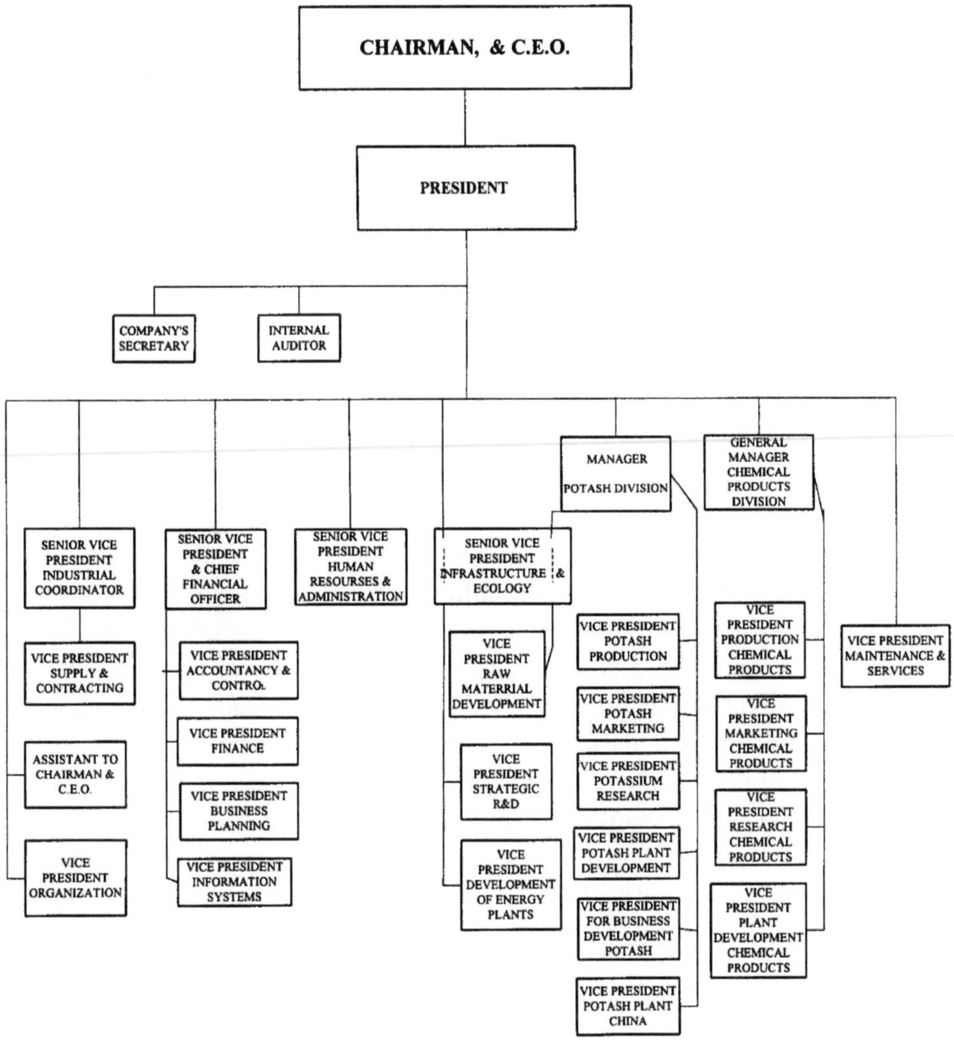


Resource Planning (ERP) system at DSW. The ERP implementation resulted from the Pandora's box that was opened following the realization that the information infrastructure and the IS applications at DSW were not Y2K compliant.

## CASE DESCRIPTION

### Decision Making Leading to ERP Implementation

To assess resource implications of converting all IS applications to comply with Y2K, the IS group prototyped Y2K conversion of one functional application. Based on this prototype, they concluded that Y2K conversion of all existing applications would cost more than $1.5M and would still leave DSW with legacy systems that revolve around a legacy DBMS.

The VPIS thus began to seriously consider avoiding Y2K conversion altogether and, at the same time, shifting from a "make" strategy to a "buy" strategy (Applegate et al., 1999.) As a subscriber of reports from META Group and GartnerGroup (1994-1996), he had already been exposed for a number of years to the growing popularity of the "buy" strategy, in general, and ERP, in particular. He was well aware that a number of software vendors offer ERP systems that integrate, on the basis modern database-management technology, a collection of modules for managing such functions as production, inventory, procurement, accounting, sales, marketing, and personnel. Although some vendors have originally developed their ERP products for mainframes, most made the transition to state-of-the-art client-server architectures.

The VPIS was also aware that some DSW competitors had already implemented ERP versions adapted to the chemical industry a few years earlier. Yet, until faced with the Y2K problem, he thought that ERP was intended either for larger and more global organizations than DSW was at the time, and/or for organizations whose IS applications, unlike DSW's, were fragmented islands of automation. Although module fragmentation was not one of DSW's problems, DSW could use an ERP package since it grew and became a multi-firm global firm.

Technologically, since none of DSW's existing applications were state-of-the-art, the VPIS expected ERP systems to be significantly superior in terms of speed and reliability. He also was aware that the core database, which was already stretched to the limit in terms of size and response time and was no longer being upgraded or even supported, was not suitable for handling globally distributed databases. Moreover, unlike ERP systems, each of the existing applications at DSW had its own user interface, making it inconvenient for users to move between applications and difficult for IS professionals to maintain data security as the number of users increased.

After becoming aware of ERP technical benefits, including speed, reliability, database distribution, convenience, and data security, the VPIS, VP of Strategic R&D, and their team began to consider the organizational implications of ERP. They soon learned that since ERP encompasses the whole organization at all sites, an ERP implementation can provide an opportunity for reengineering major organizational process and for achieving better business outcomes over time as a result of more informed managerial decision making processes.

Because of the perceived technical and organizational benefits of ERP, the VPIS and the VP of Strategic R&D began to view the turn of the millennium as an opportunity rather than a threat. They thus proposed to top management of DSW and its parent company, Israel Chemicals LTD, to avoid a Y2K conversion altogether and implement instead a Y2K-compliant ERP system that would facilitate:

1) Renewing the hardware at DSW, from host-centric to network-centric and the software infrastructure at DSW, from outdated to modern;
2) Supporting the transition of DSW from single-firm to multi-firm structure and from local to global operations; and
3) Using the ERP implementation project at DSW as a pilot from which other members in the Israel Chemicals LTD group can learn.

Once top management approved the ERP implementation, a decision was made to abandon all upgrading of existing systems and to put on hold plans for developing new applications. For example, although the need for plant-maintenance automation was already acknowledged prior to the decision to move into an ERP environment, and even though it was already budgeted at $1M, actual analysis, design, and development were suspended.

The search process that followed, for software and hardware, ended in choosing the R/3 ERP software package from SAP, the Oracle DMBS software from Oracle, and the server hardware from HP. The fit between R/3 and DSW's needs was excellent for the following reasons: a) SAP met the main functional requirements of DSW because its chemical-industry version of R/3 had already quite a number of installations worldwide. b) According to reports from META Group and GartnerGroup (1994-1996) that the VPIS has read, the larger and more financially robust SAP was more likely to survive than its competitors. c) SAP was holding the biggest market share in the world, in general, and in Israel, in particular. d) SAP gained considerable experience with and has been considered especially suitable for international multi-firm enterprises. e) SAP was investing more in R&D than competitors. f) R/3 was localized in terms of language, currency, and regulations of many countries, including Israel, where SAP partnered with the ATL LTD, an experienced Israeli vendor with whom DSW has had good relations.

Toward the end of 1996, the resolutions made by top management at DSW guided the implementation project so as to increase the prospects of reaching successful completion of the R/3 implementation by January 1, 1999. Oriented towards thorough project and risk management, these resolutions aimed at carefully controlling project size, technology, and structure (Applegate, et al., 1999.)

In order to put an upper lid on project size, DSW chose to concentrate on implementing the functions which were already automated at DSW (financial accounting, material management (inventory, purchasing and MRP), controlling, treasury and human resources, rather than implement all R/3 modules. The only exception was plant maintenance, whose long-overdue automation was already budgeted for $1M and then suspended following the decision to implement ERP. About half way into the R/3 implementation project, management decided to implement the R/3 plant-maintenance module as well, having received assurances that this could be done within the deadline for an additional $70,000.

Another critical decision was to avoid parallel implementation and its associated interfacing efforts, opting instead for the more risky "big-bang" transition to R/3. Since IS professionals within DSW were unfamiliar with R/3 technology, management realized that outsourcing the big-bang implementation project could reduce project risk (McFarlan & Nolan, 1995.) By contracting IBM as the implementing vendor, DSW was able to overcome the pressing need for unique R/3 skills and knowledge.

Given the wide organizational scope of the R/3 implementation, including the need to deal with recent mergers and acquisitions, top management was determined to increase the structure of the implementation project and to facilitate smooth upgrades to new R/3 releases in the future (Bancroft, et al., 1998.) To achieve these goals, DSW resolved to avoid as much as possible both customization of R/3 to DSW needs, on the one hand, and process reengineering, on the other. As a rule, management was more willing to abandon old DSW processes and regulations, which could not be supported by R/3 without customization, than to permit modifying R/3 to these processes. Only under such circumstances, even though project policy discouraged process reengineering, there was willingness on the part of management to reengineer new R/3-supported processes and replace the old unsupportable ones.

## The SAP/R3 Implementation at DSW

The outsourcing contract with IBM was signed on the last day of 1996 and the implementation project started on April 1, 1997. Since then, the VPIS reported once a month to the board of directors about progress made. The CEO was committed to the implementation project and regularly briefed managers at all levels about his vision and expectations. Members of top management became very involved in the implementation project and each was asked to assume responsibility, as a top user, for implementing modules within their functional area. To make sure that implementation ended on time and within budget, the project was broken up into milestones. Since meeting milestone goals and target dates was deemed critical, any delay needed top management approval.

Measures were taken to facilitate prompt decisive action. A steering committee was formed to set priorities, to handle conflict resolution throughout the project, and to promptly respond to problems. For each functional area, a joint team of three, composed of the respective top user (Pliskin & Shoval, 1989), an R/3 expert from IBM, and a systems analyst from DSW, was created. The team, assigned with responsibility for part of the implementation project, was put in charge of choosing among R/3 processes and reports and setting priorities. Since a significant portion of top-user time was allotted to the implementation, they developed familiarity with the new environment and, later on, became very effective in providing the initial response to problems that emerged. IS professionals were constantly briefed with respect to progress as well as with respect to difficulties and ways to get around them. They were encouraged to report any concern to the respective top manager to insure a fast solution.

As planned, R/3 modifications were limited to the absolute minimum and permitted under exceptional conditions and only with formal CEO approval. The formal procedure, instituted to discourage R/3 modifications, was practiced throughout the project, despite a number of unsuccessful attempts to eliminate it, especially during transition between CEOs.

Consistent with the initial intention to limit reengineering to a minimum, only a few processes were reengineered. The reengineered service entry process, for example, was perceived by users to be "the right thing to do" because it was dictated by R/3. They thus willingly adopted it before noticing that, because the new more reliable service entry process was 50% to 60% faster, a few jobs had to be eliminated. In any case, none of the employees whose jobs were eliminated was asked to leave DSW. Instead, they were transferred to other departments that were hiring at the time.

It is noteworthy that DSW, which prior to the ERP implementation opted to custom-make IS applications, wished to gain as much as possible from its decision to buy software and, at the same time, increase prospects for success. In other words, DSW was willing to abandon customization as much as it could, putting a lid on modifications and willing to force modification avoidance except under special circumstances. In an organizational culture that values labor relations, a threat of labor unrest in the form of a strike presents such special circumstances. The union was therefore consulted whenever avoiding R/3 modifications impacted compensation. In at least one case, the union's disapproval of modification avoidance forced an R/3 modification and labor relations were not hurt. It is noteworthy however that the union was willing, in most cases, to go along with and accept the implications of modification avoidance.

Additional initiatives paralleled the ERP conversion. Personal computers and terminals were upgraded. The network was stabilized to prevent down times. Data quality was addressed (through conversion, improvement, and creation). Even though time was short, specially-hired trainers joined top users and systems analysts, who participated in the implementation project intensively (Pliskin, 1989), to train, in four parallel classrooms, 400 trainees, including end users, systems programmers, operators, systems analysts, and programmers. Each trainer was put in charge of specific modules. Trainees were encouraged to come to the classroom for extra practice. A flexible and responsive computerized help-desk was staffed with individuals trained to either respond promptly or to refer swiftly to another person.

R/3 went into production at DSW on July 1, 1998, six months ahead of schedule and without exceeding the $4.95 million budget. In the beginning, the VPIS met with the IS professionals on a daily basis to air out problems. As the need to do so diminished, meeting frequency went down. As of January 1, 1999, all planned modules are working. The number of users, low- and mid-level managers, is 600 and growing.

It is noteworthy that both IS professionals and users cooperated with the ERP implementation and no resistance was observed. Although nobody in the company is willing to bet on the reasons for the lack of resistance, some speculations have been brought up. IS professionals were assured by the VPIS that their skills would be upgraded to state-of-the-art technologies through massive training and none would be fired. Users were led to believe that the R/3 implementation provided an opportunity for DSW to have modern information systems and provided with enough training to alleviate any fears about working with the new software.

As anticipated, the ERP implementation provided DSW with the opportunity to renew the hardware, from host-centric to network-centric, and to modernize its software infrastructure. Gradually, R/3 will be implemented at branches of DSW worldwide, helping DSW with the transition from single-firm to multi-firm structure and from local to global operations. The ERP implementation project at DSW served as a pilot, and the same ERP infrastructure is expected to make its way to other sister companies of DSW in the Israel Chemicals LTD. Implementation of R/3 by another Israel-Chemicals member is already approaching completion faster and cheaper than at DSW, under the leadership of an IS professional from DSW.

In retrospect, SAP R/3 has also provided DSW with some tangible savings: the number of pages printed per month was reduced by 80% from about 25000 to about 5000, because of the better ability to query online instead of printing reports. The number of shifts for server operations has gone down from 3 to 1.5 per day. With all batch processes substituted by on-line ones, information provision has improved in the sense that the raw data is now more up to date. Thus, for example, it is now possible to know in real time (as opposed to twice a week before ERP) what the real inventory levels are and, therefore, DSW is saving money on inventory without hurting production in any way. Because of the uniformity of screens for different modules, it is now possible to easily carry a transaction from one module to another (e.g., from a costing screen to a contractor screen). Another improvement has to do with the way materials are ordered from the warehouse. Until the ERP implementation, precious time was wasted when workers came to pick up materials without verifying availability beforehand. Under SAP R/3, pickup from the warehouse is permitted only after availability is verified through remote inquiry. DSW's employees accepted the R/3 process without resistance, despite failure to introduce a similar ordering process under the legacy systems in the early 1990s.

On the down size, there has also been some deterioration in information provision, especially for top managers who were accustomed to using the Commander Executive Information System. In some respects, they now have less access to information than before because Information-Center tools are not yet effectively integrated with SAP R/3. Even for lower levels of management, information provision is in some cases poorer than before because the design of several processes and procedures has been proven inappropriate. Some information provision processes now take longer and require navigation among a series of several screens whereas beforehand each of the same processes took only a single screen to complete.

In sum, DSW has already reaped substantial benefits from the strategic move to ERP. The hardware has been renewed. The IS applications and DBMS have been upgraded to Y2K-compliant and fully integrated functional modules, with uniform and smooth transitions among them. The highly needed yet missing plant-maintenance module has been implemented. DSW processes have improved and become more efficient. Having implemented a multi-company solution, organizational learning can now take place at other DSW and Israel Chemicals LTD locations. IS professionals at DSW, whose skills have been upgraded significantly, can rely upon complete documentation and apply the same set of standards and tools, including ad hoc drill-down capability and advanced quality-assurance tools.

## CURRENT CHALLENGES/PROBLEMS FACING THE ORGANIZATION

The long list of benefits and the current perception of ERP success within DSW stand in sharp contrast to the growing number of horror stories about failed or out-of-control ERP projects (Davenport, 1998.) Against this contrast, it is of interest to consider the starting conditions, goals, plans, and management practices that may have increased the chances of success at DSW.

Changes in the competitive environment in the 1990s pushed top management at DSW to approve funding for the ERP implementation in order to support company growth and restructuring from single-firm to multi-firm. The ERP implementation served strategic DSW goals such as improvement of business results, technology replacement, and reduction in the total cost of technology ownership. The ERP implementation has also created a platform for reengineering business processes in the future and for integrating the supply chain. ERP plans were well thought of, arguing a general business case in addition to a technical case. The fact that benefits have been reaped despite the limited scope and minimal modifications suggests that there was, to start with, a good fit between R/3 and the needs of DSW.

Management practices contributed to success as well. Contractual arrangements with vendors worked well and good working relationships were maintained within the implementation teams. Project management adhered to the following principles: the partial scope of the implementation was not changed during the project, except for adding the maintenance module, software modifications were avoided as much as possible; and sufficient investment was made in testing, data conversion, and user training. Even though a change of guard took place in the IS organization a few years prior to the ERP implementation, the fact that the new VPIS was not a newcomer to DSW contributed to the stability of the IS governance and experience. In retrospect, if given the opportunity to rethink management practices, the only thing DSW might have done differently is to let the IS professionals and top users in development teams work full time on the project away from DSW's premises.

It is important to acknowledge that no external events and changing conditions beyond DSW's control worked against the implementation either: financial conditions were good, there was no turnover of key personnel, and vendors neither overstated their expertise nor went out of business.

The IS group is dealing with problems and challenges by working on continuous improvement to the R/3 environment. One problem is that, probably due to the decision to opt for a big-bang implementation with only minimal reengineering and software modifications, some cumbersome and unfriendly work processes exist following the R/3 implementation. To correct this and allow DSW to better take advantage of what the R/3 environment can offer, DSW processes will be gradually reengineered and R/3 software modifications will be permitted. Thus, many efficient and effective processes that R/3 can support, but were not implemented so far, will eventually be introduced to DSW.

Another problem, which has resulted out of the decision to limit the volume of routine reporting, concerns poor design of some routine R/3 reports and user demands for additional reports. Work is ongoing to alleviate this problem—and poorly designed reports, especially those consumed by outside parties, are being redesigned. In addition, new additional reports are being planned, including control reports that are needed to support newly introduced procedural changes.

Finally, some DSW functions are not yet supported by R/3. To meet this challenge, implementation of additional R/3 modules is being planned, including project management, marketing, and production-planning. Also in the planning are systems to support senior-level decisions, such as a data warehouse and an executive information systems.

## FURTHER READING

Allen, B. (1998). Year 2000: Impact on the Healthcare Industry. *IS Audit and Control Journal*, (1), 11-13.

Ang, S. & Straub, D. W. (1999) Production and Transaction Economics in IS Outsourcing: A Study of the US Banking Industry. *MIS Quarterly*, 22(4), 535-554.

Benjamin, R. I. & Levinson, E. (1993). A Framework for Managing IT-Enabled Change. *Sloan Management Review*, (Summer), 23-33.

Henderson, J. C. & Venkatraman, N. (1993). Strategic Alignment: Leveraging IT for Transforming Organizations. *IBM Systems Journal*, 32(1), 4-16.

Jenkins, H. W. (1997). Turns Out the Year 2000 Problem is Just Beginning. *Wall Street Journal*, (May 6), A23.

Mensching, J.R. & Adams, D.A. (1991) *Managing an Information System*. New York, NY: Prentice Hall.

Pliskin, N. & Shoval P. (1987). End User Prototyping: Sophisticated Users Supporting Systems Development. *DATABASE*, 18(4), 7-17.

Rockart, J. Earl, M. J., and Ross, J. W. (1996). Eight Imperatives for the NEW IT Organization. *Sloan Management Review*, (Fall), 43-55.

## REFERENCES

Applegate, McFalan, & McKenney (1999). *Corporate Information Systems Management*. Boston, MA: Irwin McGraw-Hill.

Bancroft, N.H., Seip, H., & Sprengel, A. (1998). *Implementing SAP R/3: How to Introduce a Large System into a Large Organization*. Greenwich, CT: Manning.

Davenport, T.H. (1998) Putting the Enterprise into the Enterprise System. *Harvard Business Review*. (July-August), 212-131.

GartnerGroup Reports (1994-1996).

Kappleman, L., Editor (1997). *Year 2000 Problems: Strategies and Solutions from the Fortune 100*. London: International Thompson Publishers.

McFaralan, F. W. & Nolan, R. L. (1995). How to Manage an IT Outsourcing Alliance. *Sloan Management Review*, (Winter), 9-23.

META Group Reports (1994-1996).

Pliskin, N. (1989). Human Resource Management: Guidelines for Effective Introduction of Microcomputer Technology. *Journal of Information Systems Management*, 6(2) 51-58.

Pliskin, N. & Shoval P. (1989). Responsibility Sharing Between Sophisticated Users and Professionals in Structured Prototyping. *Information and Software Technology*, 31(8) 438-448.

Violino, R. (1997). Year 2000 Getting Down to the Wire. *Information Week*, (February 10), 14-15.

## Chapter 10

# A Three-Tiered Approach to Global E-Commerce: Experiences of Nu Skin International

David Paper
Utah State University, USA

Ken Tingey
Opennet Corporation, USA

*Access to the world enabled by the Internet facilitates internationalization as never before. However, lack of a coherent global Internet strategy can relegate any company to a strictly provincial "neighborhood" status. Globalization strategies and tactics should therefore be of central concern to all enterprises. To research the strategic issues involved in Internet-based globalization, we embarked on a case study. Our goal is to explore how the Internet and its related technologies can serve to help organizations better deal with the challenges of conducting global business. Our research enabled us to identify a set of heuristic "rules of thumb" that might be used to support Internet-based globalization efforts. In our study we discuss the many challenges to establishing successful global enterprises. We then introduce a model for understanding global business requirements in the e-commerce age. We conclude by analyzing a case study to initially validate our theoretical model and summarize our findings.*

## INTRODUCTION

Many organizations look to the Internet as an instrument to support their global aspirations, to allow them to enter new markets, to extend their brands and offerings

Previously published in *Cases on Worldwide E-Commerce: Theory in Action*, edited by Mahesh Raisinghani, Copyright ©2002, Idea Group Publishing.

geographically, and to use increasingly pervasive Web access to enter global markets. The hope is that instant worldwide access to information via the Internet can serve as a global "calling card" in consumer and business-to-business environments alike. The current global climate of peace and openness appears to be conducive to global business. The Cold War is over. China is moving toward being an "open" trading partner with the world (Weeks, 2000). Moreover, access to the Web is relatively inexpensive as compared to the past.

The Web appears to be the new vehicle to global business exchange. It seems that Web merchants are satisfied if they can get credit card money verified in real-time while leaving other business processes to a complex maze of emails, sticky notes, and off-system paper trails. However, basic business principles should still apply in e-commerce. Successful global e-commerce requires more than a Web site and simple data interchange functionality. The way of the Web has been to slide through fulfillment requirements. Few Web-based tools provide much in the way of inventory control, for example. As a result, organizations face a continental divide between whatever systems they have used in the past and new Web-based technologies. ERP, the hoped-for "do all" and "be all" of the mid-1990s, didn't meet expectations (McNurlin, 2001)—the long-term implications of which are still to be determined in the marketplace.

The question confronting large-scale organizations around the world is how these two technological worlds–the Web and internal enterprise computer systems–can and should come together. The Web is of necessity directed outward at customers and markets and collaborative partners. Traditional enterprise systems, from ERP to older legacy systems, are by their very nature mostly inwardly directed, focusing on the need to coordinate employees and manage materials according to financial, tax, and operating imperatives.

Furthermore, immediate access to the world as enabled by the Internet forces the internationalization issue as never before. Lack of a coherent global Internet strategy can relegate any company to strictly provincial "neighborhood" status. In the face of such challenges, globalization strategies and tactics should be of central concern to all enterprises. We therefore embarked on a study to explore how the Internet and its related technologies can serve to help organizations better deal with the challenges of conducting global business. Moreover, we want to identify a set of heuristic "rules of thumb" that might be used to support Internet-based globalization efforts. We begin by discussing the challenges to global enterprises. We then introduce a model for understanding global business requirements in the e-commerce age. We continue by analyzing a case study to initially validate our theoretical model.

# BACKGROUND

### Agency and Delegation Challenges For Global, Networked Enterprises

Global business opportunities did not start with the Internet. Drucker (1973/1993) outlined a mature international business environment decades ago. He drew attention to important strategy and delegation issues that continue to elude executives and managers of international organizations. Drucker believes that global enterprises cannot be managed wholly at home or abroad because top executives do not have the requisite local knowledge to make all decisions and oversee all projects. Furthermore, locals tend to optimize solely on their national or regional operations, many times causing sub-optimization of the overall company effort. We believe that the answer lies in an enterprise approach. The question remains whether the Internet and its related technologies can serve to support the global organization as a holistic enterprise. We believe that the existence of the Internet provides an opportunity for organizations to greatly expand their business initiatives.

Global e-commerce introduces massive delegation problems and broad-scale agency challenges in spite of the open, friendly nature of the Internet. Success in international business requires that organizations groom their message to match preferences and characteristics of local audiences. As such, people (agents) with knowledge and experience in local markets and local customs must be enlisted in the cause, armed with knowledge of and commitment to the primary mission of the organization (Victor, 1992). However, success in foreign markets may be stymied by behavioral challenges such as pecking orders set up between managers at headquarters and those in local markets restricting effective cooperation, delegation, and integration.

The business model at headquarters tends to follow established patterns related to the country of origin and past experience. Hence, financial procedures, models, marketing, sales, and associated systems will naturally follow edicts dictated by the home office. Production, manufacturing, and sourcing procedures are typically established to meet specific requirements of the home market based on standard products and processes. Global organizations also attempt to establish norms and procedures for employment and outsourcing requirements based on known patterns.

To extend a traditional business model to world markets, home office agents and foreign agents must engage in a comprehensive study of the existing model to apply it to new and different environments (Victor, 1992). Delegation and the associated power given to agents must therefore be carefully balanced to match headquarter objectives to local ones.

Matching corporate strengths with global requirements, however, can prove unproductive due to a number of reasons. The sheer number of agents increases so dramatically that communication difficulties arise due to complexity alone (Collins, 2001; Sklar, 2001). Diversity is also greatly increased due to different cultures. Diversity opens such problems as collective action due to the difficulty of establishing consensus and cooperation.

Obstacles to collective action come from a tendency by agents to maximize local self-interest (Kiewiet & McCubbins, 1989). As such, agents often behave "in ways that are inimical to the interests of the community as a whole." Coordination may also suffer where "[agents] become uncertain as to which strategies other members will pursue, and coordination may never be achieved" (Kiewiet & McCubbins, 1989, pp. 1-2). Traditionally, collective action is often achieved through manifestations of leadership on the part of managers (Drucker, 1998). Digital systems can remove communication barriers and assist agents at all levels if such systems are strategically deployed.

## Three-Tiered Approach to Global Business Requirements

St. John and Young (1999) distinguish between two kinds of global companies. First, the "global enterprise" attempts to find economies of scale by placing standardized products into local markets. This reduces the need for delegation because decisions come mostly from a centralized authority. Agents are less likely to have high degrees of autonomy. Second, the "transnational firm" has dispersed specialized national units that provide differentiated market offerings through integrated worldwide operations. Delegation is widespread and agents are given high degrees of autonomy in their local markets. Consistent with Drucker (1973/1993), St. John and Young underscore a need for complex and flexible systems to support unpredictable requirements of global systems that must be responsive to local markets.

The "global" model is differentiated from the "transnational" model as delegation is considered from an entirely different perspective. With a global model, delegation is minimized. Local authorities are charged with deployment of the office model with little modification, coupled with a general understanding of the overall global system. Local agents follow home office directives with little room for discretionary decisions. In the transnational model, business and systems policies are highly delegated to local agents, as they are in the best position to understand diversity in language, currency, and customs. Thus, agents have a lot of discretionary power.

Walton (1995) discusses the value of the transnational model, that is, one that allows a "structurally fluid and ever-changing" paradigm to meet market demands.

> Take the example of a high-technology company moving toward globalization. Initially, the challenge was stated as a financial goal: increasing nondomestic revenues. Despite a lot of attention and specific financial targets, progress was slow and inadequate. Executives concluded that the strategy could not be enacted with the current [global] organizational structure and that the company needed to move to a transnational organizational architecture ... empowering local managers was the implementation strategy ... the old structure was unaware of, or filtered out, the regional limitations it had placed on people and market development (Walton, 1995, p.125).

Even before the Internet, Castells argued in favor of a transnational model that he termed the "network enterprise."

> Under this strategy [transnational] ... companies relate to a variety of domestic markets ... rather than controlling markets from the outside, they try to integrate their market shares and market information across borders ... in the old strategy, foreign direct investment is aimed at taking control ... Under the [transnational] strategy, investment is geared toward the construction of a set of relationships between companies in different institutional environments. Global competition is greatly helped by "on the spot information" from each market, so that designing strategy in a top-down approach will invite failure in a constantly changing environment and with highly diverse market dynamics (Castells, 1996, p. 165).

Information technology plays a major role in facilitating the transnational model, as it enables a flexible, adaptive model to actually work (Castells, 1996). "With adequate levels of [local] information and resources [organizations can] handle errors better than fragmented, decentralized networks, provided they use adaptability on top of flexibility" (Castells, 1996, p. 166). Castells argues that the organization must integrate the logic of the corporate system with the business process logic of the organization into a digital, networked format so that managers can obtain and use information to meet the unique needs of the organization in its various markets. However, there may be a "reverse effect" in terms of agency and delegation levels. Central agents may want to delegate powers to the local agents, but local agents may not elect to accept them. For delegation and agency to work, local agents must use their agency powers to generate desired objectives typically localization, market adjustments, and lateral cooperation (Nadler and Tushman, 1997). Nadler and Tushman also promote information technology as a means of accomplishing this objective.

> Rather than seeking designs that emphasize coordination and control, what organizations need now is speed, innovation, customer focus, and radically improved productivity... information technology ... makes it possible for companies to make timely information available ... no matter where [agents] are located. Not only does information technology demolish traditional constraints of time and geography; it enhances collaboration and teamwork, eliminates the need for entire levels of bureaucracy ... the innovative use of teams [allow] people to use their collective knowledge, judgment, skill, and creativity to perform a variety of jobs and functions, rather than just one, in concert with their colleagues (Nadler and Tushman, 1997, p 9).

Organizations must also take into account localization issues such as currency, taxes, culture, etc. (Sklar, 2001). By all rights, conditions will not be constant across-the-board (Cannella & Monroe, 1997). However, the transnational model provides delegation powers to make decisions at the local level and agents are given latitude in establishing and executing their own objectives. Delegation and agency powers give local managers the ability to interact more effectively with their customers especially if they are "natives" to the culture. Our theory focuses on the agency and delegation levels with networked enterprise systems because we believe that localization issues cannot be controlled from headquarters. Therefore, our model controls for these factors. However, financial standards, general company policy, and production techniques can be better controlled centrally.

Organizations may demonstrate elements of the global and transnational models depending on levels of delegation, flexibility, and adaptability. A more realistic scenario would account for some combination of the two—a hybrid model in which some system elements retain centralized features, while other factors are open to local interpretation and implementation. While organizations may prefer one of the three—global, transnational, or hybrid—we posit that they should be expected to exhibit features of all three. The effects of Internet technologies on variations of these models are of particular interest to organizations wishing to use the Internet to extend their missions worldwide.

The "global" model is characterized by limited agency with minimal delegation in which local agents are wholly responsive to home office mandates. Agents are given very little autonomy and are charged with carrying out orders. Decisions are made at the top and "trickle" down to the bottom. Agency initiative is not considered important. Agents in the "transnational" model are trusted to use their initiative to coordinate activities at the local level and make lateral connections as deemed appropriate (i.e., to be proactive). Cross-functional and cross-geographic interaction is encouraged. The home office realizes that complexity alone minimizes its ability to control the business at the local level and has come to feel comfortable with control mechanisms—largely embedded in the information architecture of the firm (Castells, 1996). Authority is granted to agents to make decisions that cannot be made effectively in a top-down paradigm. Delegation is maximized in transnational environments wherever possible.

The "hybrid" model is a compromise. Agents are recognized (by the home office) for special skills, perspectives or implementation ideas. Agents are given authority to make decisions within limited functional areas based on customs and localization factors. However, delegation between the home office and local organizations is limited to agreed-upon boundaries. Agent initiative is valued, but tempered by careful planning and analysis of each market and each business function by the home office.

Our model is not intended to be comprehensive. We are now focusing on agency and delegation because these factors emerged from our case study. Other factors that require future consideration may be compensation structures, corporate cultures, cultural factors in the home office and in locations throughout the world where the organizations function (Cannella & Monroe, 1997; Zacharakis, 1997).

# MAIN THRUST OF CHAPTER

## Research Focus

To explore the nature of agency and delegation problems faced by global competitors where technology plays a major role, we embarked on a case study of Nu Skin International (NUS), a global direct sales leader. To collect data, we interviewed managers responsible for overseeing global implementation of Nu Skin's systems. We also corresponded with our contacts by telephone and email. We maintain current relationships with our contacts. Our theoretical model (Figure 1) helped provide a theoretical lens to frame the case. We looked at Nu Skin's global systems development strategy (given its relationship with the company's Internet and global systems strategy), the company's

*Figure 1: Three-tier networked enterprise model of global e-commerce*

| | Agency | Delegation |
|---|---|---|
| **Global** | Responsive | Minimized |
| **Hybrid** | Negotiated | Situational |
| **Transnational** | Proactive | Maximized |

intensive use of ERP systems (SAP, in particular), its commitment to Internet technologies, and its continued dependence on legacy systems of various types. Case analysis provided valuable information to aid in understanding the requirements of global enterprise competitiveness in the era of the Internet.

## Nu Skin

Nu Skin is committed to providing compelling business opportunities and superior products and services. It is a leader in the direct selling industry. For more than 15 years, it has extended its business model and products around the globe, identifying and capitalizing on important demographic and business trends. Nu Skin global sales are at nearly $900 million a year. It does business in 31 countries and has been an active global competitor for over a decade. The organization oversees the activities of over 500,000 independent distributors, who may choose at any time to participate in business opportunities in any region and country in the world through use of a fully integrated global structure that existed long before the Internet became a public phenomenon. Every month, the company calculates and remits bonuses and other rewards to its distributor force that entails tens of millions of dollars of disbursements in a country/currency mix that is based on rules, regulations, and incentive structures of tremendous complexity.

To manage such requirements, Nu Skin has brought together technological resources, people, and information systems on a global scale. As an early implementer of advanced computing tools, Nu Skin has been on the forefront in the use of distributed systems, global networking and communications technologies, large-scale, multilingual database implementation, integration of heterogeneous systems, and global ERP rollout.

## Key Players

Top management support is strong and responsive to the management team charged with global Internet management. That is, Boyd Blake, Joel Erickson, and Shane Moss have responsibility for all projects in this area and report to top management periodically.

Boyd Blake is the ERP manager at Nu Skin. According to Mr. Blake, Nu Skin's recent attempts at making use of state-of-the-art technologies, including the Internet, to expand globally have provided the company with early-stage knowledge and experience as to the strengths and weaknesses of available technologies and models. In Nu Skin's move toward globalization and integration, careful attention has always been given to local business and technological issues, lending an interesting view as to how a successful global competitor addresses its various markets.

Boyd Blake manages the overall ERP, but Joel Erickson oversees new ERP and Internet development. One of Mr. Erickson's main concerns is to manage Nu Skin's global Internet presence. Therefore, he is very concerned about customer perception of its Internet sites around the world. When Nu Skin customers complain about navigation problems on the Web and other unsatisfactory process issues, Mr. Erickson is charged with reconciliation. He must keep the Web customers satisfied, in addition to his non-Internet ERP development duties. He must also keep abreast of foreign Web sites using English to make sure that they adhere to Nu Skin standards.

Shane Moss leads the financial integration team in Japan. Mr. Moss exhibits some patience with the home country language in finance because there is a language problem. For example, accounts payable is all in Japanese because many clerks do not speak English. However, final reporting must be in English because this is a requirement from headquarters in the U.S.

## Issues, Controversies, and Problems

Nu Skin has high hopes with respect to its global Internet initiatives. The company plans to achieve a leadership role in what it has defined as "e-direct marketing" and "e-direct selling". With several country-specific sites on the Internet, Nu Skin plans a major effort to bring its off-line community to the Internet. As such, it hopes to provide instant access to its global tree of distributors, to achieve real-time information retrieval, and to seamlessly provide access to up-to-date inventory information around the world.

Such plans, however, do not revolve around a monolithic site—a global 'nuskin.com', if you will. Although such a site does and will continue to exist, the major global Internet effort will take place on a more loosely integrated country and regional site scenario. Although Nu Skin's e-direct plans are very ambitious technologically, it has no overt mandate for integrating diverse technology platforms around the world. As such, Nu Skin is faced with a tremendous technology integration and implementation challenge. However, the social aspects that come with such radical change may prove to be an even greater challenge.

Nu Skin has experience meeting individual local market needs while establishing a global presence. It also has a mature understanding of the capabilities and limitations of traditional ERP tools. Nevertheless, management realizes that they need to begin thinking "out of the box" because their global presence was brought together without considering the Internet's potential to integrate and streamline processes and social challenges that surface when working in a variety of different cultures. Nu Skin management is thus attempting to keep all of its options open until a specific solution set proves its merit in a given market. In this, the company's plans are similar to the "transnational" model. As we will see, however, the requirements of effective globalization do not allow for a single delegation style for the entire company because elements of a "global" model as well as elements of a "hybrid" environment exist within the overall corporate structure.

Many Internet users may perceive that a corporate Web site with language-based buttons for the non-English-inclined and perfunctory but functional links to back-end systems is all that is needed for successful implementation of e-commerce on a global scale. When a global enterprise like Nu Skin that already services hundreds of thousands of distributors establishes e-commerce strategies as Nu Skin has done, "e-anything" would seem to be an easy proposition at any level. The enterprise, after all, has an existing community and it already serves the function of a "portal", albeit largely offline. Nu Skin thus has a basis to command the kind of traffic that is the lifeblood of an all-encompassing e-commerce organization. However, successful global e-commerce requires much more than what initially may appear to be required.

Boyd Blake is the ERP manager at Nu Skin. According Mr. Blake, Nu Skin's recent attempts at making use of state-of-the-art technologies, including the Internet, to expand globally have provided the company with early-stage knowledge and experience as to the strengths and weaknesses of available technologies and models. In Nu Skin's move toward globalization and integration, careful attention has always been given to local issues, lending an interesting view as to how a successful global competitor addresses its various markets.

> I think that there is a split that depends on your market size and the growth potential in a market. I think it is different for a small market than it is for a large market. Japan is our largest market. They are very autonomous. They tell us what to do and we try to give them a framework to work with. A lot of it is driven by our divisions, as well, and not all divisions are in all countries, so as a division goes in we are very new in this division model. We struggle with "does the division own this, or does the country own this?" But if you look at a small market, like a Guatemala, they look to corporate for direction, how do we do this, what would you like us to do, how do we need to do this so that we will be successful, so that we will take off? So it's a couple of different models [large and small]. Your mid-sized markets, probably [require] a little of both, but I don't think its a "one size fits all." Not all small markets shun corporate [however]. They'd like corporate to do everything for them because they don't

> have the resources, they don't have the talent and skills, etc. (Boyd Blake, personal communication, July 14, 2000)

The state of the art for Internet globalization is typically considered to be a one-tiered approach, that is, there are really no new business models in this "new" economic arena (Collins, 2001). The typical business model extends the home country site with some language and content localization (Sklar, 2001). Information portals have introduced a high degree of sophistication in a similar manner (Meyers, 1999). Nu Skin's experience, however, points to a much more stylized approach, depending mostly on the dynamics of markets in individual countries and regions such as market size, resources, and local expertise.

### Heuristic Rules of Thumb

The first rule of thumb is that a multitiered environment can make use of e-commerce tools and technologies to improve its position in the global e-commerce arena. Not only does Nu Skin constitute an example of how companies in various industries, particularly consumer-related sectors, can globalize, but it provides a unique means of understanding relations between organizations that function on a global level and their intended audiences. In most cases, unsatisfied Internet visitors simply abandon sites when their global e-commerce experience is less than satisfactory. In Nu Skin's case, they complain to the company. Joel Erickson, says, ". . . when foreign Web sites come up in English, we hear about it very fast and work toward making our site(s) more efficient and effective" (personal communication, July 14, 2000).

The second rule of thumb is that language is critical in dealing with country-specific issues. Although Shane Moss indicates that there is some patience with home country language in finance, he remarks "we hear about it in the finance realm, but it is not as important, I don't think." He indicates that overall it becomes quite critical, even in underlying communications within the firm.

> Well, take accounts payable, for an example. That needs to be all in the Japanese language because they have many clerks that do not speak English. There's only a few that speak English, but a lot of the reporting that we provide to them comes out in English, mainly because it is required because we are a U.S., or a centralized U.S.-based company and we require that information in [English]. Most areas you are going to find are going to be in Japanese, but there are some areas where [it is not], but it is still important. It is very important to them that its in their language just as it would be for us if we were here and we had a foreign country coming in and

telling us about English. (Shane Moss, personal communication, July 14, 2000)

The third rule of thumb is that market tiers are important strategically in a global networked environment. In Nu Skin's 31 markets, there is a great deal of variation, from large-scale market penetration in Japan (one of the largest consumer markets in the world) to participation in smaller markets in Latin America and throughout the world. The model utilized by Nu Skin in Japan and in parts of Asia, where language elements are more critical and the scale of operations is substantial, would correspond to a "transnational" structure. Latin American markets, as outlined by Nu Skin representatives, would follow a "global" strategy similar in some sense to a "cookie cutter" approach. Latin regions follow the edicts of the home office more closely than larger markets because they have substantially less resources and they lack the technical and marketing expertise that larger regions like Japan have in abundance. Resource availability is a critical factor in global Internet business (Garten, 2001). Interspersed are many mid-tier markets in which significant requirements for local investment and modification of business issues may not be necessary. This type points toward a hybrid strategy. We believe that market size is a critical factor influencing choice of model. Of course, we understand that Nu Skin is only one case that cannot be generalized until we gather data from other organizations (which we are in the process of doing).

The fourth rule of thumb is that offline business activities can be successfully converted to online if the capabilities and limitations of technology and the dynamics of the market are well understood. Nu Skin's experience is beneficial in large part because we believe that the company is working to achieve the reverse of most global Internet organizations. Nu Skin's goal is to bring its world to the Internet to accomplish a number of things electronically that the company has traditionally achieved offline. Many organizations wish to use the Internet to gain access to markets in which they do not now participate (Collins, 2001). Nu Skin's experience in converting an existing population of affiliates provides an interesting kind of feedback that should help all global Internet-based e-commerce hopefuls. Basically, its approach is to adapt to the scope and requirements of local markets while working toward technology integration with its affiliates in specific functional areas.

The fifth rule of thumb is that flexibility and adaptability are critical when integrating Web and ERP technologies into plans for global business deployment. Nu Skin was not able to translate its ERP system screens into several desired languages due to overstatements of an overzealous ERP vendor. Interestingly, this may have aided its efforts at internationalization and localization at its current stage in the development of its global infrastructure. Nu Skin bought its ERP system based on representations that the technology supported double-byte characters at that time, but during later implementation activity, it discovered that the vendor's beta-stage product did not meet Nu Skin's requirements. As a result, Nu Skin did not attempt to develop multi-language functionality within the ERP structure as the company had originally planned. Fortunately, company managers feel that they now have more flexibility and adaptability in terms of translation options than if they had begun full language integration within the original ERP structure. As stated by Mr. Blake and Mr. Erickson,

> **Blake**: . . . when we bought the [ERP] system in '96, we found out very quickly that it was not easy to do that [support double-byte characters], and in some cases, not possible at all to have different combinations of languages.
>
> **Erickson**: We had to make some choices as to which languages we wanted to combine. We were at the forefront of those issues and really helped push SAP into further development and evolution to where today, they have a much better solution and [the] answer to that is much better today.
>
> **Researcher**: So, at this point, in the process of extending, or of making choices to [implement] those three or four key languages, did you find that it helped to have had experiences in working with multiple languages or did it restrict you? In other words, did the [language] choices you made restrict you as you tried to expand to other languages?
>
> **Blake**: I think it would have restricted us if we had been successful in taking the system to other markets. But it wasn't a problem since we didn't do much outside of the U.S. [using ERP] and then, when we finally did get approval to go do something in another language [Japanese] in Japan, the problems were that we didn't see anything else on the horizon past that point. So we took the path to do English and Japanese and now we can do almost anything we'd like (personal communication, July 14, 2000).

The sixth rule of thumb is that the Internet infrastructure should be integrated with head office data before connecting with other countries. In other words, it should be thoroughly tested with internal processes and systems before international deployment. With limited commitment to a fixed environment, Nu Skin is free to consider many options for Web-based, front-end integration worldwide. Nu Skin was also fortunate in that it developed its Internet infrastructure to integrate and process data very well within its U.S. headquarters before connecting its core business processes to other countries. Nu Skin had the advantage of streamlining and refining the Internet infrastructure before moving it global.

The seventh rule of thumb is that experience in global markets is critical to success when moving to a global Internet paradigm. Other than with respect to the monthly global bonus calculations and some material management functions that have always been managed centrally, offices outside the U.S. have operated more or less autonomously. By the time corporate management decided to expand its Web-based initiatives to support globalization objectives, Nu Skin was a seasoned competitor in many diverse markets. The company had already proven successful in deploying global strategies making use of various sized markets – small, medium, and large. Hence, Nu Skin developed proven business models that it was able to adapt in its efforts to incorporate the benefits of the Internet on a global scale. The major task of company executives and managers was to develop an overall globalization model that would fit the company's basic business model and that would allow the company to take advantage of the Internet phenomenon.

## Analysis of the Three-Tiered Model

The three-tiered globalization model serves as a guideline for selecting desirable markets around the world for entry and for evaluating strategies and tactics for improving performance of ongoing initiatives. Requirements brought on by internationalization are as old as commerce, of course, but the Internet's existence does affect the costs and benefits of doing business in various countries by making information available more cheaply, more quickly, and with less effort than could be achieved in the past.

At the high end, it should come as no surprise that large markets require more attention and resources than smaller locales. In the Nu Skin case, such larger markets would include Japan and Taiwan. In the latter case, local management spearheaded a separate computer environment altogether in order to meet local needs. Also, a higher level of autonomy by local representatives is warranted because of the sheer size of the markets themselves, as it is difficult to maintain a "command and control" structure in large markets (Drucker, 1973/1993).

Nu Skin supports many "global" market tiers, such as Guatemala, which, as Mr. Blake indicated, are happy to get anything and everything from the home office that they can. In the case of Nu Skin, many markets would fit into the "hybrid" category, where there are varying levels of autonomy and delegation, including Australia, New Zealand, and several Asian and European markets.

Achieving a balance between local and home office controls may be easier now that the Internet and related technologies exist (making a transnational strategy more attainable) (Castells, 1996). However, dissemination of information by means of the Internet cannot compensate for lack of skill or judgment on the part of local managers or employees. Employees who are charged to make critical localized decisions and otherwise represent the organization to local authorities, employee populations, and market players are not automatically endowed with the requisite experience and skill just because the Internet is available to them. Furthermore, a transnational strategy is likely to be beyond the reach of an organization with an Internet presence, but no other physical resources or connections in the market in question (which size tends to provide).

Medium-tier, hybrid markets may be the most difficult of the three to manage. In the face of uncertain potential outcomes, such environments may have unique requirements that levy costs in line with major markets, but with population demographics or lack of size and robustness in related industries and markets that make equivalent investment less compelling, even questionable. In other words, medium-tier markets typically cannot attract resources in similar ways as large-tier markets. Fulfillment and localization activities are determining factors as to whether participation in this type of market is worthwhile to the organization. By the same token, penetration into smaller markets may be more feasible if the organization overall is able to achieve transnational-like characteristics that allow for lower costs overall as a result of intelligent collaboration by local agents in position to see opportunities for optimizing the resources of the firm within local and regional areas. This is a major objective of Nu Skin in its efforts to centralize back-end systems, where materials must be deployed on a large scale while the company follows a three-tiered strategy in its Internet-centric distributor interface to meet the demands and characteristics of local markets.

Lightweight products of high value, with limited trade restraints and high relative prices, are ideal for medium-tiered markets because their return on investment can be considerable. Such products are more conducive to Internet-based globalization than heavy, low margin items that face local restrictions, trade barriers, and competition. If language and other localization issues are not consequential, medium-sized market opportunities can be very compelling.

The Internet globalization challenge was brought out by Nu Skin management when the company, as a matter of strategy, decided to expand on its initial U. S. based Internet experience with an integrated, worldwide enterprise network. As is the case with all companies that trade in goods, Nu Skin must function with some kind of physical presence in every market in which it conducts business. The relationship with physical and virtual worlds is where globalization models based on the Internet alone become problematic. Tangible items do not benefit from blanket exemptions from trade, customs, communication restrictions, and even gravity that are enjoyed in the virtual, networked world. In large part due to these conclusive limitations, it remains to be seen whether the Internet can revolutionize and significantly modify goods-based markets that in some cases have existed for centuries and longer.

Medium-tier markets may be most conducive to hybrid globalization systems that borrow from both global and transnational systems structures. Moderate market opportunities with low-level localization and fulfillment requirements will likely provide the basis for many global Internet success stories. Meanwhile, a similar market and localization environment with minimal dependence on physical product fulfillment is in line with the "global" strategy as outlined in our model. The ideal "global" product, of course, is anything that can be converted to digital form—words, music, movies, and documents. Such products do not face the logistics problems associated with physical goods.

**Solutions and Recommendations**

The Nu Skin experience is that a broad category of physical, personal care products breaks down to the three tiers simply as a function of the markets themselves. The same product line in Guatemala takes on very different characteristics when taken to Taiwan or Japan or the countries of the European Union. Thus, regardless of other factors that make the decision to be involved in local markets valid, three market tiers as indications of Internet strategies on a country-by-country basis are useful criteria.

Treatises on Internet commerce seldom deal with inventory control issues or materials management systems, though these factors have a great deal of bearing on whether success is to be achieved when doing business around the world. In the case of Nu Skin, the requirements of global fulfillment, forecasting inventories, and other mainstays of business—not often emphasized in the glitzy world of the Internet—occupy a major part of its efforts. Globally, the firm clearly follows a transnational strategy in the area of materials management and fulfillment. As outlined by Blake,

> We've seen a need to centralize our forecasting and get sales information from a central source, so we use SAP to do local forecasting and we do that by taking feeds from all of our different systems, orders systems, [and] sales systems around the world to give us a rough inventory picture … and I say rough, because it's rough, it's not timely, it's not integrated, but [the system works] through interfaces. We bring that data in and project our forecasting through SAP.
>
> In doing that, we have also pushed out some SAP functionality to seven markets, mainly in Asia, because those are our biggest markets, but they have SAP access. They enter goods receipts when they receive shipments, and we are in the process of trying to even upgrade what they do a little more all in an effort to give us a better more timely picture of global inventory so we can forecast better.

> So we've been through some initiatives [where] we've dealt with rolling bits and pieces of [information accessible] to those seven markets. We also embarked, about two years ago, [on] a project for Japan after we failed to put SAP in Taiwan. We were looking for what we should do next and Japan needed a new finance system. There was some thought that we ought to upgrade their inventory and could use a warehouse management system, but with the volume that we do in Japan, it became very apparent that we should put the whole package there because of the tremendous ordering volume … and so the ordering piece at that time started to become separated and we started to look at SAP as more of a fulfillment vehicle. We embarked on a project to do part of the inventory management [with] the full suite of financials in Japan (Boyd Blake, personal communication, July 14, 2000).

Even now, the relationship between these two worlds—the Web and the front-end world (as seen by outsiders) along with ERP and operating requirements behind the scenes necessary to make it come together right—is far from fixed. Apart from the marketing concepts and the promotional representations of these various technologies, the efforts of those charged with making them work are critical to sustained competitiveness. As stated by Nu Skin's Blake,

> … Our new wave or next generation of systems is being driven by the Web and e-commerce. We would like everything to be available via the Web, especially our ordering piece and what the distributor sees. We would like to push our ordering out to the Web from the call center model. So you can see even a bigger break between our ordering system, our back-end fulfillment system, and our commission system [although this] is still the core functionality that we've had in the past… But we've seen a break away from using it as everything and you'll see more of a break as we take a Web e-commerce solution out worldwide, or at least we think we will take it worldwide. At least that's the plan now. The plan is to have it live in the U.S. towards the end of this year. Japan and Taiwan and a couple of other countries already have their own solutions, but the plan is to replace it with this, so we are very similar to where we were four years ago saying that SAP would be our ERP around the world. Now, we are saying [that] this fancy front-end e-commerce solution [is our Web strategy and] we are going to [it] take worldwide.
> I think the two projects parallel each other and I think for some of the same reasons we didn't do SAP everywhere, I doubt [that] we will do this e-commerce solution everywhere. But that's kind of the

picture of what we have around the world and how we got there (Boyd Blake, personal communication, July 14, 2000).

Back-end, front-end, market tiers and local requirements are the elements of a successful strategy, whether for local or global e-commerce efforts. Although organizations may carve out capabilities for their existing, local business efforts without much effort, they may find that extending these capabilities to a global sphere is an onerous task—even if underlying products and services are otherwise well suited to global requirements. Figure 2 depicts our theoretical model with revisions taken from Nu Skin's experiences with global electronic commerce.

Small-scale markets are more conducive to a "global" strategy because in these environments, corporate leaders can exert higher levels of control. Small markets have less resources and typically less expertise. They are also easier to control because the absolute number of employees, facilities, complexities, and technologies is much lower than large markets. In addition, IT capabilities do not have to be as sophisticated in a localized environment because the number of clients and processes is much reduced. In contrast, large-scale markets are almost impossible to control, making them much more conducive to "transnational" strategies. IT capabilities in transnational settings must be many times more sophisticated than required in "global" environments in order to be responsive to the various players. IT resources must exhibit higher levels of integration to allow corporate executives and managers to assimilate what is happening in the disparate markets being served. Integration within small markets is much less complex, of course. Medium-sized markets are also more difficult to manage—not unlike large industrial and consumer settings. In many cases, intermediate environments are not large enough to warrant investment on the scale of large markets, but they are still much more complex than small markets (making hybrid strategies very problematic).

The Nu Skin experience is that there are no easy fix, off-the-shelf solutions. Its efforts to make the Internet work in support of its other globalization imperatives are ongoing and are not conclusive. Rather than demonstrating characteristics of either "global" or "transnational" globalization models, Nu Skin exemplifies elements of both, along with "hybrid" combinations. Thus, due to the complexities of the markets served by the organization and the differing requirements and capabilities brought on by the Internet and other information technologies, Nu Skin is adapting to the Internet by assimilating various globalization models between the home office and different parts of the world. Nu Skin is experimenting with various tiers in the context of market size and IT configurations, not unlike our theoretical model.

*Figure 2: Revised three-tier networked enterprise model of global e-commerce*

| | Agency | Delegation | Market Nature | IT Capabilities |
|---|---|---|---|---|
| **Global** | Responsive | Minimized | Small-scale | Localized |
| **Hybrid** | Negotiated | Situational | Mixed | Mixed |
| **Transnational** | Proactive | Maximized | Large-scale | Integrated |

# FUTURE TRENDS

The requirements of operating on a global scale are multiplied in large part by capabilities and demands brought on by computerized networks—most notably the Internet. "Going global" requires that organizations master the art of localization. Globalization by commercial enterprises and other institutions is a critical delegation and agency problem. Authority must be delegated in order to enter the global marketplace, but not too much, and not in all areas. Use of digital, networked systems shows great promise in this vein, but there are significant limitations that make globalization and support of attendant systems-based delegation awkward and inefficient. One such limitation is the technology itself. For instance, Asian countries must use 'double-byte' files because of the numerous characters inherent in languages like Mandarin Chinese. This causes difficulties when transforming, storing, and translating information between languages.

Drucker and others outlined two delegation models. Typically, these models represent opposite ends of the spectrum. The "transnational" model tends to push decisions out to people in the various countries and markets in which business is conducted. Limited delegation (and limited customization to meet local differences) is represented by the "global" model, in which the home office structure is duplicated as widely as possible. Herein, we have identified a "hybrid" globalization option, where a balance is struck between transnational and global systems structures. As such, empowerment is encouraged for many operational activities while strategic issues, like company standards for excellence in customer service, adopts a global decision-making approach.

# CONCLUSIONS

We have expanded our hybrid structure to outline a globalization model for organizations in the Internet age that incorporates the three-tiered market structure with delegation, agency, market nature, and IT capabilities factors. This model's primary purpose is to provide a vehicle to assist business decision makers and organizational theorists in overcoming the global "collective action" problem faced by organizations as they attempt to make use of the Internet and e-commerce technologies to penetrate markets on a global basis.

In the case of Nu Skin International, the process of sorting out local, home country, and global issues is a challenging task, managed in tune with market conditions and other factors on a case-by-case basis. As a result, the company exhibits features of all three models. In countries and regions where high growth has been achieved or is imminent, more latitude is granted to local managers and systems administrators—following a "transnational" mode. In certain cases, technology acquisition decisions have been altogether localized for improved efficiencies. A delegation strategy that requires proactive behavior on the part of local agents and high levels of interactivity in order to achieve transnational performance should not be a separate activity from the global mix of the overall organizational activities. That is, it should not be an island apart from the rest of the enterprise. Language requirements play an important role in the process, but logical imperatives play a part as well.

Central to Nu Skin's concerns is a need to function on a centralized, "global" basis for managing products and materials. The company tends to accomplish this task in a "transnational" manner, however it attempts to coordinate activities in a way that takes local and regional issues into account. A further central management requirement of its business is to oversee compensation disbursements globally due to the integrated nature of its global distributor corps. Home office personnel manage this task exclusively—truly a function that follows the "global" systems model by not delegating at all.

Centralization at Nu Skin is tempered in many cases by a process of delegation that allows for differences in market conditions, management expertise, and strengths and weaknesses of technologies that are available to the firm and its subsidiaries around the world. Agents, depending on resources and expertise, are given autonomy in many activities as long as their actions do not sub-optimize central business objectives. The Internet is a powerful tool for the firm, but not a definitive answer to all of its systems problems. The Internet does facilitate communications, but it cannot "magically" make business processes more effective. Transforming offline processes to the Internet is a major challenge that will take Nu Skin many years to master. As one example, the simple function of verifying inventory at the point of sale is a challenging task when extended around the globe. Technology alone will not solve the problem. The Internet must become even more central to strategic plans. The three-tiered, market-based delegation and agency process—coupled with centralization of certain specific, worldwide tasks—is therefore a fundamental tool to enable Nu Skin to develop more cogent plans for Internet globalization.

Our next task is to further refine our theoretical model by exploring agency theory in more depth (with more cases) and revisiting our Nu Skin contacts. We have already sent drafts of this chapter to Nu Skin as they requested. In the future, we hope to be able to gather more relevant and timely data to test the model and assist Nu Skin management in its efforts to meet its global objectives.

## REFERENCES

Cannella, A. A. and Monroe, M. J. (1997). Contrasting perspectives on strategic leaders: Toward a more realistic view of top managers. *Journal of Management*, 23(3), 213-238.

Castells, M. (1996). *The Information Age: Economy, Society, and Culture: Vol. I. The Rise of the Network Society*. Malden, MA: Blackwell.

Collins, P. (2001). Internet business executive conference. *Management Services*, 45(3), 6-11.

Drucker, P. F. (1973/1993). *Management: Tasks, Responsibilities, Practices*. New York: HarperBusiness.

Drucker, P. F. (1998). The next information revolution. *Forbes ASAP*, August, 1-3. Retrieved September 27, 2000 on the World Wide Web: http://www.forbes.com/asap/98/0824/046.htm/.

Garten, J. E. (2001). The Internet. *The Journal of Business Strategy*, 22(2), 6.

Kiewiet, D. R. and McCubbins, M. D. (1989). The spending power: Congress, the president, and appropriations. *National Science Foundation Grant SES 8421161*. Washington, DC: National Science Foundation.

McNurlin, B. (2001). Will users of ERP stay satisfied? *MIT Sloan Management Review*, 42(2), 13-14.

Meyers, J. (1999). Mission accomplished. *Telephony*, 237(5), 18-21.

Nadler, D. A. and Tushman, M. L. (1997). *Competing By Design: The Power of Organizational Architecture*. New York: Oxford University Press.

Sklar, D. (2001). Building trust in an Internet economy. *Strategic Finance*, 82(10), 22-25.

St. John, C. H. and Young, S. T. (1999). Coordinating manufacturing and marketing and internal firms. *Journal of World Business*, 34(2), 109-128.

Victor, D. A. (1992). *International Business Communication*. New York: HarperCollins Publishers, Inc.

Walton, A. E. (1995). Generative strategy: Crafting competitive advantage. In Nadler, D. A., Shaw, R. B. and Walton, A. E. (Eds.), *Discontinuous Change: Leading Organizational Transformation*. San Francisco: Jossey-

Bass Publishers.

Weeks, A. M. (2000). PRC trade strong in first half. *The China Business Review*, 27(6), 34-35.

Zacharakis, A. L. (1997). Entrepreneurial entry into foreign markets: A transaction cost perspective. *Entrepreneurship: Theory & Practice*, 21(3), 23-40.

## Chapter 11

# Growth and Consolidation in the Spanish-Speaking E-Commerce Market

Roberto Vinaja
University of Texas-Pan American, USA

*This case aims to analyze, in some detail, the major challenges in the widespread adoption of electronic commerce in the Spanish-speaking population. The case also provides a general overview of related issues in global e-commerce, specifically: language, localization, currency, cultural difference, export controls, payment methods, taxation issues, consumer protection, and legal issues. The case includes a description of the strategies followed by companies entering the Latin American market in order to illustrate some of the major cross-border issues. The case clearly exemplifies how localization involves a considerable financial investment and commitment. The chapter illustrates that while Latin America initially attracted many investors by offering one of the world's fastest growing online populations, the market was not large enough to accommodate all the new entrants.*

## A GROWING MARKET

Analyst studies in 1999 pictured Latin America as one of the world's fastest growing online populations; Internet access was growing faster in Latin America than in other regions of the world (Vidueira, 1999). The number of new websites in Spanish was rapidly increasing. This explosive growth rate was attracting investors and newcomers into the online Spanish-speaking market (Vidueira, 1999).

Previously published in *Cases on Worldwide E-Commerce: Theory in Action,* edited by Mahesh Raisinghani, 

All new entrants were looking after a share of the "potentially" huge Latin American market (Fattah, 2000b). Latin America seemed an attractive market with huge growth potential (Anonymous, 2000c). More and more U.S. e-tailers were offering Spanish versions of their own websites, while others were establishing alliances with companies with existing websites in Spanish in an effort to reach the online Spanish-speaking market.

Many U.S. companies, especially those run by U.S. Latinos, were targeting Spanish speakers. Forrester predicted the development of more ethnic portals from expatriates who retain ties to the motherland, Americans who opt out of using English as their primary language, and residents of emerging Internet markets (Mand, 1999).

Many U.S. e-tailers were realizing the revenue potential from Latin American markets (Slover, 2000). Many U.S. companies were implementing dynamic plans to direct their e-commerce efforts to reach Hispanic or Latin-American Internet users (Anonymous, 2000e; Oldham, 2000; Werner, 2000). 1-800-Flowers.com inaugurated a Spanish version of its site called 1800-LasFlores.com. Sears was also pursuing the Spanish-speaking community and opened its Spanish-language site in early 2000 (Rutledge 2000).

According to e-Marketer, the overall Internet Hispanic population is around 14 million, and almost half of those are Hispanics in the U.S. (Zoltak, 2000; Azios, 2000). StarMedia estimated the worldwide online Hispanic community at 20 million, about 5% of the overall potential market.

## THE HISPANIC POPULATION IN THE U.S.

According to the 2000 census data, there were approximately 34 million Hispanics in the U.S. In addition:

- The Hispanic group was the fastest growing group in the U.S. growing at a yearly rate of 2%.
- The Hispanic population was expected to double by 2025 (Trujillo, 1998).
- Hispanics would become the largest minority in the U.S. by 2010.
- Hispanic youth would be the largest non-white population in the U.S. by 2025.
- In 1998, the Hispanic purchasing power within the United States was over $383 billion.
- There was a 50% gap in home computer ownership between the general population and middle-class Hispanics (Trujillo, 1998).
- 19.4% of Hispanic households owned a PC, compared to 40.8% of White households (Anonymous, 1998).

- 27% of Hispanics were planning to buy a computer in the short term.
- 20% of U.S. Hispanics had Internet access at work.
- 75% of U.S. Hispanic households with incomes above $75,000 owned computers, of which half had Internet access.
- About half of households with incomes between $35,000-$74,999 owned computers, and a quarter had Internet access (Trujillo, 1998).
- In 1999, Internet access among US. Hispanics was around 4.5 million users.
- 50% access the Internet (includes any location).
- 23.4% of Hispanics not accessing the Internet cited monthly service costs as the major obstacle.
- From the bilingual Hispanic group, 47% of those who own a computer also had Internet access.
- The estimated PC purchases by U.S. Hispanics in 2000 were more than 1.5 million, equivalent to the whole U.S. college student market (Rutledge, 2000).
- Approximately 61% of US Hispanics online made a purchase in 1999, and 74% of those purchasers accessed the Internet on a daily basis (Rutledge, 2000).
- 12% of U.S. Hispanics have bought something over the Internet (Zbar, 1999).
- U.S. Hispanic shoppers were predominately male; the average age was 32, and the average annual income was $51,600.
- Younger Hispanics (13-17) tend to use the Internet more for chatting, music, etc., while those age 18-35-plus use the Internet for information and news, etc. (Anonymous 2000d).
- The average shopper bought six items and spent an average of $547 during 1999.
- 86% of the shoppers bought music and 45% bought books (Rutledge 2000).

CAG (Cultural Access Group) claimed, based on a study conducted in 2000, that the Hispanic Internet was becoming as important as Spanish language TV, radio, and print as an advertising medium.

## THE HISPANIC POPULATION IN LATIN AMERICA

- The Latin American market as a whole had 500 million potential Internet users and $2 trillion in GDP (Shetty, 2000).
- Internet use in Latin America was growing at a fast rate. The number of Latin America Internet users grew from 4.8 million in 1998 to 10.6 million in 2000 (2% of the population) and was expected to grow up to 19.1 million users by

2003 (Disabatino, 2000) or up to 38 million in 2003 according to the most optimistic estimates (Oberndorf 2000).

- Jupiter forecasted up to 66.6 million (12% of the population) in 2005 (Ebenkamp 2000). Jupiter report was based on number of Internet users, not households, because Latin Americans were more likely to access the Web at kiosks, a friend's house, at Internet cafes or at computer rental shops.
- The e-commerce market in Latin America was expected to grow from $167 million in 1998, up to $8 billion by 2003 (Oberndorf, 2000, Disabatino, 2000) of which more than $6 billion would be business-to-business e-commerce. More optimistic estimates forecasted $82 billion by 2004, out of a global Internet economy that would reach nearly $7 trillion by 2004. Brazil will register $64 billion in online sales, and Argentina, $10 billion (Anonymous, 2000c).
- According to Jupiter Communications, Latin America's online sales would reach $8.3 billion by 2005 (Ebenkamp 2000). This was an optimistic estimate that assumed that companies would find solutions to current impediments.
- The majority of Internet users were in Brazil, Argentina, and Mexico.
- Internet penetration in Latin America was only about 2 to 3% (Fattah, 2000b).
- The number of Internet users in Mexico was 500,000 and 350,000 for Colombia (Zoltak, 2000).

## ANALYSIS OF CHALLENGES

As electronic commerce in the region was growing in popularity, cross-border issues in Latin America were critical factors in the success of online ventures. Despite the expected growth, entrants faced many obstacles. Some of the obstacles for the expansion of Internet access in Latin America were:

- poor PC penetration and low Internet usage (Graves and Nucete 2000);
- a nascent and not highly reliable delivery infrastructure;
- low credit card ownership among consumers (Ebenkamp 2000);
- weak credit card processing infrastructure;
- high costs of access;
- antiquated back-office computer systems;
- inefficient distribution networks;
- expensive phone charges;
- widespread credit card fraud (Petersen, 2000a);

- tariffbarriers;
- highshippingcosts;
- distrust of mail-order shopping;
- preference to pay cash (Ebenkamp 2000);
- distrust of online transactions; and
- the need to handle multiple currency payments.

In the U.S. the major barriers for Hispanic widespread use of the Internet were the price of computers, high monthly Internet access fees, and lack of familiarity by the less acculturated Hispanics

Making e-commerce happen in Latin America would require overcoming some serious obstacles (Petersen, 2000c). Major challenges needed to be addressed before the Latin American market potential could be capitalized. Computer ownership and Internet usage was highly correlated with income. However in Latin America, relatively few people own computers. Although a growing number of Hispanics were online, only a small percentage was purchasing online. (Ebenkamp, 2000).

Financial-related websites were facing the challenge that access to financial information had always been restricted (Schmerken, 2000). The logistics infrastructure and the postal service were less than 100% efficient, and order fulfillment was really challenging. Just as in the U.S. the most successful players would likely be retailers that use the Web as an extension of their stores (Ebenkamp, 2000).

Given the current obstacles, Internet companies moving into the Latin American online market had to implement several strategies in order to be successful. Some of the strategies included adapting Websites to cultural differences, the use of localization services for translation of Websites into new languages (instead of using machine language translation) (Dodd and Graves, 1999), protecting brands via local domain registration, understanding local business customs, and implementing payment mechanisms in multiple currencies.

## Language

Some companies mistakenly thought of Latin America as a homogeneous region; nothing was so far from the truth. Although all countries (but Brazil) shared a common language, Spanish was spoken in so many different ways across the continent. The Latin America population was composed of many nations with different backgrounds and cultural values. Some U.S. retailers had committed the big mistake of just mimicking their English Websites and were offering a translated carbon copy of their English language Websites. The Latin American market had very unique characteristics, and in order to conquer the market, companies needed to implement new strategies and creative business models. There was no cookie cutter solution for the region. Indeed, Latin America was not one country and one language, but many countries and two languages (Schmerken, 2000). A report by CAG showed that English was the preferred language for U.S. Hispanics, Spanish was preferred only when U.S. Hispanics visited Latin American Websites. However, older U.S. Hispanics were more concerned about content in Spanish, and they thought that more Websites in Spanish were needed. By 2000, only about 2% of the Internet content was in Spanish (Shetty, 2000). The same reports showed that U.S. Hispanics teenagers tended to use the Internet more for chatting, and music, while the rest of the U.S. Hispanic population used the Internet for information and news. Recent studies showed that U.S. Hispanics refer to Websites from their home country when accessing information in Spanish. Web portals such as StarMedia and Terra were taking this fact into account when offering content. Websites were targeting audiences of each nation and culture.

Companies entering the Latin American online market were aware of language differences when building their websites. Many businesses today are involved in Electronic Commerce. It is important to understand that communication can be a major barrier in regards to cross-border negotiations among countries. It is also very important to provide information in the local language. Although, English is the major language of business communication in the world, many local Latin American customers would not understand any other language than that of their native country (Fridman, 2000). Websites offering information in the local language were having a selling edge over the competition not using the local language. Consumers were more likely to buy from a Web site with content in the local language. Every nation in Latin America had special characteristics and had to be considered individually. The Latin American market was culturally rich and diverse; in addition to Spanish many dialects were still in use. Winning the battle for the Latin American market meant winning a series of battles in each country. A multilingual Web site with icons that feature different language options would cater to more markets.

However, offering multiple versions of the same website was an expensive alternative. Major problems experienced by startups were the costs of creating multiple websites, and customizing/adapting to different cultural tastes and suit language variances. Latin American startups were very expensive to run because they were organized as micro-multinationals. The cost involved with maintaining multiple regional offices and developing separate flavors of a website was tremendous.

## Legal Issues

Economic activities need regulations and laws; however, most legislation was only applicable within the borders of every specific country (Mosquera, 1999). There are many legal issues that needed to be considered (Engler, 1999). Some of the legal issues included customs, laws and regulations, privacy, copyrights, jurisdiction issues, export/import regulations, and compliance with intellectual property, cryptography and security contracts (Drake, 1998). Only a few countries had established committees or initiatives for the development of domestic and global legislation to facilitate global electronic commerce.

## Currency

Currency varied from country to country. Local currency included pesos, bolivares, quetzales, cruceiros and sucres, to name a few. Because of currency devaluation, some country's currency may be of a higher rate or lower rate than the hosting website's currency. Because currency exchange rates were different, the value of the currency could increase or decrease in different countries. Purchasing in different Latin American currencies was a potential barrier, as exchange rates could literally change every minute. However, major credit card companies operating in the region provided their customer with automatic currency exchange services for multiple currencies.

## Infrastructure

The region's IT infrastructure varied from nonexistent to rudimentary to adequate. Many regions in Latin America were still lagging behind in terms of infrastructure. However, major telecommunications providers were reluctant to invest in less profitable parts of the continent and preferred to focus on closed loops to selected cities. For example, 70% of the overall connectivity market constituted traffic between Brazil and Argentina, and many companies were focusing their efforts in this region.

International connectivity was a critical factor to provide Internet access for the nascent Latin America's e-commerce economy. However, a major hindering block in Latin America was the high cost of international half circuits. Telecom Argentina charged $71,351 per month for 2Mbps leased lines. Chile's Entel charged $26,600 and Brazilian Embratel charged $20,109 per month. One of the reasons for these high tariffs was that while many new ISPs had been entering the region, the supply market was highly concentrated (Shetty, 2000).

There were several obstacles to achieving universal coverage for businesses all over Latin America. An alternate solution to the current obstacles was the use of the Wireless Application Protocol (WAP) to enable Internet access through wireless devices.

### Cultural Differences

Companies entering a new market must get to know the culture of that country. Countries may have diverse views on e-commerce regulations. A few countries in Latin America did not have defined electronic commerce initiatives, and therefore they lacked a established body of regulations. Credit card penetration was extremely low. The estimated card ownership was as low as 10% in some regions. Concerns over credit fraud and a cultural preference for price haggling and face-to-face transactions were prevalent (Pereiera, 1999). According to Francisco Ramirez, sales director with Mexico City reseller Getronics: "Mexicans do not even make purchases by phone because merchants insist on verifying credit card signatures. It's not a technology issue, it's a cultural issue."

Building customer loyalty and trust in the region would require additional efforts. Many Latin American customers were not used to catalog shopping, store returns or exchanges. Many Latin Americans were also distrustful about the security of online transactions and were reluctant to buy products they had never seen or touched. In order to overcome this attitude, many companies were exploring alternative payment methods and others were implementing alternative Web currencies and extensions of their traditional services (Ebenkamp, 2000).

In the Latin culture, community features such as chat rooms, personal pages, email and shopping were strongly emphasized. The average Latin American user spent a considerable amount of time chatting or sending e-mails (Fattah, 2000a). According to Fernando Espuelas, founder of StarMedia, there is a difference in Internet utilization in Latin America. Whereas the U.S. user has a greater focus on information retrieval, the Latin American users places more emphasis on communication.

### New Initiatives

Many obstacles had hindered e-commerce in Latin America; however, the situation was beginning to improve. Many countries in Latin America, such as Mexico, Brazil and Argentina, were opening their economies and markets to the new globalization trend and facilitating the development of global e-commerce.

Electronic commerce in Latin America was expected to rise after a slow start. Although Latin America was still lagging behind North America in technology, many Internet trading partners were investing in the region in key technologies, such as phone lines, computers, Internet hosts, and cell phones.

The trend of increasing local content and regulatory cooperation would pave the way for economies of scale that would drive the growth of e-commerce in the region. Optimistic forecasts expected a continuous drop in computer prices, the proliferation of satellite and wireless technology, a growing number of free Internet service providers, the issuance of new e-commerce legislation, and new developments in telecommunications (Gower, 2000). One of the benefits of Latin America as an emerging market was that entrants were able to learn from success and failure stories from their experienced counterparts in more mature markets.

Many analysts expected the index of PCs per head of population to grow. Many ISP were trying to provide alternative and affordable options for Internet cccess. The number of Web-based applications was also increasing at a fast rate (Shetty, 2000).

There were many projects underway. Approximately, 15 new fiber projects were close to completion. Major projects totaling 170,000 kms of cable were under construction. The new improvements in infrastructure would likely decrease the price of Internet connectivity.

According to the Americas Telecommunications Indicators 2000, released by the ITU, in some of the large markets, the incumbent carriers had anticipated the Internet growth and captured a large share of Internet subscribers. For example, in Argentina, Telecom Argentina's Arnet and Telefonica Argentina's Advance shared 43% of the Internet subscribers in the country. In Chile, incumbents CTC's and Entel's Internet service had left only a 5% market share to independent ISPs. Likewise, in Mexico, Uninet, the official ISP of Telmex had taken 50% of the access market and in Venezuela, CANTV Servicios had 35% of the market.

Many local governments were encouraging Internet service providers and telecommunications companies to offer Internet access at affordable prices and, if possible, offer free access (Oberndorf, 2000). The Chilean government mandated interconnection among Internet service providers and a national peering exchange was established so that Chile's bound Internet traffic would stay within the country, rather than traveling through outside networks before reaching a neighboring city. Other countries were implementing innovative measures to provide Internet access to the general population. Colombia's Minister of Communications, Claudia de Francisco Zambrano became a champion of universal Internet access through an elaborate program called Comparatel. Plans were to take the Internet to all Colombian municipalities, the poorest of which would receive free access. In addition, the Colombian Ministry of Education was planning to wire 2,000 schools. The plan was ambitious and many analysts were skeptical about the cooperation of the national carrier, EBT (Shetty, 2000).

Despite the major challenges, many improvements, mostly resulting from the ongoing privatization of telecommunications were underway (Pereiera, 1999). The number of Spanish language Internet sites was constantly growing, and the number of Spanish-speaking users was also increasing.

## MAJOR PLAYERS AT THE END OF 2000

Many Web portals were offering localized content and local Web brands for every specific country. Some portals were targeting the U.S. Hispanic population, others focused on the Latin American population and still others were trying to reach both groups. Each company was trying to target Latin America in many different ways. Many sites were targeting specific niches among U.S. Hispanic Internet users and those in Latin America (Zoltak, 2000). Each Web site was trying to differentiate their style and offerings, both in terms of content and community. Some portals specialized in content, while others were emphasizing community.

Table 1 lists the major players and whether they were targeting the U.S. Hispanic population, the Latin American population or both. Table 2 indicates which sites emphasized content versus those emphasizing community features.

### Starmedia.com

StarMedia was founded in 1996 by Fernando Espuelas, a former AT&T executive and Jack Chen, a securities analyst (Katz, 1999). Starmedia was one of the first pioneer Latin portals (Malkin, 1999). Based in New York, StarMedia Network started with $80 million in private financing. The company's IPO in May 1999 raised $110 million. The company had over 750 employees and operations in Argentina, Brazil, Chile, Colombia, Mexico, Puerto Rico, Spain, Uruguay, Venezuela and the United States. Starmedia was mainly a portal with a few incursions in the ISP business. The traffic at Starmedia.com was estimated at 2.1 billion page views in the first quarter of 2000. StarMedia had a Pan-regional approach, targeting Spanish and Portuguese speakers worldwide. The company operated a network of sites in virtually all Latin American countries, including several in Brazil. The main focus of Starmedia was the Latin American market, but it was also targeting the U.S. Hispanic market. Starmedia was one of the pioneers and offered community features both in Spanish and Portuguese. Starmedia specifically targeted users in major population capitals in Latin America and the U.S. The content was customized to every country.

Partnerships with advertisers and merchants was an important factor of Starmedia's business model (Fattah, 2000a). Most revenues at Starmedia came from advertisement. The site had about 150 advertisers, including Ford, Chrysler, General Motors, Intercontinental Hotels and Lufthansa (Mand, 1999). Starmedia acquired several local companies in 1999, including LatinRed, one of the largest Spanish-language online communities, OpenChile, a Chilean portal, Zeek!, a Portuguese site directory, and AdNet, a Mexican portal. Starmedia also owned several other media properties, such as Pidemasonline.com,

*Table 1: Major players who targeted the US Hispanic population, the Latin American population, or both*

| STRATEGY | COMPANIES | DESCRIPTION |
|---|---|---|
| OPENING A SUBSIDIARY | YAHOO! | Yahoo! entered the market by opening a subsidiary. |
| ACQUISITIONS | STARMEDIA<br><br>TERRA | Starmedia made a series of strategic acquisitions.<br>Terra also acquired local companies in several countries |
| JOINT VENTURE | AOL | AOL formed a 50-50 joint venture with Venezuelan Cisneros group. |
| MERGERS | TERRA | Terra merged with U.S. Lycos. |
| PARTNERSHIPS | MICROSOFT TELMEX | Microsoft and Telmex joined efforts to launch T1MSN |
| NICHE | QUE PASA | Que Pasa was mainly focused on the U.S. Hispanic population. |
| GROWING OVERSEA CUSTOMERS | EL SITIO<br><br>YUPI | El Sitio was entering the U.S. market<br><br>Yupi was gradually targeting the Latin American population. |

*Table 2: Content vs. those emphasizing community features*

| COMPANIES | Emphasis on U.S. market | Emphasis on Latin America | Emphasis on content | Emphasis on community |
|---|---|---|---|---|
| AOL | X | X | X | |
| EL SITIO | | X | | X |
| QUE PASA | X | | X | |
| STARMEDIA | X | X | | X |
| T1MSN | | X | X | |
| TERRA | | X | X | |
| YAHOO! | | X | X | |
| YUPI | | | | |

launched in conjunction with Pepsi-Cola, Cade, Guia SP, and Paisas.com. Starmedia also operated StarMedia Mobile, its wireless division, and StarMedia broadband, its broadband services arm. Starmedia had strategic partnerships with several companies, including Netscape Communications, RealNetworks, Billboard, Dell, Reuters, eBay, NBC, Hearst Communications and Fininvest (Fattah 2000a).

StarMedia was recently faced with a major challenge to its position in the U.S. as Spain-based Terra Networks made a strategic move by acquiring the U.S. portal, Lycos.

### Terra.com

Terra Networks S.A. was a publicly traded company and its major stockholder was Telefonica de Madrid, Spain, with portals and ISPs around Latin America. Terra's Web site, opened in 1997, was offering a balanced combination of customized content and connectivity features. Terra's Web site offered links to customized pages for every country it served. An audit conducted in the fourth quarter of 1999 estimated Terra's traffic at 1.2 billion page views.

Terra built a strong position in 1999, thanks to 2 million ISP customers in the region and to a $100 million IPO in late 1999 (Fattah, 2000a). Terra was expanding its market share in Latin America by acquiring Internet service providers in Brazil and Mexico (Ewing, 2000). Terra bought Infosel, its Mexico unit for $280 million.

Terra had just acquired Lycos in May, 2000 for $12.5 billion. Lycos was the fourth-largest portal in the U.S. and is a portal in 25 countries, through 65 sites in 13 languages. The combination Terra-Lycos would potentially reach 91 million users in 40 countries. (Disabatino, 2000). Terra was attracted to Lycos because of Latin America's proximity to the U.S. and the large Spanish-speaking U.S. population. The Lycos purchase would help Terra reach 30 million U.S. Hispanics (Fattah, 2000a). Terra-Lycos would be based in the U.S. and Bob Davis, former Lycos CEO would continue performing the same function.

### Yupi.com

Yupi was a private company based on Miami Beach, Florida founded in 1996 by Colombian-American Carlos Cardona (Vidueira, 1999). Yupi started as a Spanish-language search engine on the AltaVista network (Robinson, 1999). Yupi was a leading search engine in Spanish offering community links and related content. Yupi's traffic was estimated at about 143 million page views in February 2000. Yupi featured special separate Websites focused on health, kids, and auctions.

### Quepasa.com

The Hispanic portal QuePasa was initiated in 1998 and was based in Phoenix, AZ. QuePasa raised $48 million and surged from $12 to $19.75 a share on its first day of trading (Vidueira, 1999). It achieved 30 million visits at the end of 1999. Quepasa was targeting specifically the U.S. Hispanic population. The Web site provided two versions, one in English and one in Spanish. Content was focused on Hispanic pop starts and leaders in the U.S. and news generated in the U.S. Quepasa emphasized content over community options.

QuePasa.com launched in 2000 its own shopping pages targeting the U.S. Hispanic consumer market. The shopping site, available in both

English and Spanish, featured merchandise from popular sites such as Amazon.com, eToys, Dell Computer, Barnes and Noble, Staples, JCPenney and other merchants in 13 product categories. QuePasa also entered a partnership with MapQuest to offer maps and door-to-door driving directions in both Spanish and English. This strategy was intended to drive traffic to the brick-and-mortar stores of QuePasa's online partners. QuePasa had also recently signed famous pop star Gloria Estefan as its spokesperson (Zoltak, 2000).

### El Sitio

El Sitio, established in 1997, was based in Buenos Aires, Argentina. El Sitio was the first company from Argentina to go public on the Nasdaq. El Sitio started as a content provider, but gradually offered Internet features after acquiring several fiber network companies in Brazil, Argentina and Colombia (Fattah, 2000a). Nevertheless, the ISP unit was not considered the core function. El Sitio reached 300 million page views in the last quarter of 1999. El Sitio offered content in both Spanish and Portuguese. El Sitio focused on community aspects rather than content. El Sitio also bought DeCompras, a privately held Mexican online retailer for $44.7 million in stock and cash. El Sitio's increased its efforts to appeal to the U.S. Hispanic population. El Sitio signed on Sammy Sossa, a famous major league baseball player, as its spokesperson in the U.S. and developed a new site: Sammysossa.com targeting U.S. Hispanic sports fans.

### T1MSN

The richest man in the world, Bill Gates, joined efforts with Carlos Slim, president and CEO of Latin America's largest telecommunications company, Telmex, to develop a Web portal in Latin America (Fineren, 2000). Telmex and Microsoft combined their expertise into a Spanish-language Internet portal aimed at Latin Americans. Telmex and Microsoft were equal partners in the venture and according to the agreement, Microsoft would offer Spanish-language content through its MSN Web service. In addition, MSN would give Spanish-speaking subscribers bundled software services: free e-mail through Hot Mail, instant messaging and Web publishing services. All 1.8 million Hot Mail registered users in Mexico were turned into users of T I MSN. The portal got a guaranteed captive audience because Spanish-language subscribers to Telmex's Prodigy and Microsoft's MSN.com were automatically forced into the portal. The site reached 2.3 million unique users only two months after launch (Fattah, 2000a). The portal also included online shopping and access to other Microsoft services (Rutledge, 2000).

### AOL

Internet service providers such as America Online (Willoughby, 2000) were also beginning to target Hispanic Americans (Gonzales, 1999). AOL offered $575 million of stock to the public to feed its expansion in the Latin American Internet market (Goldsmith et. al., 2000). AOL took its first big step into the Latin American market by launching a Brazilian portal in partnership with the Cisneros Group, a Venezuelan media conglomerate (Fritsch, 1999). The Cisneros Group invested $100 million to fund a 50-50 joint venture with AOL to bring AOL to Latin America.

### Yahoo!

Yahoo! was also offering local guides. For example, Yahoo! Argentina was designed specifically for users in Argentina and individuals with special interest in the country. Yahoo! also developed an agreement with Hispanic Television Network, a Hispanic television media company. Yahoo! would provide Internet broadcasting solutions for Hispanic Television Network through Yahoo! Broadcast. Yahoo also unveiled Camp Yahoo! in Spanish, a Spanish-language Internet education initiative.

## THE FUTURE

The projected growth in Latin America initially attracted many investors, but the market was not accommodating all the companies looking for business. Despite the projected growth, pioneer companies had not posted a profit yet. The Latin American market was fragmented and overcrowded, and shares prices were dropping. Analysts expected that the numerous ISP players would consolidate to only a few major players via mergers, acquisitions and strategic partnerships, a trend that would gradually change the current landscape as foreign Internet competitors came to the Latin American market (Shetty, 2000). Many analysts predicted that troubled companies would become takeover targets (Druckerman, 2000a). Rumors about acquisitions were generated on a daily basis, followed by denials of the companies involved.

According to analysts, after the drop of the NASDAQ index in the U.S., the volume of investments in Latin American Internet companies was expected to dramatically decrease (Reuters, 2000). The plummeting NASDAQ index and the subsequent tightening of funds for Internet ventures were accelerating the pace of consolidation in the market. Many start-up ventures, including Yupi, had delayed their plans for initial public offerings.

Luis Mario Bilenky became president of StarMedia. Francisco Alberto Loureiro, formerly with AOL, became its chief operating officer (Petersen,

2000a). StarMedia posted losses in the first and 44 million for the second quarter of 2000 because of increased product development costs and marketing expenses. Star Media's shares dropped from a high of $70 in July 1999 to $15 in August 2000 (Petersen, 2000c). In the last quarter of 2000, StarMedia announced it was cutting 125 of its 850 positions, in a company-wide restructuring (Rewick 2000). This was called a proactive cost-cutting measure aimed at making the company profitable once more. There were some rumors that StarMedia would not rule out the possibility of a merger with larger U.S. Internet players. Terra had just established a stronghold in the U.S. Hispanic market after acquiring Lycos (Palatnik, 2000). El Sitio was hit by the collapse in the stock market and was facing major problems in its acquisition and expansion efforts (Fattah, 2000a).

QuePasa's stock was recently being traded at 33% below its first-day closing price. QuePasa.com laid off two-thirds of its staff as it recorded disappointing third-quarter results. Its shares fell 24%. QuePasa, Phoenix, said it was firing 38 of 58 employees in an effort to reduce spending, as it continues its search for a buyer. It also posted a third-quarter loss of $7.9 million, down from $8.3 million a year ago, and disclosed that at the end of September it had only $9.5 million in cash.

The chaotic situation implied that only a few companies would survive—the ones with the most resources to support the initial losses, and those patient enough to wait the years it would take to make a profit. Although industry forecasts indicated an expected growth, the future of the Spanish-speaking e-commerce market was still uncertain.

## REFERENCES

Anonymous (1998). Blacks, Hispanics still own far fewer PCs than Whites: Study. *Jet*, August, 94(12), 39.

Anonymous. (2000a). El Sitio expands by acquiring Mexican online retailer. *New York Times*, March, C.4.

Anonymous. (2000b). AOL Latin America plans stock offering of up to $575 million. *The Wall Street Journal*, January, B8.

Anonymous. (2000c). Hypergrowth for e-commerce? *Futurist*, September-October, 34(5), 15.

Anonymous. (2000d). The wired Hispanic market. *Direct Marketing*, June, 63(2), 54-56.

Anonymous. (2000e). MasterCard continues tradition of innovation with first Spanish-language Web site by major payment card brand. *Business Wire*, December 19.

Azios, D. A. T. (2000). Hi-tech latinos. *Hispanic*, April, 13(4), 60.

Disabatino, J. (2000). US, Latin America blending e-commerce. *Computerworld*, May, 34(22), 45.

Dodd, P. and Graves, L. (1999). Globalization, overcoming challenges of geography and language. Vision report, February, *Jupiter Communications*. Retrieved on the World Wide Web: http://www.jup.com/sps/research/reportoverview.jsp?doc=sos99-14.

Drake, W. J. (1998). Toward sustainable competition in global telecommunications: From principle to practice. *International Lawyer*, August 23.

Druckerman, P. (2000a). Latin Web firms venture out to markets in Spain, US-Hispanics abroad offer quick route to retail revenue. *The Wall Street Journal*, April, A23.

Druckerman, P. (2000b). Latin American Web concerns struggle to stay in business. *The Wall Street Journal*, October 9.

Ebenkamp, B. (2000). Manana's opportunities. *Brandweek*, February, 41(9) 26.

Engler, N. (1999). Global e-commerce, how products and services help sites expand worldwide. *Information Week*, October 4. Retrieved on the World Wide Web: http://www.informationweek.com/755/global.htm.

Ewing, T. (2000). Deals & deal makers: Softbank aiming at online market in South America. *The Wall Street Journal*, January, C18.

Fattah, H. (2000a). Livin' e-vida. *Brandweek*, July, 41(27), IQ46-IQ54.

Fattah, H. (2000b). Latin crowd. *Mc Technology Marketing Intelligence*, August, 20(8), 36-44.

Fineren, D. (2000). Microsoft and Telmex plan a latin region Web portal. *New York Times*, March, C.4.

Fridman, S. (2000). *Global E-commerce Requires More Than A Foreign Language Website*, January. Retrieved on the World Wide Web: http://www.bizreport.com/news/2000/01/20000124-11.htm.

Fritsch, P. (1999). AOL takes major step in Latin America-US firm joins local group to launch Web site aimed at Brazilian users. *The Wall Street Journal*, November, A24.

Goldsmith, C., Boston, W. and Druckerman, P. (2000). You've got time Warner!—World looks much smaller from abroad-foreign media, Web firms see AOL deal sparking copycat acquisitions. *The Wall Street Journal*, January, B13.

Gonzales, E. J. (1999). Latinoamerica.com. *Hispanic*, March, 12(3), 34-38.

Gower, M. (2000). Helping emerging markets thrive in the digital age. *Business Mexico*, September, 10(9), 54.

Graves, L. and Nucete, V. (2000). No easy victories in Latin America's projected $8.3 billion online commerce market. *Concept Report, Jupiter Communications*, March 15.

Katz, I. (1999). La vida loca of a latin Web star. *Business Week*, November, (3653), 196.

Malkin, E. (1999). The Web's southern frontier. *Business Week*, December, (3661), 120.

Mand, A. (1999). Missing the target. *Brandweek*, February, 40(5), 65-68.

Mosquera, M. (1999). Lawmakers offer bills to spur e-commerce. *TechWeb News*, May. Retrieved on the World Wide Web: http:// www.techweb.com/wire/story/TWB19990506S0024.

Oberndorf, S. (2000). Going worldwide via the Web. *Catalog Age*, February, 17(2), 41-42.

Oldham, C. (2000). Internet marketers search for way to reach latino audience on Web. *The Dallas Morning News*, November 29.

Palatnik, M. (2000). Nasdaq composite's drop speeds Latin American Internet M&As. *Dow Jones Newswires*, May 12.

Pereiera, P. (1999). E-business washes into Latin America. *Computer Reseller News*, December, (873), 5, 12.

Petersen, A. (2000a). Opening a portal: E-commerce apostles target Latin America, but it's a tough sell-limited use of the Internet, lack of venture capital stymie effort in Mexico-Big US players take aim. *The Wall Street Journal*, January, A1.

Petersen, A. (2000b). StarMedia lures executives, including one from AOL, to lead its operations. *The Wall Street Journal*, January, B23.

Petersen, A. (2000c). StarMedia net loss widened in quarter, but beat estimates. *The Wall Street Journal*, August 8.

Reuters (2000). *Internet en A. Latina no atraera grandes inversiones en 2001*. Feb. 7.

Rewick, J. (2000). Starmedia sets layoffs in effort to reach profitability by 2002. *The Wall Street Journal*, September 14.

Robinson, E. (1999). Salsa beat: Meet Latin America's Net Bets. *Fortune* 140(7), 346-348.

Rutledge, K. (2000). Increased interest speaks well for Spanish language Websites. *Discount Store News*, May, 39(9), 25-27.

Schmerken, I. (2000). Financial site serves up a Latin beat. *Wall Street & Technology*, February, 18(2), 52.

Shetty, V. (2000). Latin lessons. *Communications International (London)*, May, 27(5), 42-45.

Slover, P. (2000). Web companies cater to Hispanic market with Spanish-language site. *The Dallas Morning News*, September 28.

Trujillo, S. D. (1998). Opportunity in the new information economy: Technology, the great equalizer. *Vital Speeches of the Day*, June, 64(16), 490-492.

Vidueira, J. (1999). Rocketing to cyberspace. *Hispanic Magazine*, July-August.

Werner, H. M. (2000). International: Bank of America equity builds IT services co. for Latin America. *Venture Capital Journal*, June, 39-40.

Willoughby, J. (2000). Offerings in the offing: AOL ole? *Barrons*, July, 80 (31), 31.

Zbar, J. D. (1999). Powering up Internet en Español. *Advertising Age*, November, 70(49), S2.

Zoltak, J. (2000). The Web goes south. *Billboard*, June, 112(25), LM1-LM3.

## Chapter 12

# ICT Diffusion and Strategic Role within Italian SMEs

Giacomo Buonanno, Stefano Gramignoli, Aurelio Ravarini and Marco Tagliavini
Università Cattaneo, Castellanza (VA) – ITALY

Donatella Sciuto
Politecnico di Milano, Castellanza (VA) – ITALY

*Information System literature is rich in studies concerning the impact of Information and Communication Technologies (ICT) on organizations. This chapter, however, focuses on a specific context: small and medium size enterprises (SMEs). It presents an up-to-date picture of the ICT employed and of the activities ICT support within Italian SMEs. Moreover, it provides results regarding the relationship between ICT and SME strategy. Data were collected through a survey on IT managers of 143 SMEs placed in Northern Italy. Research results highlighted relevant differences on how small and medium size organizations employ ICT in order to influence or support their strategy.*

## INTRODUCTION AND PRIOR RESEARCH

In the '80s, Information and Communication Technology (ICT) evolution mainly affected large companies, able to invest a large amount of their resources in Information System (IS) development. Large information centers and specialized staff were required to effectively manage the ICT introduction and development in these companies.

The initial advantages which promoted ICT introduction in many companies, were mainly related to the productivity improvements. However, in the following years the impact on company strategies was theoretically proposed and empiri-

Previously Published in *Challenges of Information Technology Management in the 21st Century*, edited by Mehdi Khosrow-Pour, Copyright © 2000, Idea Group Publishing.

cally tested (Porter, 1985; Wiseman, 1985; King, 1994; Willcocks, 1997; Kemerer, 1998).

As the broadness of ICT impact begun to expand beyond the boundaries of large enterprises, the academic world has been increasingly interested in the phenomenon of ICT adoption and use by SMEs, recognizing their peculiarity with respect to larger implementations. During the 90's a lot of research outlined the increasing use of ICT within SMEs:

- a survey over 228 Canadian SMEs presented a positive relationship between ICT adoption and organizational development (Raymond, 1992);
- a study on a sample of 201 British small enterprises with under 10 employees reported that approximately 50% had at least one personal computer (Bums, 1992);
- a research carried out in Japan (Japan Small Business Research Institute, 1995) pointed out that more than two thirds of the surveyed SMEs already owned one or more computers, and were working to reach higher levels of computerization, with a further 16% still in the planning phase of giving the business a superior information orientation.
- the few research in Italy, showed that ICT support to business activities is on average lower when compared to other European countries (ASSINFORM, 1997); nevertheless the diffusion of IS-related assets appears widespread even within small and very small Italian enterprises (Bartolozzi, 1995).

Other studies, however, outlined that ICT adoption and management amongst SMEs have not been exempt from problems:

- from technological point of view, the lack of technical expertise seems to limit technical specification and selection policies (Monsted, 1993; Schleich, 1991), leading to hardware and software inadequacy (Cragg, 1995).
- from organizational perspective, Cragg (1995) reported that management paid insufficient attention to the Information System: this can be related to the small size (or even absence) of IS staff, possibly combined with the lack of ICT consultants (Soh 1994; Palvia, 1995; Zinatelli, 1995). As a consequence, IS of small businesses show fail in providing adequate support to many decision activities (Lai, 1995; Lang, 1997).

It is worth noticing that such a rich literature, nevertheless, has given scarce consideration to the analysis of the strategic role ICT could play into SMEs. Therefore, while relationship between ICT and business strategy represent a well-known issue, there is little evidence that ICT can influence or even support SME strategies.

## RESEARCH MODEL

Because of the wide impact of ICT in organizations, the ICT utilization issue can be analyzed from a number of different points of view. The first aim of the research presented in this paper is to provide an up-to-date picture of technologies employed by Italian SMEs with a specific focus on the activities they support.

Nevertheless a correct interpretation of collected data requires the preliminary identification of the importance given by the company to ICT management. Therefore the first research question investigated deals with an organizational issue:

RQ1: which are the organizational roles devoted to the IS management within SMEs?

The second step of the research was focused on collecting information about technology adoption and its impact on the organization:

RQ2: which kind of hardware equipment is adopted by SMEs?

RQ3: which activities are more often supported by ICT among SMEs and which are the IS performances (efficiency and effectiveness) in supporting them?

When considering such rapidly evolving technologies it is fundamental not to limit a survey to current data, but also to outline which changes are expected to take place in years to come. A second set of relevant information was thus collected in order to answer to the following question:

RQ4: which is the expected development of ICT among SMEs?

Finally, a precise aim of this study was to fill the highlighted lack of research in the field of ICT impact on company strategies. Specifically, the following issue was investigated:

RQ5: which is the strategic relevance ascribed to ICT by top management and which is the actual impact of ICT on SMEs strategy?

As a consequence, the research goal was to survey the most complete range of ICT impacts on an organization (Figure 1).

*Figure 1: The research model*

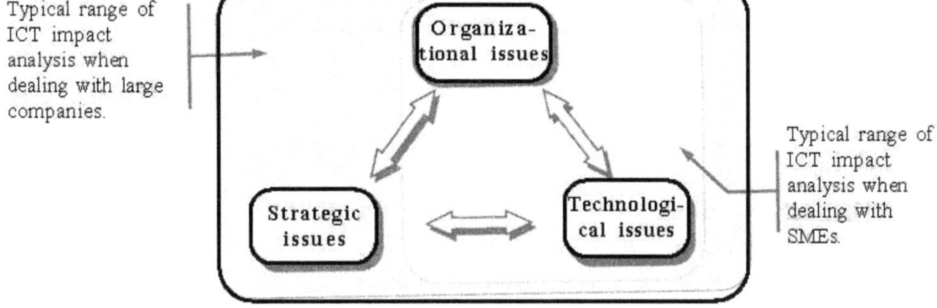

# METHODS

### The Instrument

The analytical phase of the research was carried out by mailing a simple questionnaire to 1354 Italian SMEs located in the province of Varese, in Northern Italy. The IT manager, or, whenever such role was not present, the man in charge of computer system development, was asked to answer to 63 closed questions, organized in 7 sections according to the research model:

1. basic company information (number of employees, industry, turnover of past three years);
2. IS staff (internal staff or external consultants, full-time or part-time);
3. hardware equipment;
4. software applications;
5. activities supported by ICT;
6. annual costs and investments to manage the IS;
7. impact of ICT investments on main business leverages and strategies.

Besides mailing, a few direct interviews were held to increase the answer rate and to improve the statistical significance of the sample: 130 SMEs returned the questionnaire spontaneously, 13 were directly contacted by the authors.

### The Sample Groups

In order to accomplish a more detailed analysis, respondents have been classified by size and by activity sector.

Referring to the current definition of SMEs provided by the European Community (DG XII's Press Releases, 1996), *small* enterprises has been identified by a turnover lower than 7 million ECU, while *medium* enterprises by a turnover greater than 7 million ECU and lower than 40 million ECU.

Moreover, respondents were gathered according to the activity sector code ATECO91 (AT91 code) (NACE, 1990; ISTAT, 1991): in order to perform a significant analysis, seven macro-classes were identified, corresponding to larger sets of similar businesses:

- metal and mechanical industry (including AT91 codes from 26 to 29)
- paper industry (AT91 codes 21 and 22)
- textile industry (AT91 codes 17 and 18)
- chemical and plastic industry (AT91 codes 24 and 25)
- other manufacturing firms (AT91 codes from 26 to 30)
- distribution (AT91 code 51)
- services (AT91 codes 40, 63, and 74)

## Statistical Validity

The statistical validity of the sample has been preliminarily verified in terms of number of participants versus the number of contacted companies[1].

Secondly, the Chi Squared test (Berenson, 1989) has been used to test the independence of the sample from the population of SMEs located within the province of Varese. The test has been applied to four representative variables: company size, number of branch offices, export quota and industry.

As shown in Table 1, each Chi Squared value is lower than the maximum level, allowing to extend research results to the universe of SMEs placed in the surveyed area.

*Table 1: Chi Squared test verifying the independence of the sample from (a) the company size, (b) the number of branch offices, (c) the export quota and (d) the branch of industry (with the exception of those companies belonging to the "other" sector).*

| Size | Population distribution | Sample distribution |
|---|---|---|
| Medium | 28.4% | 35.0% |
| Small | 71.6% | 65.0% |
| Total | 100.0% | 100.0% |
| Independence | Max level | Sample level |
| $\chi^2$ Test | 3.84 | 3.54 |

(a)

| Number of branch offices | Population distribution | Sample distribution |
|---|---|---|
| One | 83.4% | 83.6% |
| More than one | 16.6% | 16.4% |
| Total | 100.0% | 100.0% |
| Independence | Max level | Sample level |
| $\chi^2$ Test | 3.84 | 0.18 |

(b)

| Export Quota | Population distribution | Sample distribution |
|---|---|---|
| High | 23.3% | 22.9% |
| Small | 38.2% | 37.9% |
| Medium | 14.8% | 15.7% |
| Only domestic mkt | 23.7% | 23.6% |
| Total | 100.0% | 100.0% |
| Independence | Max level | Sample level |
| $\chi^2$ Test | 7.81 | 0.10 |

| Homogeneous Macro Sectors | Population distribution | Sample distribution |
|---|---|---|
| Paper | 5.4% | 3.6% |
| Chemical & plastic | 12.8% | 11.4% |
| Distribution | 2.0% | 2.1% |
| Services | 5.0% | 8.6% |
| Metal & mechanical | 29.5% | 29.3% |
| Other manufacturing | 10.2% | 14.3% |
| Textile | 35.0% | 30.7% |
| Total | 100.0% | 100.0% |
| Independence | Max level | Sample level |
| $\chi^2$ Test | 3.84 | 0.18 |

## The Economic Context

The northern part of Italy has always been characterized by a great number of SMEs, which represent its economic base. Most of the companies have always been characterized by the wish to keep their independence within the economic context; therefore, nowadays SMEs set is very diversified, with many different companies providing a very heterogeneous range of products and services.

In the 60's the surveyed area was characterized by manufacturing companies, mainly belonging to the textile industry. More recently, their diffusion has significantly decreased with respect to the service sector, although still remaining the primary leverage of the local economy (ISTAT, 1951-1991).

At the same time, both small and medium size companies were affected by the reduction of the average number of employees, determined by different reasons. Most of the small firms (often owned and managed by a single family) decided spontaneously not to expand their business in order to be able maintain the complete control of the company (Preti, 1991; Bramanti, 1995). On the other hand, ICT adoption and Business Process Reengineering (BPR) allowed medium companies to reduce the number of workers (mainly those committed to low-level tasks, easily replaced by automatic control machines) simultaneously improving the productivity.

In fact, the introduction of ICT caused not only a decrease in manpower: it arose as well a more efficient way of producing goods, processing and exchanging information (UnionCamere, 1996). From the organizational point of view, it also determined the need of new and more specialized roles (Earl, 1995; Earl, 1994; Feeny, 1997, Gramignoli, 1999). These requirements have sometimes induced delaying ICT investments within companies experiencing problems in recruiting people with specific competencies.

## RESULTS

After an overall description of the sample, next sections will present research results providing answer to the research questions shown above.

### Basic Information on the Sample

The heterogeneity of the answers suggested to evaluate them by dividing the sample in two separate subsets, according to the company size: Figure 2 and Figure 3 show in detail the composition of the sample.

Small companies activity sectors are shown in Figure 3(a), while medium ones are detailed in Figure 3(b). Small companies average year of establishment is 1964, while it is 1946 for medium ones. 56% of small companies and 79% of the medium ones serve an international market.

It is relevant to underline that the average income per employee within medium companies is 1.7 times higher than the one of small companies, thus motivating even more a separate analysis of the two subsets.

### RQ1: Information System Staff

The presence of internal staff dedicated to the business IS has been evaluated by separating technical roles (i.e. development and maintenance) from manage-

*Figure 2: Subdivision of the companies included in the sample. (a) Subsets based on the company size. (b) Subsets based on seven main activity sectors.*

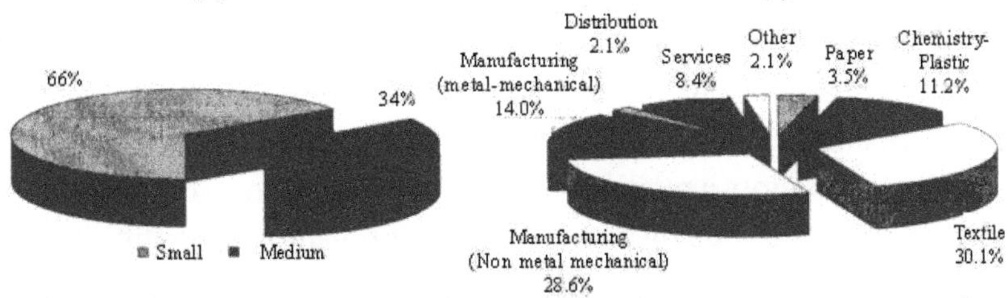


*Figure 3: Subdivision of (a) small and (b) medium size companies by sector.*

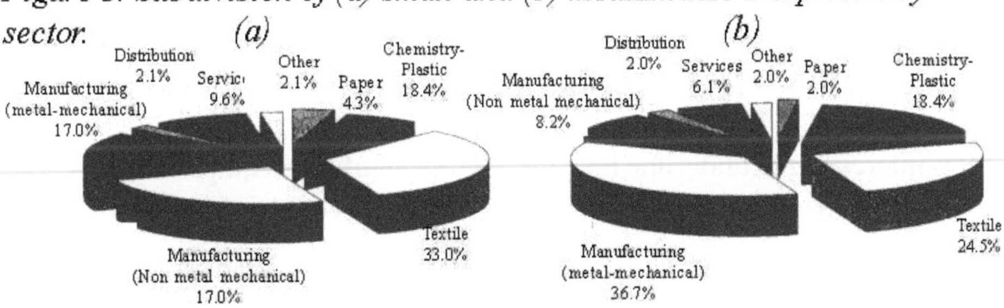


ment ones. The share of respondents having at least one employee dedicated to the IS development and maintenance is about 65%; 50% of companies make use also of external consultants. Conversely, while 17% of the companies do not care about the IS management, at least one third of the sample makes use of both internal and external resources to effectively drive the business IS.

Obviously, small size companies have a lower number of IS dedicated employees, but even a lower impact of external consulting (on the average they require external intervention 7 days per year against 28 days per year required by medium size companies).

*Figure 4: diffusion of hardware equipment within the surveyed companies.*

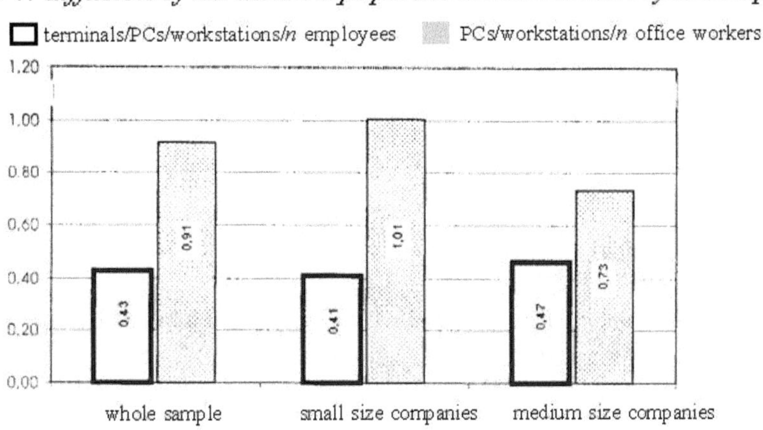

## RQ2: Hardware Equipment

On average, the surveyed companies present rich hardware equipment: almost one seventh of companies owns a mainframe, while more than a half has a minicomputer. However, almost all the mainframes are adopted by medium size organizations, while distribution of minicomputers is quite uniform between medium and small companies.

Research results show a relatively exiguous presence of workstations, which seems to contradict the wide diffusion of CAD application: a possible reason can be recognized in the difficulties encountered in the designers' re–training to these new tools.

Personal Computers are evenly spread over all the companies, with a higher ratio of desktop computers per employee in small companies with respect to medium ones.

It is interesting to underline how the average hardware equipment is comparable with the figures coming from other European countries (ASSINFORM, 1997).

## RQ3: IS Support to Company's Activities

The most widespread application concerns accounting, as shown in Figure 5: it is adopted by 86% of all companies (possibly most of the remaining 14% companies outsource such a task), and 34% of respondents consider this application profitable for the company.

Although CAD and production management applications are considered capable of differentiating a company from its competitors, their diffusion in the sample is lower than any reasonable prediction. CAD applications, especially, present a very low adoption rate.

The surprising low importance assigned to payroll management has its roots in the frequent outsourcing of this activity, mainly due to the complex and frequently changing related normative, which requires a frequent maintenance of the related software. Moreover, the smaller the size of the company, the more convenient it is outsourcing this activity.

Inventory and storage management is highly automated: at least 7 companies out of 10 adopt software tools, and about 27% of the whole sample regard it as convenient.

## RQ4: ICT Adoption Trends

As shown in Figure 6, more than one third of the companies are already connected to the Internet, while almost another third expects to be connected in the

*Figure 5: Company activities supported by software applications and related importance.*

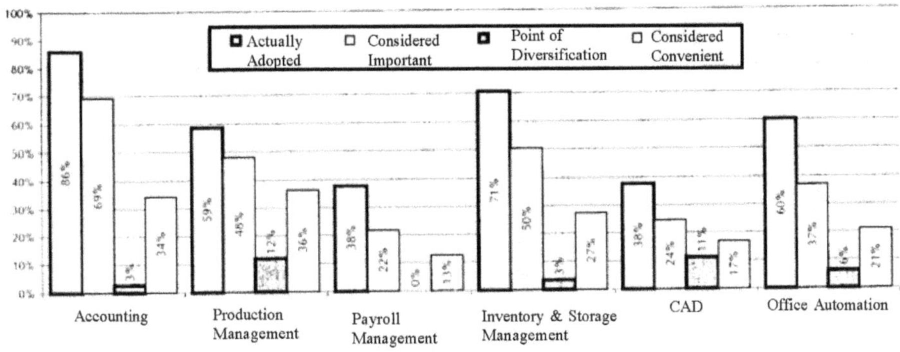


*Figure 6: Current diffusion of ICT and foreseen evolution within the next three years. In each column the base portion represent the current diffusion, the intermediate refers to next year, the top portion is related to next three years. In each group the first column represents small size companies, the second refers to medium size companies, the third shows the situation of the whole sample.*

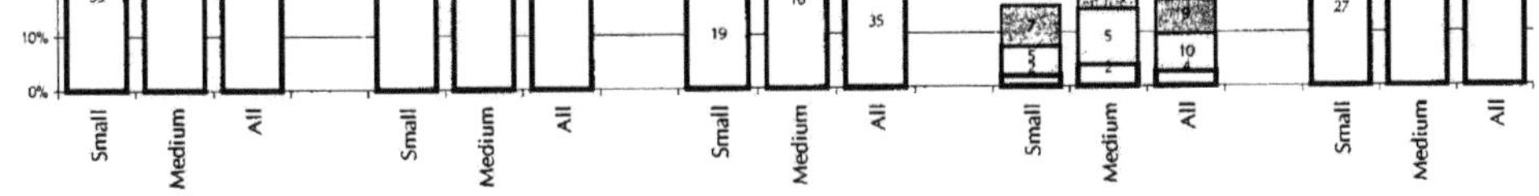


next three years. Most of these companies foresee the adoption of electronic mail (e-mail) and the World Wide Web (WWW) services.

On the other hand, intranets have a very limited diffusion: only 3% of the sample is already making use of an intranet, but its adoption is supposed to multiply six times in the next three years, reaching a diffusion of about 18%. This percentage becomes 27% when considering only those companies that currently have a connection to the Internet, thus showing that current assets in Internet related technologies are expected to generate a good return, and lead to new investments in the same field.

Electronic business applications are taking their very first steps. Companies' web sites are mainly devoted to advertisement and distribution of commercial information, whilst involving only marginally other functions.

Direct connections are mainly adopted by medium size companies, which use them to connect different locations of the same company or to allow a more direct interaction with their partners.

Small size companies highlight a limited interest in sharing information with external partners. On the other hand, they claim to increase next year investment in Internet related technologies showing an increasing awareness of the opportunity to extend commercial interactions, i.e. their range of activity.

Particular classes of applications, such as groupware or Decision Support Systems, are only used in few medium size companies, and their expected growth in next three years is very limited.

## RQ4: ICT Investments

The analysis of ICT investments has been carried out according to a classification based on three levels (Figure 7). There is a clear difference between small and medium size companies: the majority of small companies, on average, invest for less than 25.000 Euro (therefore occupying only the first two levels). On the contrary, medium size companies are more uniformly spread over all three levels, thus showing an aptitude toward higher investments (possibly inferring a positive relationship between company size and absolute value of IS budget).

The difference appears particularly evident when referring to personnel expenses, confirming that larger companies contrive to allocate higher resources to the IS management.

## RQ5: The Role of ICT within SMEs

All SMEs involved in this work were asked to evaluate the impact of ICT investments on company performance, as reported in Figure 8. In particular, each company was asked if ICT investment had an impact on costs, process and product quality, efficiency, diversification, and market share and whether such impact was positive or negative.

It is worth noting that even if the results for diversification and market share factors are similar, actually they do not refer to the same companies. Hence, we cannot imply that a positive impact on the market share may be induced by the positive impact on diversification.

Small companies are characterized by a stronger perceived impact on costs and product quality, while medium ones prevail on process quality.

These results evidence a gap between the awareness of SMEs and the actual ICT influence as shown by collected data. When answering a direct question in the questionnaire, SMEs asserted that ICT influences their strategic choices. However, a whole comprehensive analysis of the results contradicts such assertion, pointing out the weaknesses of the SMEs entrepreneurial culture (such as reluctance towards ICT investments, difficulties in the evaluation of their profitability, lack of knowledge on long term strategic objectives) as reported by literature.

*Figure 7: ICT investments in last three years. In each column the base portion represents the percentage of companies investing less than 5000 Euro, the intermediate portion refers to an investment between 5000 and 25.000 Euro, while the top portion is related to investments higher than 25.000 Euro.*

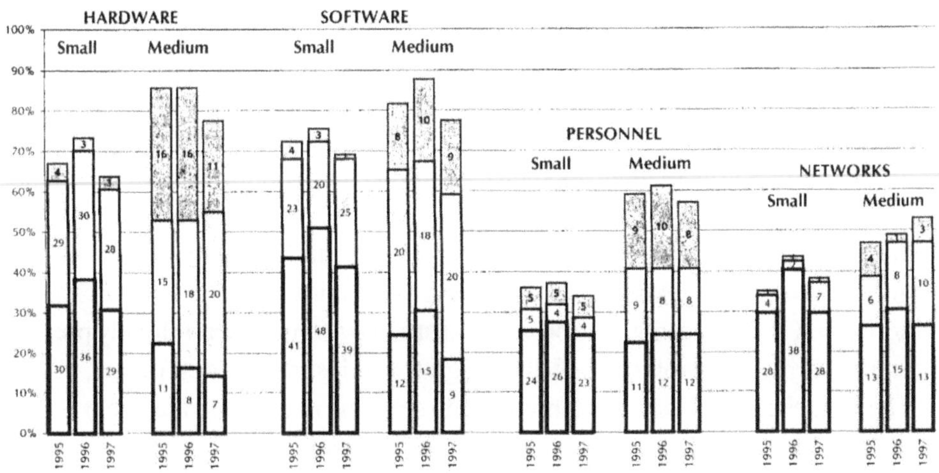


*Figure 8: Evaluation of the impact of ICT investments on the company results. In each column the base portion represents the percentage of companies declaring a positive impact, the intermediate portion (when present) refers to negative impact, the top portion refers to companies declaring no impact. Columns do not reach 100% because some companies did not answer to these questions.*

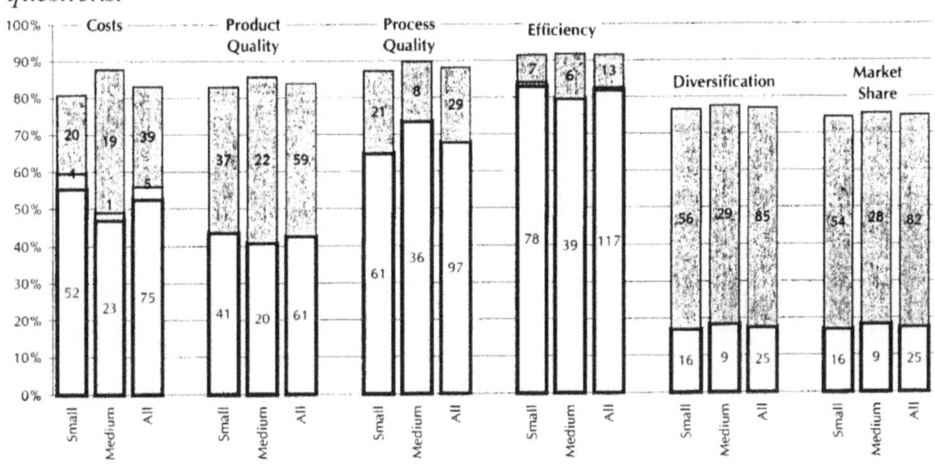

In order to clarify the ICT strategic role within SMEs we found useful to compare research results with the characteristics of the economic environment.

a) Production oriented software applications are used by most of the companies: it is inferable an influence of the strongly product-process oriented culture, typical of most enterprises in this geographical area. Nevertheless, office automation tools are playing an increasingly important role, possibly as a consequence of the easily measurable efficiency improvement that can be gained by using these applications.
b) The average ICT equipment reflects the caution characterizing investment choices of SMEs. Even so, it should be noticed that respondents predict an inversion of this trend: they expect to raise their ICT investments in the next three years. This change is going to be strongly influenced by the prevention policy for the Y2K bug. However, even small companies forecast a wider impact of ICT on their business (mainly due to adoption of Internet related technologies) and the whole set of SMEs recognizes such technology as an important tool to support both customer/provider relationships and internal processes. Such evolution may influence the long lasting *product-oriented* culture and move it towards a more *customer-oriented* one. Moreover, the fast investment growth characterizing ICT may smooth the progress of small systemic areas (already present in the Varese area) towards a fully integrated industrial district. In fact, ICT is flexible, cheap and with a potentially high strategic impact and may help SMEs in leaving behind the obstacles preventing a profitable integration.
c) SMEs traditional opposition against external control determines reluctance towards external financing (e.g., banks). Whenever the introduction of an innovation has to be evaluated, this determines a too scrupulous (and expensive) risk analysis, possibly leading to dangerous bottlenecks. This is particularly true when dealing with intangible ICT investments. The difficulty in assessing the expected profitability of ICT investments is confirmed by the strong aptitude towards investments in data processing compared with the lack of decision support tools, which require a more complex value analysis.
d) The exploitation of ICT investments for strategic purposes requires a preliminary effort, consisting in defining a competitive strategy, planned by the top management and shared within the company (Banker, 1993; Bigdoli, 1997; Knights, 1997). The lack of such a long-term vision could compromise the big potentiality deriving from the strategic use of ICT. The influence of past ICT investments on most performance indicators (as pointed out by the research) represents a stimulus toward a better awareness, thus a constructive attention, of the ICT strategic role.

By comparing results concerning the support of ICT to strategy and operations is possible to outline very different conclusions for companies of different size.

Small companies seem to be aware of the quick development and diffusion of ICT and recognize its importance for an active presence in the world market. The remarkable automation of information flows support the thesis that small companies have already crossed mere utilization of ICT for efficiency purposes. On the other hand, their investment policies show contradictions, highlighting lacks of planning activities. CAD applications, for instance, seem under–used even within contexts where design activities play a strategic role. Moreover, Internet adoption often seems to be chosen to emulate competitors more than to support the company activities that actually need it.

Nevertheless, the weight of production management applications within such a heavily product-oriented context and the perspectives of Internet technologies use suggest that small companies are already using ICT as a competitive weapon. It is therefore difficult to predict in how many small enterprises ICT will actually maintain a strategic role, while experiencing managerial and organizational deficiencies. In fact, IS development in a strategic direction must imperatively be supported by a clear competitive and environmental analysis on middle and long-term horizons.

Within medium enterprises, the awareness of the role ICT will play in the next future coexists with a high organizational-managerial coherence, typical of larger companies. As a consequence, medium size enterprises seem to be much more prepared to sustain an evolutionary development of the ICT role, more and more oriented to support the competitive strategy.

## CONCLUSIONS

While previous literature on ICT impact is mainly focused on large companies, the analysis presented in this paper is concentrated on SMEs. Results show that there is a remarkable difference between small and medium size companies: both are aware of ICT importance, but medium ones seem to have more resources to spend on ICT adoption and management. On the contrary, small companies are characterized by a limited technical knowledge: this seems to be the major barrier to maximize the profitability of ICT adoption.

Computer based tools are used by all the examined companies and their diffusion confirms a high (and still growing) degree of ICT adoption: e.g., the ratio between computer systems and employee is twice the Italian average and remains in line with the most industrialized European countries. Hence, we can argue that SMEs in Northern Italy have a high potential regarding ICT adoption and deployment: both managers and owners feel that ICT will play a strategic role in the future of these enterprises.

Nevertheless, an effective connection between ICT investments and the strategic control of the enterprises is still lacking: ICT adoption is often considered just an operational cost, without any clear understanding of its strategic implications. A profound connection between the company strategic orientation and the ICT investment policy might improve the effectiveness of such investments, even though in many cases the problem seems related to the awareness of the company strategic course. Other forces driving this evolution could be the succession phase that many enterprises are going through and the growing managerial presence (especially in medium size companies).

Our future work will be mainly focused on a more detailed analysis of small companies, trying to monitor their evolution in a rapidly evolving context. On one hand, ICT cost and complexity are continuously reducing, thus making ICT introduction cheaper and easier. On the other hand, a delay in an aware strategic use of ICT could cause small companies losing their ability to compete, and even compromising their future.

## ENDNOTE

[1] The minimum size *n* of a sample, necessary to obtain a *D* maximum error in the *Eg* % of the cases, can be assessed through the formula: n= [N•Eg•P'•(1-P')]/[(N-1)•D2+(Eg2•P'+(1-P'))], where N is the dimension of the population and P' is the percentage of the *n* elements which own the characteristic under analysis (Giardina, 1990).

## REFERENCES


ASSINFORM (1997). Rapporto Assinform sulla situazione dell'informatica in Italia, in Italian: Report on the Information Technology use in Italy.

Banker, R., Kauffman, R. and Mahmood, M. (1993). *Strategic Information Technology Management: Perspectives on Organizational Growth and Competitive Advantage*, Hershey, PA: Idea Publishing.

Bartolozzi, P., and Garibaldo, F. et al. (1995). Lavoro creativo e impresa efficiente, ricerca sulle piccole-medie imprese in Italian: Creative working and the efficient firm: a research upon small-medium enterprises, Ed. Ediesse, Roma, Italy.

Berenson, M. L., and Levine, D.M. (1989). *Basic Business Statistics: concepts and applications*, NJ: Prentice-Hall.

Bigdoli, H., (1997). *Modern Information Systems for managers*, San Diego, CA: Academic Press.

Bramanti, A. and Odifreddi, D. (1995). Lo sviluppo delle "aree avanzate": apertura e identità nei sistemi economici territoriali : Varese come caso nazionale, (in Italian: "Advanced areas" development: openness and identity within economic territorial systems: Varese as a national case study) Ed. Angeli, Milano, IT.

Bums, T., and Hatter, W. (1992). Attitudes to PCs - General Public and Very Small Businesses, (research study conducted for IBM UK Limited), MORI, London.

Cragg, P.B., and Zinatelli, N. (1995). The Evolution of Information Systems in Small Firms, *Information and Management*, 29, July, pp. 1-8.

DG XII at the European Commission - A new common definition for SMEs, Press Releases, 13.02.1996, http://europa.eu.int/comm/dg12/press/1996/pr13-2bi.html (October, 14th 1998).

Earl, M. J. (1995). The Changing Shape and Skills of the new IS Function, London Business School, Centre for Research in Information Management, WP 95/3.

Earl, M. J., and Feeny, D. (1994). Is Your CIO adding Value? *Sloan Management Review*, Spring.

Feeny, D. F., Abs, V., Millie, E., Minter, A., Selby, C., and Williams, J. (1997). Defining New Skills and Competencies, Oxford Institute of Information Management, Research Paper, Templeton College, Oxford.

Giardina, B. (1990). Statistica per aziende e ricercatori in Italian: Statistics for firms and researchers, FrancoAngeli, Milano.

Gramignoli, S., Ravarini, A., and Tagliavini, M. (1999). A Profile for the IT Manager within SMEs, *Proceedings of the ACM International Conference SIGCPR 1999: Managing Organizational Knowledge for Strategic Advantage: The Key Role of Information Technology and Personnel* (April, 1999), New Orleans, Louisiana, USA.

ISTAT Istituto Nazionale di Statistica (1991). Classificazione delle Attività Economiche in Italian: Activity Sectors Classifications, Istat, Roma.

ISTAT Istituto Nazionale di Statistica (1951-1991). Censimento generale dell'Industria, del Commercio, dell'Artigianato in Italian: Census of the Economic Activities in Italy.

Japan Small Business Research Institute (1995). The New Challenge, Chapter 5: Introduction of information technology for the purpose of rebuilding corporate operational management systems.

Kemerer, C. F. (1998). *Information Technology and industrial competitiveness: how IT shapes competition*, Boston, MA: Kluwer Academic Publishers.

King, W. R., and Thompson-Teo, S. H. (1994). Facilitators and inhibitors for the strategic use of information technology, *Information & Management*, 27, pp. 71-87.

Knights, D., Noble, F., and Willmott, H. (1997). We should be total slaves to the business. Aligning Information Technologies and Strategies, in Bloomfield B.P., Coombs R., Knights D. and Littler D. (Eds.), *Information Technology and Organizations: Strategies, Networks and Integration*, Oxford University Press, Oxford.

Lai, V. (1994). A survey of rural small business computer use: success factors and decision support, *Information & Management*, 26, pp. 237-304.

Lang, J.R., Calatone, R.J., and Gudmundson, D. (1997). Small firm information seeking as a response to environmental threats and opportunities, *Journal of Small Business Management*, January, 35(1), pp. 12-18.

Monsted, M. (1993). Introduction of Information Technology to Small Firms: a Network Perspective, *Entrepreneurship and Business Development*, (Ed.) Klandt, H., Avebury, Aldershot.

NACE Rev. 1 (1990). L. 293 of 24/10/1990, in CEE Regolament 3037/90.

Palvia, P. et al. (1994). Determinants of computing in very small businesses, *Information & Management*, 27, pp.161-174

Porter, M.E., and Millar, V.E. (1985). How information gives you competitive advantage, *Harvard Business Review*, July-August, pp. 149-160.

Preti, P. (1991). L'organizzazione della piccola impresa in Italian: The Small Enterprise organization, EGEA, Milano, IT.

Raymond, L., and Bergeron, F. (1992). Personal DSS success in Small Enterprises, *Information & Management*, 22, pp. 301-308.

Schleich, J.F., Carney, W.J. and Boe, W.J. (1990). Pitfalls in microcomputers system implementation in small business, *Journal of System Management*, 41.

Soh, P. P. C., Yap, C. S., and Raman, K. S. (1994). Impact of consultants on computerization success in small business, *Information & Management*, 22, pp. 309-319.

UnionCamere Lombardia (1996). Gli investimenti delle imprese manifatturiere lombarde nel 1994, in Italian: Investments of manufacturing enterprises in Lombardia in 1994, Quaderno 1/96.

Willcocks, L., Feeny, D., and Isley, G. (1997). *Managing IT as a Strategic Resource*, London: McGraw-Hill.

Wiseman, C. (1985). *Strategy and Computers: Information Systems as Competitive Weapons*, Homewood, IL: Dow Jones Irwin.

Zinatelli, N., Cragg, P. B., and Cavaye, A. L. M. (1995). End User Computing Sophistication and Success in Small Firms, *European Journal of Information Systems*, 5.

## Chapter 13

# Critical Strategies for Information Systems Development Projects: Perceptions of Developers in Korea

Chung Kim, Dane Peterson, Jerry Chin and Tonya Barrier
Southwest Missouri State University

## INTRODUCTION

Despite enormous progress in the methodologies and technologies used in the development and implementation of Information Systems (IS), IS failure continues to plague the efforts of many companies. According to a recent report, up to 90 percent of all IS projects fail to achieve their goals, 80 percent are late and over-budget, and 40 percent are abandoned (Clegg et al., 1997). Since information systems are becoming increasing critical for the strategic operations of organizations, the high rate of failure for IS projects is of great concern to both organizational management and IS professionals.

Developing and implementing successful international IS have proven to be even more difficult due to such factors as the diversity and disagreement among the participants (Mantelaers & Van Den Berg, 2000). However, with an increasing number of firms expanding into international markets, the need to develop and implement IS which cross national boundaries and span diverse cultures is rapidly growing (Palvia, 1998). Adding to the importance of developing successful international IS is the realization that they constitute an essential tool for assisting organizations in coping with the increased competition associated with global markets. Thus, there is an urgent need to acquire knowledge about whether the factors found to be critical for developing successful IS in one country are also applicable to developing international IS involving countries with different cultural norms.

This study examined the views of IS developers from (South) Korea regarding factors perceived to be critical in IS development. There are a couple of reasons why it may be important to explore the views of IS developers from Korea. First, Korean is an economically strong and technologically advanced country (Palvia, Palvia, & Zigli, 1992). Thus, Korean firms actively markets goods and services in a number of other countries. In addition, there are a large number of joint ventures and mergers among companies that link Korea with other nations. A second reason why an examination of the views of IS developers from Korea may be of importance is that the Korean culture differs in number of ways from countries with a Western culture (Hofstede, 1980). Cultural differences are often cited as a reason that IS development in a global environment may differ significantly from the methods that have typically been proven to be successful in the Western culture (Shore, 1998; Shore & Venkatachalam, 1994). In addition, cultural differences have particularly proven to be a major obstacle in global IS development when Asian offices are involved (Burnson, 1989). Thus, an examination of the views of IS developers from Korea regarding the factors considered to be critical for IS success should provide valuable information to IS managers engaged in the development of international IS involving Korea and possibly other nations in East Asia.

## CRITICAL FACTORS FOR IS SUCCESS AND FAILURE

Since Rockart (1979) highlighted its usefulness, the concept of Critical Success Factors (CSFs) has found widespread application in IS planning and implementation of various types of IS. However, the broader literature on IS implementation indicate that there may be many inhibiting factors which play a key role in IS failure. It has been suggested that there is a need to identify both CSFs and Critical Failure Factors (CFFs) (Williams & Ramaprasad, 1996). Otherwise, a disproportionate amount of attention may be devoted to enhancing factors only to discover that there are inhibiting factors that prevent IS success.

Earlier studies suggested that the reasons for IS success may not necessarily coincide with the reasons for IS failure (Lyytinen, 1988). In other words, CSFs and CFFs may not be exactly the opposites of each other so that the absence of a CSF becomes a CFF. However, many factors identified in IS failure studies are also identified as CSFs in the IS success literature. Moreover, to develop two completely different taxonomies for CSFs and CFFs could be difficult and possibly confusing (Williams & Ramaprasad, 1996).

## RESEARCH ON FACTORS CONTRIBUTING TO SUCCESS/FAILURE OF IS DEVELOPMENT

Numerous studies have been conducted to identify the factors that are critical for IS success (e.g., Ginzberg, 1981; McDoniel, Palko, & Cronan, 1993; Pinto & Slevin, 1990; Swanson, 1988). The focus of these studies was typically on enhancing factors that increase the probability of IS success. For instance, Ginzberg (1983) used a discriminant analysis procedure to examine which factors best distinguished between successful and unsuccessful projects as determined by user satisfaction. His results indicated that organizational commitment, commitment to change, and extent of project definition and planning were the best predictors of system success.

More recently, it has been suggested that the factors critical for successful IS development could be classified according to three groups (McDoniel et al., 1993). The three groups are the users, management, and IS developers. Factors related to the user include user involvement, commitment, expectation, and training. Management-oriented factors include commitment to the project, support for change, resource adequacy, and situational stability. Designer-oriented factors include designer understanding, design quality, proper techniques, and project management. To determine which of these success factors were critical, the factors were used as predictor variables in a multiple regression analysis with user satisfaction as the dependent variable. The results indicated that the significant predictors of user satisfaction were user expectation, situational stability, and project management.

In another example of an attempt to identify the important success factors, Pinto and Slevin (1990) developed a set of ten critical success factors. The set of factors were constructed on the basis of prior conceptual studies (Baker, Murphy, & Fisher, 1983; Cleland & King, 1983; Martin, 1976). The set of ten factors included (1) project mission, (2) top management support, (3) project schedule/plan, (4) client consultation, (5) personnel selection/training, (6) technical tasks, (7) client acceptance, (8) monitoring and feedback, (9) communication, and (10) trouble-shooting. Pinto and Slevin found that the critical factors for IS success varied across the project's life cycle. Planning factors were important during the initial stages of project development, but were overtaken by tactical issues as the project progressed through its life cycle.

Other investigations have approached the issue of identifying the critical factors from a IS failure perspective. These studies have examined unsuccessful IS projects in search of key factors that are likely to contribute to the failure of IS projects (e.g., Ewusi-Mensah & Przasnyski, 1991; Flowers, 1997; Jiang, Klein, Balloun, 1998; Liebowitz, 1999; Lyytinen, 1988). As an example, Ewusi-Mensah

(1997) examined abandoned IS projects and found that poorly defined project goals, lack of project team experience and expertise, lack of project control, inappropriate technology, lack of top management involvement, and escalating project costs were among the major reasons for IS failure. In another investigation of CFFs, Flowers (1997) reported that IS failure could be attributed to poor reporting structures, poor consultation, over-commitment, changing requirements, project timetable slippage, underestimated complexity, inadequate testing, and poor training.

## PURPOSE OF THE STUDY

Although numerous studies have investigated potential CSF's or CFF's, there are apparently no studies that examined both types of factors in the same study. One objective of this study was to examine if there are any differences in the relative importance of the factors in terms of how they contribute to the success or failure of IS projects. One possibility is that the factors believed to be most important for IS success may be same as the factors, which if not present, would most likely cause IS failure. Alternatively, IS developers may view some factors as important contributors to IS failure, but view the same factors as having little relevance for contributing to the success of IS projects.

Another goal of this investigation was to compare the results obtained in this study involving the perceptions of IS developers from Korea with the results of prior studies conducted in the U.S. The results of this study should make it possible to determine if the views of IS developers from Korea regarding the critical IS development factors are similar to those reported in studies involving IS developers from the U.S. Thus, the present study addressed the following four research questions:

1. What are the most important factors for IS *success* as perceived by IS developers in Korea?
2. What are the most important factors for IS *failure* as perceived by the developers in Korea?
3. Are some factors viewed as more likely to contribute to IS success than to contribute to IS
failure?
4. How similar are the perceptions of IS developers from Korea regarding the critical factors in comparison to studies examining the views of IS developers from the U.S.?

## RESEARCH METHODOLOGY

Based on prior studies examining factors that contribute either to the success or failure of IS projects, this study identified 18 potential factors that may be considered critical. These factors are controllable and are also applicable to IS developers in international environments. The 18 factors are listed in Table 2. The failure factors are expressed as the absence or insufficient condition of the success factors (i.e., insufficient user participation, lack of top management support, etc.).

The survey was pre-tested with several faculty members employed with a university in the U.S. It was then pilot-tested with IS developers from two companies located in the U.S. The questionnaire was translated into the Korean language by a faculty member at a Korean University and pre-tested by several graduate students and faculty members working at the same school. Finally, a back-translation was performed on the questionnaire by a faculty member from the U.S.

IS developers evaluated the importance of each of the 18 items displayed in Table 2 on a seven-point scale ranging from (1) no contribution to (7) very high contribution for project success. In addition, for a second set of 18 failure factors, each respondent evaluated the items on a seven-point scale from (1) no contribution to (7) very high contribution in terms of contributing to IS failure. The order in which the items were presented on the survey was randomly determined.

*Companies.* The data for IS programmers were obtained from ten organizations. The type of organizations included five in consulting/computer service companies, most of which belonged to a parent conglomerate company. Four companies were in manufacturing and the remaining company was in the oil industry. Four of the companies had less than 500 total employees, three had between 500 and 1,000 total employees, one company had between 1,000 and 5,000 total employees and the remaining two companies had over 5,000 total employees. The number of IS employees ranged from 20 to over 250. The IS budget was under one million for one company, between 1.1 and 5 million for six companies, between 5.1 and 10 million for one company, and over 10 million for two companies.

*Respondents.* The surveys were randomly distributed to IS developers in the ten organizations. All respondents receiving the survey completed the questionnaire. The number of

IS developers surveyed in each organization ranged from four to 29, with a total of 127 usable surveys. Table 1 presents a demographic profile of the survey respondents. As can be seen in Table 1, the majority of the respondents were males (90.6%). Most of the respondents had at least a college degree (83.5%), but only 10.2 percent had a graduate degree. The average age of the respondents was 32.24 with a standard deviation of 4.42 years. The average number of years of experience in the IS field was 6.25.

*Table 1: Profile of Respondents*

| Variable | Frequency | Percentage |
|---|---|---|
| Gender | | |
| Male | 115 | 90.6% |
| Female | 12 | 09.4 |
| Education | | |
| High School | 6 | 04.7% |
| Some technical training | 15 | 11.8 |
| Bachelor's degree | 93 | 73.2 |
| Master's degree | 13 | 10.2 |
| Major in College | | |
| Computer Science | 54 | 42.5% |
| MIS or CIS | 6 | 04.7 |
| Business | 5 | 03.9 |
| Others | 62 | 04.9 |
| Job Title | | |
| VP/Director | 1 | 00.8% |
| IS manager | 3 | 02.3 |
| Supervisor/project manager | 13 | 10.2 |
| Programmer | 55 | 43.3 |
| Analysts | 3 | 02.4 |
| Programmer/analyst | 40 | 31.5 |
| Other | 12 | 09.4 |
| Organizational Position | | |
| General Manager | 4 | 03.1% |
| Deputy General Manager | 4 | 03.1 |
| Manager | 18 | 14.2 |
| Assistant Manager | 55 | 43.3 |
| Employee | 46 | 36.2 |
| Age (mean) | 32.24 | |
| Years in Organization (mean) | 4.87 | |
| Years in the Industry (mean) | 6.25 | |

## RESULTS AND DISCUSSION

The 18 factors are arranged in Table 2 according to the perceived importance of the factors for contributing to IS success. The sample means and standard deviation for the success factors are presented in the middle two columns of Table 2. As illustrated in Table 2, user participation, clearly stated objectives, team member commitment, and top management support were viewed as the four most important factors for IS success. These results are generally consistent with most studies conducted in the U.S. in terms of the factors believed to be critical for IS success (Schmidt, Lyytinen, Keil, & Cule, 2001).

The last two columns of Table 2 summarize the descriptive statistics for the 18 failure factors. The results indicate that three of the same four factors, user participation, clearly stated objectives, and top management support, are viewed as the factors most likely to contribute to IS failure. Thus the factors viewed as most important for success are generally also the factors, when absent, that are perceived as most likely to contribute to IS failure.

As illustrated in Table 2, the success factors were rated higher in importance for success than the failure factors were for contributing to failure. For example, the mean for user participation was 6.07 for success and 5.89 for contributing to failure. This would seem to support the assumption that developers tend to think in terms of the factors that contribute to the success of a project and are less inclined to think in terms what might lead to IS failure (Lyytinen, 1988).

*Table 2: Critical Factors Ranked by Importance for Success*

| Critical Factors | Success Ratings | | Failure Ratings | |
|---|---|---|---|---|
| | Mean | s.d. | Mean | s.d. |
| 1. (insufficient) User participation in the project | 6.07 | 1.056 | 5.89 | 1.248 |
| 2. (Lack of) Clearly stated objectives | 6.06 | 0.994 | 5.38 | 1.291 |
| 3. (Lack of) Team member commitment | 5.95 | 0.836 | 5.02 | 1.501 |
| 4. (Lack of) Top management support | 5.91 | 1.066 | 5.57 | 2.069 |
| 5. (Lack of) Project leader's project monitoring/control | 5.84 | 0.929 | 5.17 | 1.320 |
| 6. Project leader's (in)experience | 5.80 | 0.882 | 5.63 | 1.302 |
| 7. (Ill)Alignment of project and corporate goals | 5.71 | 1.032 | 5.05 | 1.201 |
| 8. Use of (in)appropriate technology | 5.69 | 0.965 | 4.98 | 1.391 |
| 9. (Lack of) Detailed project plan | 5.65 | 0.979 | 5.19 | 1.283 |
| 10. (No attempt to) Reengineering business processes | 5.65 | 1.028 | 5.18 | 1.712 |
| 11. (Im)Proper project scope | 5.54 | 0.963 | 5.20 | 1.235 |
| 12. (In) Adequate training for team | 5.47 | 0.9128 | 5.18 | 1.746 |
| 13. (Lack of) Project leader's feedback to team | 5.45 | 0.902 | 4.85 | 1.228 |
| 14. (Lack of)Team member's self-control on the progress | 5.41 | 1.084 | 4.67 | 1.440 |
| 15. Utilizing an (in)effective methodology | 5.41 | 0.958 | 4.59 | 1.444 |
| 16. Team member (in)experience | 5.36 | 1.010 | 4.65 | 1.269 |
| 17. (Little use of) Utilizing a prototype | 5.11 | 1.125 | 4.44 | 1.307 |
| 18. (insufficient) Peer review on project progress | 5.11 | 0.995 | 4.56 | 1.207 |

To compare the relative importance of the 18 success and failure factors, the responses for each respondent were standardize. That is, the 18 responses for the success items were first standardized for each participant in the study and then the same procedure was used to standardize the responses for each participant on the 18 failure factors. The means for the standardized responses are presented in Table 3 along with the rank order of importance for each of the 18 factors.

The value of Spearman's rank order correlation coefficient is .886, indicating that there is a strong agreement in terms of how the factors are perceive to contribute to both IS success and IS failure. The last column of Table 3 shows the results of related samples t-tests for each of the 18 factors. Only four of the differences were significant. As shown in Table 3, user participation and project scope were viewed as more likely to contribute to project failure than to IS success. On the other hand, team member commitment and use of an effective methodology were viewed as more likely to contribute to IS success than to IS failure.

## CONCLUSIONS

This study identified critical factors for IS success and critical factors for IS failure as perceived by IS developers in Korea. Of the four most important factors, three are common to both success and failure. The three factors perceived critical for both IS success and failure are:

- User participation
- Clearly stated objectives
- Top management support

*Table 3: Critical Factors: Standardized Responses, Ranks and t-tests*

| Critical Factors | Success Mean | Success rank | Failure Mean | Failure rank | t-statistic |
|---|---|---|---|---|---|
| 1. (insufficient) User participation in the project | .514 | 1 | .838 | 1 | -2.59** |
| 2. (Lack of) Clearly stated objectives | .543 | 2 | .414 | 2 | 1.10 |
| 3. (Lack of) Team member commitment | .374 | 3 | .095 | 7 | 3.08** |
| 4. (Lack of) Top management support | .297 | 4 | .381 | 3 | -.081 |
| 5. (Lack of) Project leader's project monitoring | .257 | 5 | .190 | 4 | 0.78 |
| 6. Project leader's (in)experience | .196 | 6 | .057 | 9 | 1.58 |
| 7. (Ill-) Alignment of project and corporate goals | .132 | 7 | .029 | 11 | 1.12 |
| 8. Use of (in)appropriate technology | .092 | 8 | .039 | 10 | 0.56 |
| 9. (Lack of) Detailed project plan | .036 | 9 | .175 | 5 | -1.44 |
| 10. (No Attempt to) Reengineering business | .016 | 10 | .088 | 8 | -0.88 |
| 11. (Im)Proper project scope | -.118 | 11 | .163 | 6 | -3.17** |
| 12. (In) Adequate training for team | -.203 | 12 | -.250 | 13 | 0.55 |
| 13. Utilizing an (in)effective methodology | -.213 | 13 | -.380 | 16 | 1.97* |
| 14. (Lack of) Project leader's feedback to team | -.231 | 14 | -.203 | 12 | -0.37 |
| 15. (Lack of)Team member's self-control | -.269 | 15 | -.290 | 14 | 0.27 |
| 16. Team member (in)experience | -.290 | 16 | -.330 | 15 | 0.62 |
| 17. (Little use of) Utilizing a prototype | -.545 | 17 | -.540 | 18 | -0.07 |

The finding that user participation is the most important factor is consistent with a number of studies conducted in the U.S. (Aladwani, 2000; Cleland & King, 1983; Ewusi-Mensah, 1997; Jiang, et al., 1996). User participation in the preliminary analysis stage and throughout the later stages in IS implementation is known to be critical for accurate reflection of user requirement in systems design and for user acceptance of the final system.

The clearly stated objectives factor has also been found to be important for system success in prior studies conducted in the U.S. Various authors emphasized that the objectives and the ultimate benefits of a new project should be declared at the outset of a project (Pinto & Prescott, 1990; Pinto & Slevin, 1987; Ewusi-Mensah, 1997). Similarly, top management support has been reported to be related to the success or failure of many types of IS (Aladwani, 2000; Cleland & King, 1983; Ewusi-Mensah, 1997; Jiang et al., 1996; Ginzberg, 1981; Liebowitz, 1999; McDoniel et al., 1993; Pinto & Prescott, 1990).

One factor that Korean IS developers viewed as more important than what is typically found in studies involving IS developers in the U.S. was project leader's project monitoring/control. This may be explained in terms of the high power distance culture of Korea. In a high power distance culture, top management and leaders are expected to make most of decisions and monitor project progress. Thus, it might seem reasonable to expect that IS developers in Korea would consider project leader's control as an important source for determining the success of IS projects

A factor viewed as more important for failure than for success was project scope. In order to avoid IS failure, it is important for IS developers to have an appropriate understanding of the boundaries of the project or what aspects of the system will be included in the project (Ewusi-Mensah, 1997; Liebowitz, 1999; Stone, 1997). The failure to clearly define the scope of the project may often lead to IS failure. Conversely, it is apparently unlikely that IS developers may attribute IS success to having an appropriately defined project scope, since they may assume the scope should always be appropriately defined.

In general, some of the factors viewed as least important by IS developers from Korea were the technical factor and/or tactic-related operational factors (e.g., effective methodology, utilizing a prototype, peer review, etc.). This result has also been observed with IS developers from Western cultures (Lyytinen, 1988). Apparently, IS developers view organizational factors as more crucial for IS success, possibly because they have less control over these factors making and find it more difficult to manage obstacles outside their control.

In summary, as organizations expand across borders, IS managers may need to become familiar with the different cultures, values, and strategies prevailing in other nations. IS managers also need to analyze the differences in order to develop effective IS implementation strategies for global IS environment. This study provides an initial starting point for future studies we plan to conduct on comparing various perspectives of IS developers from different cultures.

## REFERENCES

Aladwani, A. M. (2000). IS project characteristics and performance: A Kuwaiti illustration. *Journal of Global Information Management*. 8 (2), 50-57.

Baker, B. N., Murphy, D. C., & Fisher, D. (1983). Factors affecting project success. In D. I. Cleland & W. R. King. (Eds.), *Project Management Handbook*. New York: Van Norstrand Reinhold.

Burnson, A. (1989). The perils of going global. *Infoworld*, August 14, 39-40.

Clegg, C., Axtell, C., Damadoran, L., Farbey, B., Hull, R., Lloyd-Jones, R., Nicholls, J., Seell, R., & Tomlinson, C. (1998). Information technology: A study of performance and the role of human and organizational factors. *Ergonomics Journal*, 40 (9), 851-871.

Cleland, D. I., & King, W. R. (1983). *Systems Analysis and Project Management*. New York: McGraw-Hill.

Ewusi-Mensah, K. (1997). Critical issues in abandon information systems development. *Communications of the ACM*, 40 (9), 74-80.

Ewusi-Mensah, K. & Przasnyski, Z. H. (1991). On information systems project abandonment: An exploratory study of organizational practice. *MIS Quarterly, 15(1)*, 67-85.

Flowers, S. (1997). Information systems failure: Identifying the critical failure factors. *Failure and Lessons Learned in Information Technology Management: An International Journal*, 1, 19-30

Ginzberg, M. J. (1981). Key recurrent issues in the MIS implementation process. *MIS Quarterly*, 5(2), 47-59.

Hofstede, G. (1980). Motivation, leadership, and organization: Do American theories apply abroad? *Organizational Dynamics*, Summer, 42-63.

Jiang, J., Klein, G., & Balloun, J. (1998) Perceptions of system development failures. *Information and Software Technology*, 39 (14), 933-937.

Liebowitz, J. (1999). Information systems: Success or failure? *Journal of Computer Information Systems*, 40(1), 17-20.

Lyytinen, K. (1988). Expectation failure concept and systems analysts' view of information system failures: Results of an exploratory study. *Information and Management*, 14 (1), 45-56.

Mantelaers, P. & Van Den Berg, W. (2000). Transnational information systems: development and management issues. *Journal of Global Information Management*, 8(1), 34-44.

Martin, C. C. (1976). *Project Management*. New York: AMACOM.

McDoniel, P. L., Palko, J., & Cronan, T. P. (1993). Information systems development issues affecting success. *Journal of Computer Information Systems*, 34(1), 50-62.

Palvia, P. C. (1998). Research issues in global information technology management. *Information Resources Management Journal*, 11 (2), 27-36.

Palvia, S., Palvia, P., & Zigli, R. (1992). Global issues: An introduction. In Palvia, Palvia & Zigli (Eds.). *Global Issues of Information Technology Management*, Harrisburg, PA: Idea Group Publishing.

Pinto, J. F., & Prescott, J. E. (1990). Planning and tactical factors in the project implementation process. *Journal of Management Studies*, 27(3), 305-327.

Pinto, J. K. & Slevin, D. P. (1987). Critical factors in successful project implementation. *IEEE Transactions on Engineering Management*, 34(1), 22-27.

Schmidt, R., Lyytinen, K., Keil, M. & Cule, P. (2001). Identifying software project risks: An international Delphi study. *Journal of Management Information Systems*, 17(4), 5-36.

Shore, B. (1998). IT strategy: The challenge of over-regulation, culture, and large scale collaborations. *Journal of Global Information Technology Management*, 1(1), 1-4.

Shore, B. & Venkatachalam, V. (1994). Prototyping: A metaphor for cross-cultural transfer and implementation of IS applications. *Information & Management*, 27 (3), 175-184.

Stone, J. (1997). *Developing Software Applications in a Changing IT Environment*. New York: McGraw-Hill.

Swanson, E. G. (1988). *Information System Implementation*, Homewood, IL: Irwin.

Williams, J. J. & Ramaprasad, A. (1996). A taxonomy of critical success factors. *European Journal of Information Systems*, 5, 250-260.

## Chapter 14

# A Web-Based Data Warehouse Planning – A Large-Scale Project Management Information System in Taiwan

Jia-Lang Seng
National Chengchi University, Taiwan

*This chapter describes a data warehouse planning project to develop a large-scale management information system for a Taiwan government agency in managing a nationwide multi-billion science and technology project management and control process. The planning and consulting period lasted through the time period of 1997 and 1998. The outsourcing and implementation tasks are expected to continue through 2000. In this chapter, we present the framework and model of the data warehouses developed by the committee during the planning and consulting stage. Our focus is on the data design and module design for a multi-dimensional database system. The data design aims to provide an integral model for the internal and external data sources. A core view of the multi-dimensional database is developed. The module design attempts to give an analytic model of the field, division, project, and agent performance along the dimension hierarchies. Actual versus budget and actual versus forecast analysis along with the time series analysis and ad hoc analysis are supported in the systems.*

# INTRODUCTION

The science and technology project management (STPM) agency at the Ministry of Economy (MOE) in Taiwan manages a nationwide mission of multi-billion US dollar science and technology project formulation, implementation, and execution process. Starting the time period of 1997-1998 and continued through the time period of 1999-2000, STPM is undertaking a multi-year development of a large-scale project management information system in order to automate and streamline the decision making and performance evaluation process of STPM. We were on a planning and consulting committee with the mission to propose a Web-based data warehouse information system for a contemporary STPM. The goal of this project management information system is to integrate the information resources in current STPM agency and to improve the information sharing among all STPM internal and external units nationwide.

STPM office is responsible for the formulation, implementation, and execution of the national science and technology projects (STP). Project formulation involves internal administrators and external expert representatives from the industry, government, and university. Project execution requires the involvement of a large number of outside contractors such as the universities, research institutes, Taiwan companies, approximately one thousand throughout the country. Documents such as proposals, minutes, contracts, reviews, and manuals are bulky and lengthy and represent the main data input of information resources to be processed and stored. Decision support, data analysis, and management reports represent the main data output to be produced and generated.

However, under the current legacy information system, the data service performance is far below the desired acceptance level of quality and functionality. Every time when the management needs a new report, staff has to ask each of the contractors and agencies to re-prepare the original data sources in order to generate the new report. Next time when the same request comes up with a slight change, staff has to go through the same process and re-request help from the contractors and agencies. Staff has to re-enter the same data and re-design the same report with the change. The waste of time and effort is unimaginable and unacceptable in today's information technology (IT) operation. Under the legacy information system, whatever the preparation for the prior report is lost or cannot be re-used because the system is disparate, heterogeneous, and non-database-centric. Business performance of STPM is hence far below the desired acceptance level.

The key issue faced by STPM is mainly data management and design. These data issues include (1) too many redundant data entry, (2) inconsistent data content, (3) too much time-consuming report preparation and generation, (4) inefficient textual document processing, (5) no timely data output, and (6) very little data analysis toward decision making. The Legislative Yuan (similar to the Congress in the US) has pressed MOE to improve the STPM performance once for all or they have to face the pressure of more stringent budget and audit hereafter. Starting 1997-1998, STPM has determined and begun to re-engineer their project management and control process.

We approached STPM problems with a strategy of an overall information infrastructure re-design and an information intent re-development to tackle the data management and design issues. Our plan was to first conduct a large-scale data requirement, data analysis, and data design. We geared to design an analytic and decision-making oriented data warehouse architecture. We used the Web clients as the standard workstations for both internal and external offices and agencies. Databases were to be distributed over the proprietary governmental Intranet and Internet. The information infrastructure is illustrated in Figure 1. Distributed data warehouses were the chosen and adopted information technology solution. Data sources and information analysis were designed toward the satisfaction of management decision, control, and evaluation.

*Figure 1: STPM Intranet and Internet Information Infrastructure*

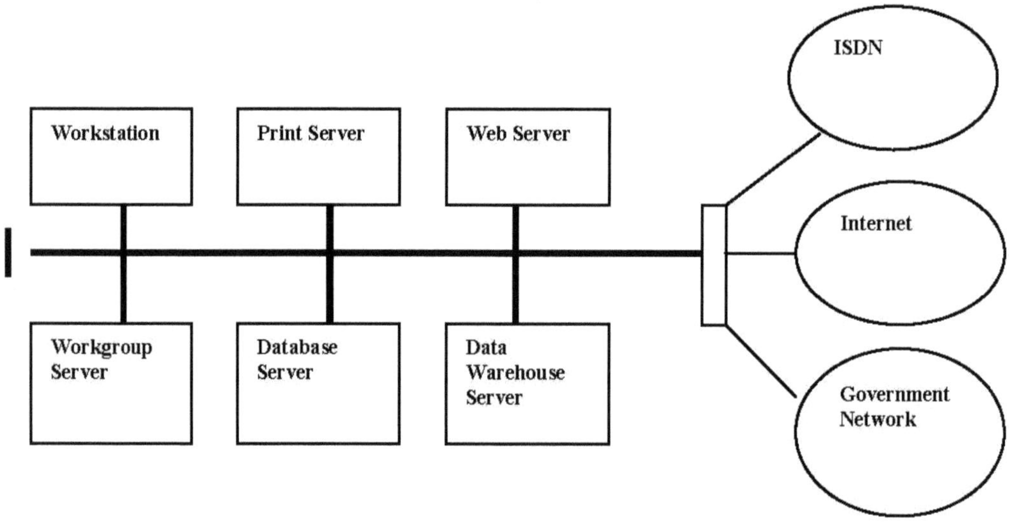


The chapter is organized as follows. Section one is an introduction to this field study. Section two is a review of the data warehouse technology. Section three describes the current STPM operation and management. Section four presents the design and development of the data warehouses of STPM project management information system. And, Section five gives a summary and future research direction to conclude the chapter.

## DATA WAREHOUSING

Data warehousing (DW) technology is defined as a collection of decision support tools to enable knowledge workers to make better and faster decisions. IDC predicts the market size of data warehousing including the hardware system, the software system, and the development tools is between $20 billion US dollars in 2000 to$200 billion U.S. dollars by 2005 (Celko & McDonald, 1997; Hurwitz, 1996; Kimball, 1996). The promises of the data warehouse technology include the following capability and capacity in order to provide and foster the above forecasted market size.

(1) Data warehouses can contain consolidated data from several operational databases over long periods of time and can be over hundred gigabytes to terabytes in size.
(2) Data warehouse workload can be query intensive with mostly ad hoc and complex query that can access millions of records and perform bulk of scans, joins, and aggregates.
(3) Data warehouse query throughput and response time is even more important than regular transaction throughput and response time.
(4) Data warehouse can be modeled as an information system that is hierarchical, dynamic, and multi-dimensional.

The above set of system characteristics is the main reason why the STPM office chose the Web-based data warehousing solution to meet the needs of their project management and control process.

### Data Warehousing Functions

Compared with a general-purpose information system, a data warehousing system is distinct and is characterized as follows (Chen, Han & Yu, 1996; Ferguson, 1997; Inmon, 1996). A data warehouse is:

(1) Subject-oriented – Query is the main subject from the user's viewpoint.
(2) Time-variant – Time is the most important dimension in data analysis.
(3) Summarized – Data is usually extracted and processed in statistic format.
(4) Projected – If and simulated query can be performed.
(5) Analytical – Both analytical and operational query can be conducted.
(6) Historical – Business transactions form the basis of data warehouse analysis.
(7) Partitioned – Any databases can be fragmented to give multi-dimensional database.
(8) De-normalized – Schema design differs from relational schema and does not require data normalization in order to reduce join operation.
(9) Multi-dimensional – Data result is extracted from multi-level and nested level of operational databases.

(10) Hierarchical – Databases are structured top down and level by level.

In addition, we describe and list a set of core features and functions provided by the data warehouse technology to further explain the promises it can bring to this project. Figure 2 shows the overall data warehouse architecture. A data warehouse system provides the following functionality.

(1) Data rollup - increases the level of aggregation.
(2) Data drill down - decreases the level of aggregation or increases details along one or more dimension hierarchies.
(3) Data slice and dice – partitions the columns of data warehouses according to the dimension of query results.
(4) Data pivot - points to and preset the dimensions of data warehouses.
(5) Data reorient - revolves the dimensions of data warehouses.
(6) Data selection - selects rows of one or multi-dimensional data.
(7) Data projection - projects columns of one or multi-dimensional data.

*Figure 2: Data Warehouse Architecture*

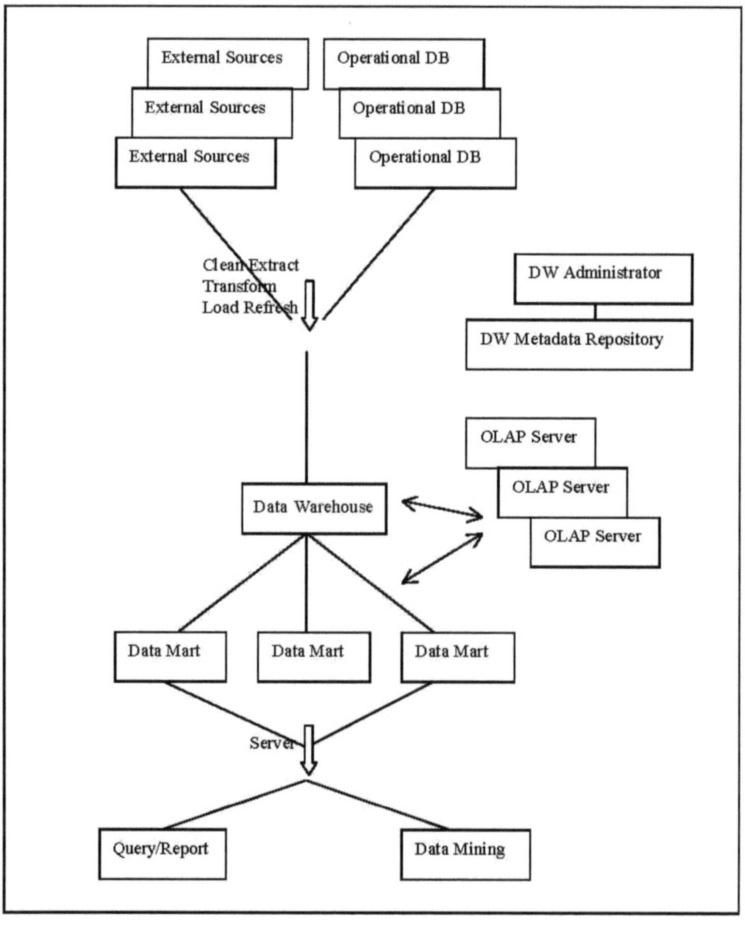

## Data Warehousing Components

In this section, we describe the main components of a data warehouse and explain how the technology can solve the fundamental data issues of STPM. A typical data warehousing system is composed of nine main components (Chen et al., 1996; Inmon, 1997). We draw a simplified system structure in Figure 3.

(1) Data cleaning is to detect any inconsistent field length, descriptions, value assignments, missing entries, and any violation of integrity constraints.
(2) Data migration is to allow the simple transformation rules to be specified.
(3) Data scrubbing is to use the domain-specific knowledge to scrub data.
(4) Data auditing is to discover the rules and relationships and signal any violation of the stated rules by scanning data.
(5) Data loading is to load rows in an incremental and piecemeal manner.
(6) Data pre-processing is to check integrity constraints, to sort, to summarize, to aggregate, to build indices, and to plan access paths.
(7) Data refreshing is to propagate updates on source data to correspondingly update the base data and the derived data stored in the warehouse.
(8) Data shipping is to take a snapshot and a mirror of a table that is updated in a remote database. After-row triggers are used to update the snapshot log table whenever the source table changes.
(9) Metadata of data warehouses includes all of the information necessary for setting up and using a warehouse such as the description of the source databases, the definition of warehouse schema, and the dimension and hierarchy of data warehouse.

*Figure 3: Data Warehousing Components*

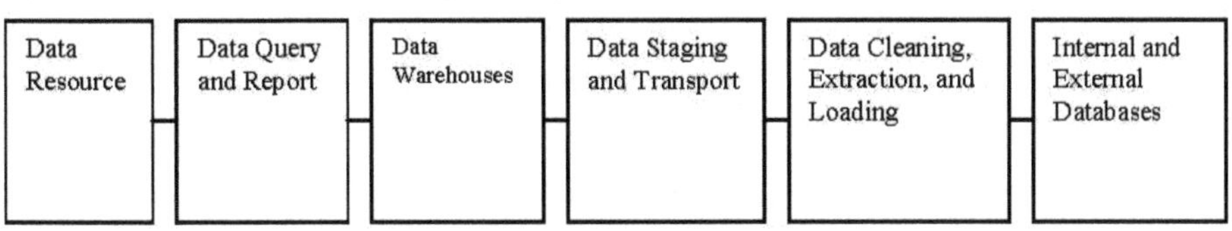


## Data Warehousing Integration

Data warehousing is an integral IT tool. In other word, integration is critical to the success of a data warehouse installation. Internal and external data sources have to be incorporated into the databases. Bulk and incremental data loading have to be supported. Heterogeneous and distributed data extraction and migration have to be performed. These integration services meet the required quality criteria of transparency, portability, scalability, and cost effectiveness. Data warehousing uses the following tools to achieve the integration goal. This set of twelve integration tools is classified into the class of data dictionary tools, the class of data design tools, and the class of data development tools. These tools are platform-independent.

(1) It provides a data repository of database schema.
(2) It provides a design facility.
(3) It provides connectivity through middle-ware such as native drivers, ODBC, and JDBC.
(4) It supports client/server architecture and Web browser/thin client, two-tier, three-tier with TP monitor, four-tier with WEB and TP monitor, and n-tier with distributed CORBA objects.
(5) It gives application development facility of its integration with object browsers, CASE tools, and GUI generators.
(6) It can be object orientation.
(7) It provides a programming language.
(8) It provides application deployment tools such as interpreters, compilers, debuggers, and interface builders.
(9) It supports performance studies on the aspects of interface processing, database access, application object processing, load balancing, and language processing.
(10) It offers component integration and Internet development capability with Active X, VBX, OLE container, OLE automation server, Java beans, Java applets, HTML/CGI generation, and distributed objects.
(11) It offers context search and GUI interface.
(12) It offers Web enablement including (a) directory services, (b) message handling services, (c) security services, (d) workflow services, and (e) application development services.

## STPM OPERATION

STPM office is responsible for the nationwide science and technology policy formulation, implementation, and execution through annual science and technology projects. Efficiency and effectiveness of the project execution has important effects on the national competitiveness. Entering the twenty-first century, Taiwan positions herself as the green silicon island servicing the global high-tech industry. She aims to focus on the high-tech research and development. Her core competency centers on the national innovation systems and high technology breakthrough. STPM plays an increasingly important role on the breakthrough. A fast growing number of private and public research and development units and agents are involved in executing these STP in order to strengthen and empower the national competitiveness. The scale and sophistication of STPM operation is therefore further complicated.

### Organization and Personnel

STPM organization is structured based on the national classification of six high-tech industry. There are five administration offices to manage the six classified industry fields in the organization. The first office is called the general service unit who is in charge of the accounting, asset, budget, legal service, personnel, and Legislative Yuan liaison service. The second office is named the planning unit who is responsible for the industry service, project planning, and international service. The third office is called the electric and electronic engineering industry unit who is in charge of the supervision of STP execution in the area of electric and electronic high tech industry on the technology transfer and diffusion. The fourth office is termed the mechanic engineering industry unit who is monitoring the STP execution in the area of mechanic engineering high tech industry on the technology transfer and diffusion. The fifth office is named the social welfare industry unit who is obliged to control the STP execution in the process of transferring and diffusing the social welfare knowledge and know-how. Each office unit is staffed with one executive, several area managers, and a small size of staff, with help of outside consultants. The director is the chief of the STPM office. Figure 4 shows the organization chart.

*Figure 4: STMP Organization Chart*

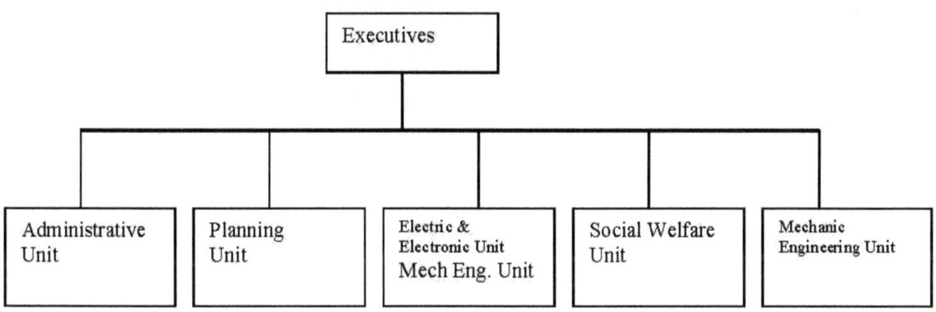


## Information Requirements

We adopted a number of data collection methods to acquire and analyze STPM system requirements. These methods included observation, meetings, interviews, and manuals. We collected these data requirements and compiled them into the following four categories.

(1) Operational Data Requirements – This category includes the non-digital, digital, formatted, non-formatted, compatible, and incompatible data sources from internal and external data sources such as agencies and companies.

(2) Environmental Data Requirements – This category includes the data sources of fax, copy, email, document, manuals, and minutes.

(3) Organizational Data Requirements – This category includes the schedule, schedule change, budget, budget change, review, and review change.

(4) Managerial Data Requirements – This category includes various formats of job descriptions, handbooks, rules, and reviews.

## STPM PROJECT MANAGEMENT SYSTEM PLANNING

In this section, we present three models we develop in our data warehouse plan. These models are named the system model, the data model, and the function model. The system model depicts the main systems of STPM project management. The data model describes the databases to be used in data warehousing. The function model details the modules to be designed.

### System Model

STPM information system is divided into three categories based on the lifecycle of the STP formulation, implementation, and execution that include the strategic, tactic, and operational stages. At each stage, STPM system is further divided into three operations such as the budgeting, contracting, and evaluating. The basic unit is an individual field project and proposal of science and technology. Each unit is built on top of the other to form a project information system hierarchy. From the planning viewpoint, the system model provides the following IT solutions. They include: (1) total information planning, (2) total information structure, (3) total information management, and (4) total information networking as shown in Figure 5.

The main functions of each system are mostly multiple on-line Web sessions. We group them into the following classes.

(1) Bulk loading of data from internal or external data sources.
(2) Incremental loading of data from operational systems.
(3) Aggregation of input data.
(4) Calculations of new data based on project management model.
(5) Time series analysis of project management performance.
(6) Queries with a high degree of complexity such as join, aggregation, and partition.
(7) Drill down through seven dimension hierarchies.
(8) Ad hoc simple and complex queries.

Below we describe and list the thirteen new STPM system modules running the above on-line sessions. We draw them in Figure 5.

(1) Long-term science and technology strategic field planning,
(2) Mid-range tactic proposal planning,
(3) Mid-range operational project planning,
(4) Mid-range budget proposal planning,
(5) Annual field budget planning,

(6) Updates to annual field budget planning,
(7) Individual operational project planning,
(8) Annual individual project budget planning,
(9) Individual project contract planning,
(10) Individual project execution,
(11) Field plan execution,
(12) Individual project execution,
(13) Field plan execution.

*Figure 5: Data Warehouse Planning Chart*

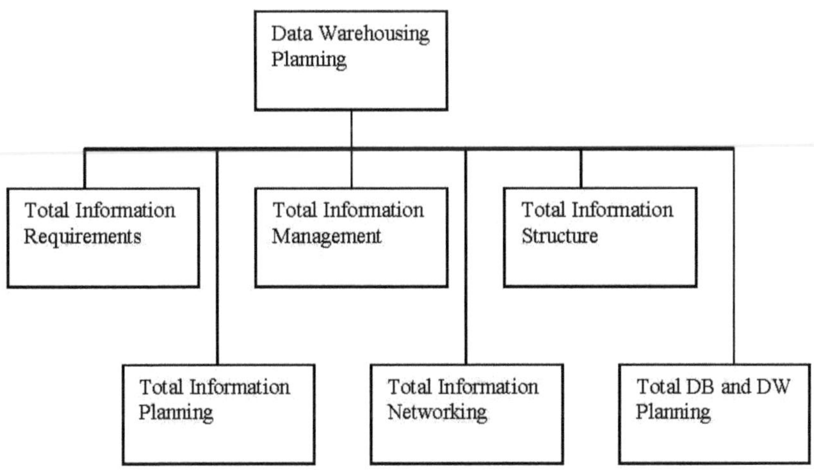


## Data Model

Data model describes both the input and output data requirements of STPM. Input data requirements come in from several data sources including the documents from each stage of the STP formulation, implementation, and execution lifecycle, such as the industry inquiry form, the science and technology field planning form, the field budget form, the field proposal form, the annual budget worksheet, the committee meeting minutes, and the individual project budget form. Output data requirements emerge from the management decision making needs and its data resources include a variety of formal and ad hoc reports including the contracts, evaluation sheet, progress report, project report, and audit report.

A multi-dimensional view of these data is to form the databases of dimension, dimension member, and hypercube hierarchy. The logical database structure of the data warehouses is made up of seven dimensions including field, division, project, agent, expert, asset, and time dimensions as shown in Figure 6. Data files of each dimension first are generated in order to initialize the data warehouse as bulk loading. Then the data files are generated in order to populate the data ware-

*Figure 6: Multiple Dimensions of Members and Hierarchies*

| | |
|---|---|
| Dimension 1: | Field |
| Hierarchy 1: | Top |
| | Subfield |
| | Class |
| | Base |
| Dimension 2: | Division |
| Hierarchy 2: | Top |
| | Subdivision |
| | Group |
| | Base |
| Dimension 3: | Project |
| Hierarchy 3: | Top |
| | Base |
| Dimension 4: | Agent |
| Hierarchy 4: | Top |
| | Base |
| Dimension 5: | Expert |
| Hierarchy 5: | Top |
| | Base |
| Dimension 6: | Asset |
| Hierarchy 6: | Top |
| | Agent |
| | Project |
| | Base |
| Dimension 7: | Time |
| Hierarchy 7: | Top |
| | Historical Periods |
| | Future Periods |
| | Base |

house as incremental loading with current updates, inserts, and deletes of project management.

Based on these data files, we developed a data warehouse architecture that is made up of a number of operational databases as shown in Figure 7. These databases are described and listed as follows.

(1) National science and technology plan database containing annual and nationwide science and technology research and development plans and proposals.

(2) Departmental science and technology plan database containing annual and major areas of science and technology research and development plans and proposals.

*Figure 7: Database Architecture*

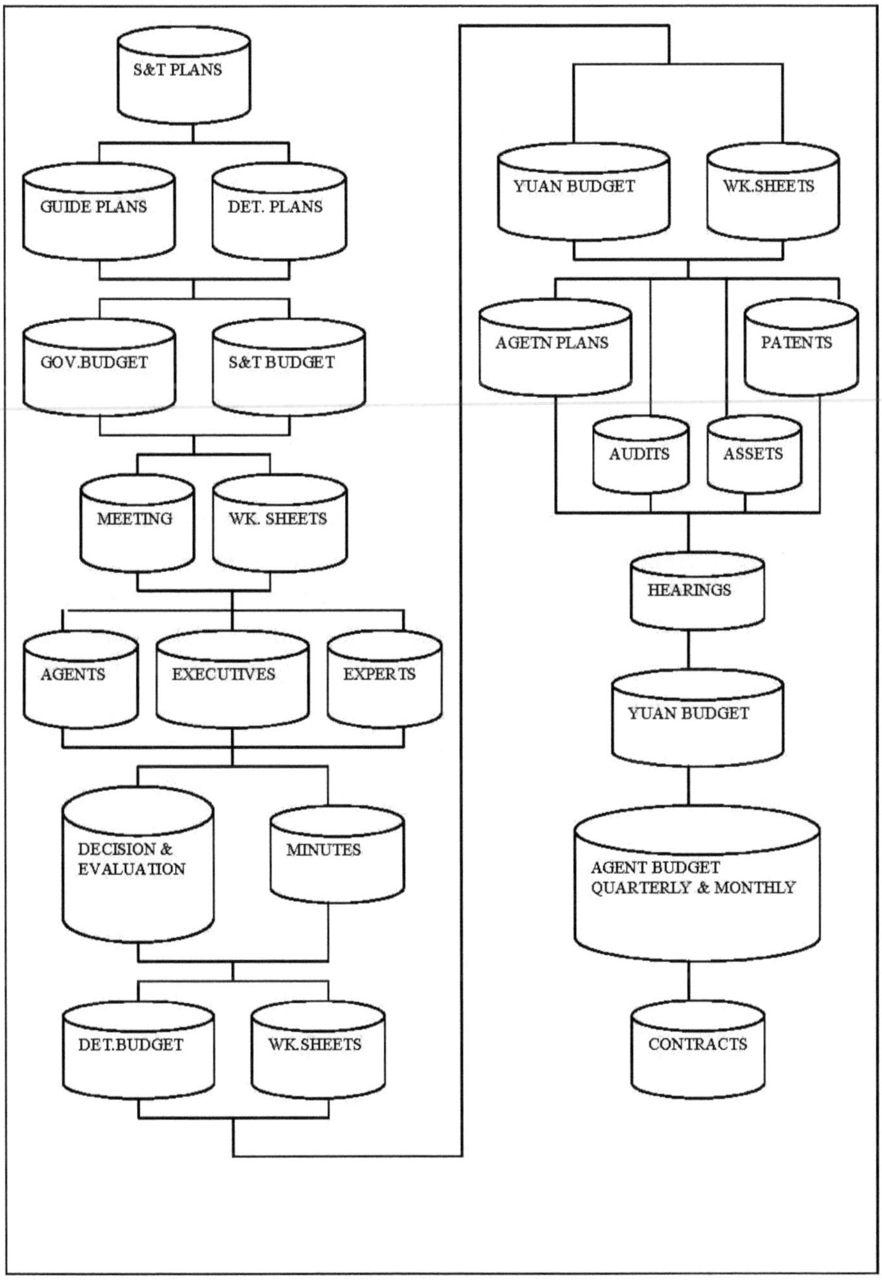

(3) Field science and technology plan database containing annual and individual field of science and technology research and development plans and proposals.
(4) National science and technology budget database containing annual and nationwide science and technology project budgets.
(5) Departmental science and technology budget database containing annual and major areas of science and technology project budgets.
(6) Field science and technology budget database containing annual and individual field of science and technology project budgets.
(7) Meeting database containing plan, budget, and audit meeting minutes.
(8) Work sheet database containing a variety of work sheets from each plan and budget meeting.
(9) Agent and contractor database containing information about the outside agencies and contractors.
(10) Expert and consultant database containing expert and consultant information.
(11) Contract database containing details of contract with each outsourcing agency.
(12) Project database containing details of individual science and technology project.
(13) Project report database containing details of mid-term and term project report from each contract.
(14) Project evaluation database containing performance evaluation and results including papers, patents, products, and science and technology transfer.
(15) Auditing report database containing outside independent auditors' reports.

## Function Model

In this section, we present the key modules of the new STPM system. Each module does the work of one system function defined in the above system model. The aforementioned data model gives the database definition of each module in order to perform the data warehousing operation. Specifically, three modules support the administrative unit to do the office automation. These include the administrative management, office management, and operational support. Two modules assist the management decision making. These include the executive decision support and budget management. Another three modules support the core STPM operation. These include the annual project management, annual project results management, and asset management. We describe and list them as follows.

(1) Administrative support module supports (1.1) information services, (1.2) personnel services and connection with external personnel services, (1.3) inquiries and reviews, and (1.4) audit and validation.
(2) Office management module supports (2.1) document management, (2.2) document development, (2.3) expert/consultant management, (2.4) news report, and (2.5) name card and contact management.

(3) Operational support module supports (3.1) legal service, (3.2) international service, and (3.3) industry service.
(4) Budget management supports (4.1) overall budget operation, (4.2) individual project budget operation, (4.3) overall budget execution, and (4.4) individual project budget execution.
(5) Executive decision support system supports (5.1) strategic planning, (5.2) policy planning, (5.3) tactic planning, (5.4) operational planning, (5.5) asset analysis, (5.6) results analysis, (5.7) performance analysis, (5.8) meeting and minutes management, (5.9) finance management, (5.10) accounting management, and (5.11) personal organizer
(6) Annual project management module supports (6.1) pre-planning, (6.2) proposal and project review, (6.3) budget review and contract assignment, (6.4) project control, (6.4.1) quarterly execution management, (6.4.2) mid-term validation, (6.4.3) project change management, (6.5) project evaluation, (6.5.1) term validation, (6.5.2) overall validation, (6.6) common project management, and (6.7) special project management.
(7) Annual project results management module supports (7.1) industry cooperation, (7.2) science and technology project results management, such as patent, journal, conference, technical report, technology diffusion, know-how import/export, industry service, extended investment, stockholders'wealth and company worth, and (7.3) science and technology project follow-up management, including industry investment, industry value, product cost reduction, man power transfer, man power import/export, human resources training, information service.
(8) Asset management module supports (8.1) asset depletion, (8.2) asset depreciation, and (8.3) asset disposition.

The actual queries are described and listed as follows and illustrated in Figure 8 as well.

(1) Field budget analysis is to analyze the budget amount and distribution varied by fields.
(2) Division budget analysis is to analyze the budget amount and distribution varied by divisions.
(3) Project budget analysis is to analyze the budget amount and distribution varied by projects.
(4) Agent budget analysis is to analyze the budget amount and distribution varied by agents.
(5) Field actual versus budget performance analysis is to analyze the difference and distribution between the actuals and budgets varied by fields.
(6) Division actual versus budget performance analysis is to analyze the difference and distribution between the actuals and budgets varied by divisions.

*Figure 8: Multiple Dimensions of Analysis*

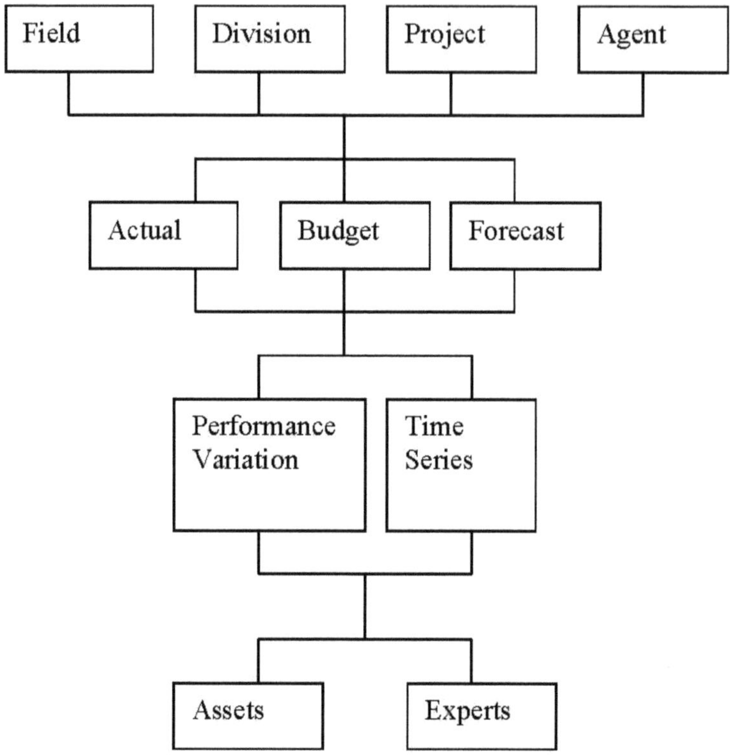


(7) Project actual versus budget performance analysis is to analyze the difference and distribution between the actuals and budgets varied by projects.

(8) Agent actual versus budget performance analysis is to analyze the difference and distribution between the actuals and budgets varied by agents.

(9) Actual time series analysis is to analyze the actual amount along the time hierarchy varied by field, division, project, and agent.

(10) Budget time series analysis is to analyze the budget amount along the time hierarchy varied by field, division, project, and agent.

(11) Forecast time series analysis is to analyze the forecast amount along the time hierarchy varied by field, division, project, and agent.

(12) Expert analysis varying by fields and projects is to analyze experts on the advisory and review committees based on expertise, fields, and projects.

(13) Asset analysis varying by fields and projects is to analyze assets acquired in each project based on accounts, fields, and projects.

(14) Agent varying by fields and projects is to analyze agents and contractors based on their characteristics, fields, and projects.

(15) Ad hoc analysis.

## SUMMARY

In this chapter, we have presented a large-scale information system planning of data warehouses for a Taiwan governmental organization's project management and control system. The presentation includes a review of the data warehousing technology, the STPM operation, the design of data warehouses, and the development of data services. We have achieved the data warehouse planning through the use of three models of information requirements. These models consist of the system model, the data model, and the function model. The system model depicts the main science and technology planning, budgeting, and auditing functions required in STPM. The data model describes the multi-database schema definition. The function model details the main modules in the support of each STPM new system function. The planning and advisory committee was working with the STPM staff to guide and monitor the insourcing and outsourcing process. As we learned that the system requirements and specifications have been given to the outsourcers. The information system is expected to implement through the year of 2000.

## REFERENCES

Anahiry, S. (1997). *Data Warehousing in the Real World: A Practical Guide for Building Decision Support Systems,* Addison Wesley.

Appleton, E. L. (1996). Use Your Data Warehouse to Compete, *Datamation,* 42(10), May 15, pp. 34-5.

Atre, S., and Storer, P. (1997). Data Distribution and Warehousing, *DBMS,* 8(10), pp.54-62.

Barquin, R., and Edelstein, H. (1996). *Planning and Designing the Data Warehouse,* Prentice Hall Co.

Berry, L. (2000). *Mastering Data Mining,* John Wiley Co.

Brackett, M. H. (1996). *The Data Warehouse Challenge: Taming Data Chaos,* John Wiley and Sons.

Celko, J., and McDonald, J. (1997). Don't Warehouse Dirty data, *Datamation,* 41(20), pp.42-44 and pp.51-52.

Chaudhuri, S. and Dayal, U. (1997). An Overview of Data Warehousing and OLAP Technology, *SIGMOD Record,* (26):1, March, pp.65-74.

Chen, M., Han, J., and Yu, P. (1996). Data Mining- An Overview form a Database Perspective, *IEE TOKDE,* (8):6, December, pp. 866-883.

Colliat, G. (1996). OLAP, Relational and Multidimensional Database Systems, *ACM SIGMOD Record,* (25):3, September, pp. 764-69.

Corey, M. (1997). *Oracle Data Warehousing,* McGraw-Hill.

Darling, C.B. (1996). Think Outside the OLAP Box, *Datamation*, 42(8), April 15, pp.88-92.

Darling, C. B. (1996). How to Integrate Your Data Warehouse, *Datamation*, 42(8), April 15, pp.94-95.

Ferguson, M. (1997). Parallel Computing and the Data Warehouse: Status Report, *InfoDB*, 9(5), pp.11-16.

Fernandez, P., and Schneider, D. (1996). The Ins and Outs of Data Warehousing, *ACM SIGMOD Record*, (25):2, June.

Groth, R. (1998). *Data Mining: A Hands-on Approach for Business Professionals*, Prentice Hall.

Hackathorn, R. (1995). Data Warehousing Energizes Your Enterprise, *Datamation*, 41(2), February 1, pp. 28-45.

Haisten, M. (1997). Information Discovery in the Data Warehouse, *InfoDB*, 9(6), pp.14-25.

Hamm*ergren, T. (1996).* Data Warehousing: Building the Corporate Knowledge Base, International Thompson Computer Press.

http://www.arborsoft.com/OLAP.html.

http://members.aol.com/fmcguff/dwmodel/dwmodel.htm.

http://www.olapcouncil.org.

http://www.tpc.org.

http://pwp.starnetinc.com/larr/articles.html.

Hurwitz, J. (1996). Preparing for the Warehouse, When You are Beginning a Data Warehouse Project, Start with the Metadata, *DBMS*, April.

Inmon, W. (1996). *Building Data Warehouse*, John Wiley Co.

Inmon, W. (1996). The Data Warehouse and Data Mining, *CACM*, (39):11, November, pp.49-50.

Inmon, W. (1997). *Corporate Information Factory*, John Wiley Co.

Inmon, W. (1999). *Building Operational Data Stores*, John Wiley Co.

Kelly, S. (1997). *Data Warehouse in Action*, John Wiley Co.

Kimball, R. (1996). Automation Data Extraction, *DBMS*, July, pp.15.

Kimball, R. (1997). Drill Down, Up, and Across, *DBMS*, 10(3), pp.14-28.

Kimball, R. (1997). Letting the Users Sleep, Part I, *DBMS*, 9(12).

Kimball, R. (1997). Letting the Users Sleep, Part II, *DBMS*, 10(1).

Kimball, R. (1999). *Data Warehousing*, John Wiley Co.

Kimball, R. (1999). *Data Marts*, John Wiley Co.

Kimball, R. (2000). *Data Warehousing Toolkit*, John Wiley Co.

Kimball, R. and Inmon, W. (1996). *Data Warehouses Toolkit*, Wiley.

Kimball, R. and Strehol, K. (1994). Why Decision Support Fails and How to Fix it, *Datamation*, June 1.

Little, R., and Gibson, M. (1999). Identification of Factors Affecting the Implementation of Data Warehousing, HICSS-32.

Mattison, R. (1996). *Data Warehousing: Strategies, Technologies, and Techniques*, McGraw-Hill.

Reeves, L. (1996). *Building A Data Warehouse for Decision Support*, Prentice Hall Co.

Simpson, D. (1997). Building Your Warehouse on MPP, *Datamation*, 42(1).

Tanler, R. (1996). Putting the Data Warehouse on the Intranet, *Internet System*, May, pp. 34-37

Van Den Hoven, J. (1998). Data Warehousing: Bringing it All Together, *Information Systems Management*, 15(2), Spring, pp.92-95.

## Chapter 15

# Behind the Celtic Tiger: Key Issues in the Management of Information Technology in Ireland in the Late 1990s

Joseph McDonagh
Trinity College, Ireland

Andrew Harbison
Ernst & Young, Ireland

## INTRODUCTION

Like most of its counterparts in the developed world, in the last ten years the Irish economy has become highly dependent on information technology (IT), both as a source of export earning and as a means for businesses enhance their competitive performance. Indeed, given its unusually high dependence on IT-based industries, Ireland is particularly exposed to the need to maximise the benefits accrued from information technology. The country is now, in absolute terms, the world's second largest supplier of software, with over 40% of all packaged software sold in Europe originating in Ireland (IDA Ireland, 1999). Nearly one third of all personal computers sold in Europe are manufactured in Ireland, with many major manufacturers having their European bases in Ireland. Ireland has also developed a thriving indigenous IT industry. Iona, Kindle and CBT are but the best known of the 600 small, medium and large Irish IT companies operating at the current time. The IDA presently estimate that, on average, Irish IT company start-ups occur at the rate of three per week (Brown, 1999).

Given Ireland's remarkable success in attracting, developing and retaining IT-specialised companies, it is surprising that no detailed analysis has ever been carried out of the IT management issues within the country. Such an analysis might be of immense use in identifying areas of concern; points where pressure may be growing in the rapidly expanding Irish IT sector. Identification of such factors may allow managers to make allowances for them, thereby lessening their effects.

In July 1998, the School of Business Studies, Trinity College, Dublin began a study of key issues in information technology (IT) management in Ireland. The work was to focus on all areas of practice in IT management in the Irish economy, and to attempt to identify trends in the field. Our study was the first large-scale examination of IT management issues in the Irish context, a smaller case based study having been carried out by Moynihan (1990). A number of studies have also been carried out in recent years in a number of different countries (Watson et al, 1997). Considerable work has been done in the UK (e.g., Price Waterhouse, 1998 (a) & (b); Galliers et al., 1994), the US (Brancheau, Janz & Wetherbe, 1996) and in Asia (e.g., Yang, 1996;), among others.

Watson et al (1997) in their comparison of a broad range of key issues surveys world - wide identified a number of issues that tended to continually be identified as key issues in IT management. In their chapter, they rank the top ten issues as described in Table 1.

*Table 1: Key Issues in IS Management: International Rankings.*

| Overall Ranking | International Issue |
|---|---|
| 1 | Strategic Planning |
| 2 | IS Organisational Alignment |
| 3 | Information Architecture |
| 4 | IS for Competitive Advantage |
| 5 | Data as a Resource |
| 6 | Human Resources |
| 7 | Security and Control |
| 8 | Integrating Technology |
| 9 | Software Development |
| 10 | The Role and Contribution of IS |

An examination of any recent work on IS management issues form any journal in recent years will reveal a similar set of results. Typically, broad managerial issues such as IS strategy development, the alignment of information systems with the organisation, and leveraging information systems for competitive advantage are high in the rankings of the most important issues. Further down the rankings will be operational and technical issues such as information systems security and maintenance, ensuring IS reliability and managing software development.

## INSTRUMENT & METHODOLOGY

The instrument chosen for the survey was a questionnaire, originally developed at Warwick Business School, which has been applied to large samples of companies, in the UK and elsewhere, a number of times already. (Parker & Idundun, 1988; Galliers, et al. 1994) The targets of the questionnaire were senior managers with responsibility for IT in larger organisations across a broad cross section of the Irish public and private sectors.

The instrument was made up of two parts. Part One of the questionnaire consisted of 21 multi-part, multiple-choice questions. Looking at personal details of the respondent, organisation-specific information, industry and sector data, and specifics each organisation's information systems / technology operations.

Part Two, the principal focus of this paper, consisted of 40 questions, covering a broad range of current issues, from Information Systems Strategy and using IS for competitive advantage to the impact on the organisation of technologies such as e-mail and intranets. Participants ware asked what the considered to be the most important and the most problematic information systems management issues facing their organisations over the previous five years, presently, and over the next five years. They were asked to indicate their views by rating each issue six times on a scale of 1 to 10, (3 times for importance, past, present and future, and 3 times for associated problems, past, present and future). High numbers represented more important or problematic issues, low numbers indicated less important / problematic issues and zeros represented "don't knows."

The questionnaire is specifically designed for companies with a significant IT infrastructure. Typically such firms tend to be quite large in terms of both numbers of employees and overall turnover.

The companies in the sample were drawn from the Irish *Business and Finance* databases. Subjects were selected subject to criteria including turnover, number of employees and sophistication of IT infrastructure. Upon completion of the selection process, questionnaires were posted to 902 public & private companies and organisations. Given the small size of the Irish economy, the companies selected made up the vast majority, if not the entire population of non-SME firms operating in Ireland. Such a sample group allowed us to minimise the possibility of sampling errors inherent in the postal survey technique.

A total of 144 usable responses were received, a 16.0% response rate. This is comparable to rates of 8 to 10% in equivalent UK studies. An identifiable bias was apparent in the responses, with larger organisations responding at a much higher rate than smaller ones. Among the largest 300 trading companies sampled, the response rate was 19.6%, among the smallest 200, less than 5%.

*Table 2: The top 10 information technology management issues identified as being important 3 to 5 years ago.*

| Issue | Important | | Problematic | |
|---|---|---|---|---|
| | Rating | Rank | Rating | Rank |
| Reliability of IS | 6.25 | 1 | 5.39 | 11 |
| Maintenance of Information Systems | 5.72 | 2 | 5.46 | 10 |
| Management of Software Development | 5.47 | 3 | 5.22 | 15 |
| Security of IS | 5.40 | 4 | 4.83 | 23 |
| Information Systems Strategic Planning | 5.39 | 5 | 4.85 | 22 |
| IT Staff Awareness of Business Imperatives | 5.29 | 6 | 5.04 | 19 |
| Limited Perception of the Role and Contribution of the IS Function | 5.28 | 7 | 5.72 | 3 |
| Communication Gap between IS and non-IS staff | 5.20 | 8 | 5.67 | 5 |
| Impact of IT on Jobs and Roles | 5.08 | 9 | 5.23 | 12 |
| Managing Legacy Systems | 4.96 | 10 | 4.97 | 21 |

*Table 3: The top 10 information technology management issues identified as being problematic 3 to 5 years ago*

| Issue | Important | | Problematic | |
|---|---|---|---|---|
| | Rating | Rank | Rating | Rank |
| IT for Process Integration | 5.88 | 1 | 4.71 | 16 |
| Management of Distributed Systems | 5.77 | 2 | 4.64 | 19 |
| Limited Perception of the Role and Contribution of the IS Function | 5.72 | 3 | 5.28 | 7 |
| Managing Data as a Corporate Resource | 5.71 | 4 | 4.80 | 14 |
| Communication Gap between IS and non-IS staff | 5.67 | 5 | 5.20 | 8 |
| Electronic Communications Infrastructure | 5.67 | 6 | 4.53 | 22 |
| Managing Access and Use of External Data | 5.64 | 7 | 4.18 | 27 |
| IT-enabled Business Process Redesign | 5.47 | 8 | 4.15 | 28 |
| Information Handling Capabilities | 5.47 | 9 | 4.31 | 25 |
| Maintenance of Information Systems | 5.46 | 10 | 5.72 | 2 |

## RESULTS AND DISCUSSION

### Past Issues.

The top 10 most important and most problematic issues over the past 5 years are summarised in Table 2 (below) and Table 3 (overleaf).

Immediately we note a difference between the Irish responses and those collected internationally (See Table 1 above). While some broader strategic questions, such as IS strategy formation, are mentioned, the issues of most interest to Irish managers, both in terms of perceived importance and difficulty, are operational in nature.

The principal issue of importance for Irish IT managers over the last five years has been the reliability of the systems they manage. This issue encompasses both the evaluation of risks associated with systems failure and the provision of suitable contingency arrangements in the event of failure, as well as other matters. The top ranking of this issue is, no doubt, to some extent due to concerns surrounding the Year 2000 date-field problem and related issues. Yet as we shall see below, the same managers predict information systems reliability will retain much of its importance beyond 2000. In terms of the difficulty of dealing with this issue, overall respondents ranked it 11th of the 40 issues studied. This suggests that while IS reliability causes some problems, it is by no means the most intractable problem facing IT managers. Closely related to the reliability question are the second ranked issue in terms of importance; IS/IT Maintenance, the third; Software Development and the fourth, Information Systems Security. Fortunately, of these only IS/IT maintenance was viewed as having been especially problematic.

All four top ranked issues are closely related to the day-to-day running of an Information System. It is clear that in the past, at least, Irish IT managers have seen their duties as more to do with the day to day functioning of the information system, including its upgrading from time to time. The only broader "strategic" issue is ranked only fifth (IS Strategy), while no other related issue appears in the top ten. The remaining issues are either technical (managing Legacy Systems), or are concerned with improving the relationship between the IT function and the organisation as a whole.

A serious dislocation between organisations' IT functions and the rest of the employee body appears to underlie issues ranked 6th, 7th, 8th and, to some extent, 9th. Over the last five years, IT managers have placed a great deal of importance on lessening the "culture" gap between these two groups. Notably, IT managers appear to believe that it was a great deal easier to induce IT staff to understand the problems of their non-IT colleagues, than vice versa. While creating IT staff awareness of business imperatives ranked only 19th in terms of difficulty, and managing the impact of IT on employees' jobs and roles 12th, lessening non-IT staff's misunderstanding of the role of the IT function was perceived as being a lot more difficult.

## Current Issues

From the perspective of those issues that are most problematic (Table 3), the issues that IT managers rate as having been the most difficult over the last five years (1st, IT for process integration and 2nd, management of distributed systems) are not seen as having been particularly important overall.

The 3rd and 5th ranked issues, we have discussed above. Altering the limited perception of the role and contribution of the IS function (#3) and closing the communication gap between IS and non-IS staff (#5) are seen as having been both important and problematic IS management issues over the last five years.

Tables 4 & 5 show the current most important and problematic issues in information management, as IS managers perceive them. It also shows the change in both average score and ranking for each issue. Notably, average ratings for both importance and associated problems increased by approximately 1.5 in comparison with the earlier period. It is observed, however, that this increase appears to be constant over all questions. In practice, it may simply reflect that the problems managers have solved already may simply seem less intractable than those they have to cope with in the present.

We see first that IS reliability has retained its status as the most important issue for IS managers, the principal change over the last five years being that it is seen to be a good deal more difficult to achieve. One might have assumed given the greater usability of modern systems, and the move over the last few years to more open systems would have made it easier to ensure the reliability of systems, yet it appears that the opposite is in fact the case. It is observable too that the maintenance of IT systems has become relatively more problematic. It would appear that the complexity of modern systems has been increasing faster that IT managers ability to cope.

Security, already important in the past, has become an even more crucial issue. More worryingly, managers now perceive that the security problem has become far more difficult to solve. As more and more vital data is stored on organisations' information systems the problem of providing adequate back up and disaster recovery has become far more complex. The advent and proliferation of e-mail and internet applications has also greatly increased the number and scale of threats facing the typical information system.

"Business-level" issues have begun to appear at the top of the importance ranking. Broad issues such as Information Systems Strategic Planning (#2), Managing Data as a Corporate Resource (#6), IT for Process Integration (#7) and IT and IS for Competitive Advantage (#10) are ranked much higher in terms of

*Table 4: The top 10 information technology management issues identified as being important at the present time.*

| Issue | Important | | Change - 5 Years | | Problematic | |
|---|---|---|---|---|---|---|
| | Rating | Rank | Rating | Rank | Rating | Rank |
| Reliability of IS | 7.99 | 1 | 1.74 | 0 | 6.11 | 5 |
| Information Systems Strategic Planning | 7.87 | 2 | 2.48 | 3 | 5.74 | 16 |
| Electronic Communications Infrastructure | 7.74 | 3 | 3.21 | 19 | 5.57 | 21 |
| Security of IS | 7.54 | 4 | 2.14 | 0 | 6.47 | 2 |
| Recruitment and Currency of in-house IS Skills | 7.35 | 5 | 2.53 | 8 | 7.08 | 1 |
| Managing Data as a Corporate Resource | 7.20 | 6 | 2.40 | 8 | 6.05 | 7 |
| IT for Process Integration | 7.16 | 7 | 2.45 | 9 | 6.04 | 8 |
| Staff Education about IT Capabilities | 7.11 | 8 | 2.32 | 7 | 6.30 | 3 |
| Exploitation of E-mail | 7.07 | 9 | 3.53 | 25 | 5.16 | 32 |
| IT and IS for Competitive Advantage | 7.06 | 10 | 2.37 | 7 | 5.69 | 19 |

*Table 5: The top 10 information technology management issues identified as being problematic at the present time.*

| Issue | Problematic | | Change - 5 Years | | Important | |
|---|---|---|---|---|---|---|
| | Rating | Rank | Rating | Rank | Rating | Rank |
| Recruitment and Currency of in-house IS Skills | 7.08 | 1 | 2.62 | 29 | 7.35 | 5 |
| Security of IS | 6.47 | 2 | 1.64 | 21 | 7.54 | 4 |
| Staff Education about IT Capabilities | 6.30 | 3 | 1.15 | 14 | 7.11 | 8 |
| Maintenance of Information Systems | 6.24 | 4 | 0.78 | 6 | 7.05 | 11 |
| Reliability of IS | 6.11 | 5 | 0.72 | 6 | 7.99 | 1 |
| Change Management Skills | 6.08 | 6 | 0.90 | 10 | 6.77 | 16 |
| Managing Data as a Corporate Resource | 6.05 | 7 | 0.34 | -3 | 7.20 | 6 |
| IT for Process Integration | 6.04 | 8 | 0.16 | -7 | 7.16 | 7 |
| Information Handling Capabilities | 6.02 | 9 | 0.56 | 0 | 6.50 | 19 |
| Management of Distributed Systems | 5.90 | 10 | 0.13 | -8 | 6.93 | 13 |

importance in comparison with five years ago. This may reflect a growing sophistication among IT managers, and a more global view of IT in the context of the entire business. Also notable is the fact that respondents believe that, relatively speaking, the management of some of these issues is still relatively straightforward. IS strategy development and information handling are perceived as being relatively more problematic than they were, but are still by no means the most difficult problems facing managers.

Managing data as a corporate resource, considered difficult in the past, has not become any easier. It has, however, become a great deal more important. It would appear from this fact that information management techniques still have not been able to catch up with the continuing development of information gathering resources. More positively, it is notable that the management of distributed systems is perceived to have become considerably easier over the last few years. Improved connectivity products and the move to "open systems" in recent years have no doubt contributed to this improvement.

Even before this study was carried out it was clear that Irish organisations were have difficulty recruiting and retaining trained IT staff. In our study, respondents ranked this issue fifth in terms of importance, but first in terms of difficulty, and by a considerable margin. The last two years has seen a considerable increase in staff recruitment from abroad, both of expatriate Irish and non-Irish IT specialists. Over 60% of graduates entering Irish third level institutions this year will be commencing courses in either business or some form of IT and the emphasis placed by respondents on staff IT education is evidence of companies turning to internal training to provide themselves with extra IT-literate personnel. (McGarvey, 1999)

### Future Issues

Tables 6 & 7 show those issues in information management predicted by Irish IS managers to be the most important and problematic in the next five years.

With a single exception, Irish IT managers believe that the issues that will be important in five years time, are the same that are important today. The most important issue is foreseen to be managing organisations' electronic communications infrastructure. This issue is closely related to the only new item in the ranking; understanding and exploiting internet opportunities. It is also reflected in the importance assigned to the management of e-mail, the relative importance of which is predicted to increase.

The survey respondents clearly see the boundaries of their organisations, and in particular their organisations' information systems, begin to blur. The responsibilities of information technology managers have shifted considerably. Over the ten years covered by the study, we see a shift of emphasis from and operations-

*Table 6: The top 10 information technology management issues identified as being important in 3 to 5 years time.*

| Issue | Important | | Change/5 Years | | Chng./10 Years | | Problematic | |
|---|---|---|---|---|---|---|---|---|
| | Rating | Rank | Rating | Rank | Rating | Rank | Rating | Rank |
| Electronic Communications Infrastructure | 8.78 | 1 | 1.04 | 2 | 4.25 | 21 | 4.62 | 37 |
| Reliability of IS | 8.61 | 2 | 0.62 | -1 | 2.36 | -1 | 6.25 | 4 |
| Security of IS | 8.54 | 3 | 1.00 | 1 | 3.14 | 1 | 6.85 | 2 |
| Information Systems Strategic Planning | 8.35 | 4 | 0.48 | -2 | 2.96 | 1 | 5.56 | 13 |
| Managing Data as a Corporate Resource | 8.31 | 5 | 1.11 | 1 | 3.51 | 9 | 5.65 | 11 |
| Understanding and Exploiting Internet Opportunities | 8.25 | 6 | 2.19 | 24 | 6.08 | 33 | 5.05 | 27 |
| IT for Process Integration | 8.20 | 7 | 1.04 | 0 | 3.49 | 9 | 5.25 | 22 |
| Exploitation of E-mail | 8.12 | 8 | 1.05 | 1 | 4.58 | 26 | 4.87 | 34 |
| Staff Education about IT Capabilities | 8.06 | 9 | 0.95 | -1 | 3.27 | 6 | 6.20 | 5 |
| Management of Distributed Systems | 7.95 | 10 | 1.01 | 3 | 3.31 | 9 | 5.09 | 26 |

*Table 7: The top 10 information technology management issues identified as being problematic in 3 to 5 years time.*

| Issue | Problematic | | Change/5 Years | | Chng./10 Years | | Important | |
|---|---|---|---|---|---|---|---|---|
| | Rating | Rank | Rating | Rank | Rating | Rank | Rating | Rank |
| Recruitment and Currency of in-house IS Skills | 6.95 | 1 | -0.13 | 0 | 2.49 | 29 | 7.71 | 16 |
| Security of IS | 6.85 | 2 | 0.39 | 0 | 2.03 | 21 | 8.54 | 3 |
| Maintenance of Information Systems | 6.26 | 3 | 0.02 | 1 | 0.80 | 7 | 7.49 | 21 |
| Reliability of IS | 6.25 | 4 | 0.14 | 1 | 0.86 | 7 | 8.61 | 2 |
| Staff Education about IT Capabilities | 6.20 | 5 | -0.10 | -2 | 1.05 | 12 | 8.06 | 9 |
| Change Management Skills | 6.15 | 6 | 0.07 | 0 | 0.97 | 10 | 7.65 | 18 |
| Impact of IT on Jobs and Roles | 5.97 | 7 | 0.25 | 11 | 0.73 | 5 | 7.15 | 25 |
| IT and IS for Competitive Advantage | 5.85 | 8 | 0.16 | 11 | 1.22 | 20 | 7.92 | 11 |
| Flexibility of IS | 5.79 | 9 | -0.06 | 4 | 0.74 | 9 | 7.68 | 17 |
| Measuring the effectiveness and productivity of the IS Function | 5.72 | 10 | -0.14 | 2 | 0.68 | 10 | 6.86 | 30 |

based transaction processing focus, towards a much greater focus on information provision and management.

The shift from the situation where IT simply supported the business process, to one where it is an inherent part of product delivery is reflected in a number of the predicted rankings. Managing IT to provide competitive advantage to the organisation, and delivering information systems that will be flexible enough for the new operating environment are predicted to become relatively problematic over the next few years.

Not all has changed however. Even after the Year 2000 problem has passed, IS reliability and security are predicted to remain important issues in IS management. Nor is the difficulty of dealing with such matters predicted to ease to any great extent.

Likewise, IS strategy development, and managing IS for process integration are predicted to remain important issues. Managers predict that strategy development will marginally increase in difficulty over the next five years. Conversely, process integration through information technology is predicted to become a great deal easier, as new products and techniques become available. Similarly the management of distributed systems is expected to lose nothing in importance over the next five years, but is predicted to become a great deal easier from the perspective of management.

Surprisingly, the recruitment of IS staff is predicted to drop from the list of most important issues facing IS managers (to $16^{th}$ in the ranking), although the problems associated with recruiting new IS staff are not expected to ease at all. Meanwhile, the training of IS staff internally retains much of Its importance as an issue, while retaining many of its associated problems. These results are difficult to interpret. They may suggest that organisations intend to place much more emphasis on internal training, making external recruiting less of an issue.

## SUMMARY AND CONCLUSIONS

The issues identified in the study can be classified into four main areas of concern:

- *Technical and Operational Issues.* These are concerned with the effective running of information systems, and would be closest to the general perception of the "traditional" role of the IT manager. Typical among this group would be questions of information system security, reliability and maintenance. In the past managing legacy systems was a problem in this area. In the present and future, questions of distributed system management, electronic communications infrastructure and internet use are considered important.

IS reliability is seen as a critical issue which is difficult to manage, despite all the efforts of system developers over the last decade. Nor is it seen as an issue that is likely to go away. Information Systems security is another key issue, which is likely to cause serious problems for managers into the distant future

- *Strategic and Planning Issues.* These are by no means as important to Irish managers as they are to their counterparts in other countries. IS strategy formation is considered a key issue, and there is a belief that it is likely to become more difficult in the future, but for the moment it is not seen as especially problematical. Irish managers have learned to see data as important corporate resources, and have begun to appreciate the difficulties in managing that data.

There is evidence that Irish information technology managers are increasingly inclined to see the softer non-technical side of their discipline, but there appears to still be a strong orientation to the more mundane technical aspects of the job.

- *Relationship with the Organisation.* There is a perception that, in the past, a communications gap existed between organisations' IS functions and the organisations themselves. The results suggest that most managers believe that this problem has been, to a significant extent, overcome.
- *Recruitment of IS Staff.* This issue has become, by a considerable margin, the most difficult for Irish organisations. At present it is seen as key issue in information technology management, but is predicted by many managers to decrease in significance as other sources of IS personnel are identified.

## REFERENCES

Brancheau, J.C., Janz, B.D. & Wetherbe, J.C. (1996). Key Issues in Information Systems Management: 1994-95. SIM Delphi Results, *MIS Quarterly* 20 (2) June.

Brown, E. (1999). Irish Green, *Forbes*, April 19.

Galliers, R.D., Merali, Y. & Spearing, L. (1994). Coping with information technology? How British executives perceive the key information systems management issues in the mid-1990s, *Journal of Information Technology* 9, pp.223-238.

IDA Ireland web site at *www.idaireland.com/is.html.*

Moynihan, T. (1990). What Chief Executives & Senior Managers Want from their IT Departments, *MIS Quarterly* 14 (1).

Parker, T. & Idundun, M. (1988). Managing Information Systems in 1987: The Top Issues for Managers in the UK. *Journal of Information Technology*. 3, pp.34-42.

Price Waterhouse (1998) (a) UK Financial Services IT Survey 1998 and (b) UK Government IT Survey 1998 Price Waterhouse Ltd., London.

Watson, R.T., Kelly, G.G., Galliers, R.D., & Brancheau, J.C. (1997). Key Issues in Information Systems Management: An International Perspective, *Journal of Management Information Systems* 13 (4).

Yang, H.L. (1996). Key information management issues in Japan and the US, *Information & Management* 30 (5).

## Chapter 16

# Turning E-Commerce Theory into Action in Ireland: Taming the Celtic Tiger

Ira Yermish and Dale A. Bondanza
St. Joseph's University, USA

## INTRODUCTION

E-Europe is not a place or a group of countries but the philosophy of an emerging region within the scope of a technological revolution. E-Europe can be viewed from many perspectives including Euro currency makeup, European Union membership, or technological contributions to the region. Ireland, the "Celtic Tiger," presents an interesting opportunity of entrepreneurial activity. As Evans and Wurster (2000) have pointed out, the traditional value chains may be redefined. The availability of the Internet as an enabling communications technology has markedly changed the possibilities for countries far from the centers of economic power. On the other hand, the realities of the Internet have been obvious. Porter (2001) and Kanter (2001) have identified many of the realistic strategies for traditional and new businesses. Yet there will always be room for new ventures provided, of course, that entrepreneurs understand these issues and this challenging environment. Joshi and Yermish (2000) have summarized the challenges facing the entrepreneur in this kind of environment.

This case will examine a possible strategy for entrepreneurial consultants seeking to provide expertise to rising e-businesses. Readers of this case will be encouraged to explore the issues further, filling in the missing details and developing a full-fledged business plan for action for a pair of young entrepreneurs.

Previously published in *Cases on Worldwide E-Commerce: Theory in Action,* edited by Mahesh Raisinghani, Copyright ©2002, Idea Group Publishing.

First, they will examine the demographics and technology issues facing Ireland. This will provide the appropriate background for the second section where they address the issues of the entrepreneur seeking to start an e-commerce consulting practice in Ireland. In the final section, they will suggest additional steps to be taken. Readers will be faced with evaluating these suggestions in light of the background information and any additional research that they may undertake.

## SETTING THE STAGE

Colleen O'Conner and Sean Kelly are Irish citizens in their late twenties. Both of them just completed an MBA program in the United States after their undergraduate program in business in a university in Cork, Ireland. They have the entrepreneurial spirit and see opportunities to convert their new knowledge into a business venture. O'Conner and Kelly both have excellent communications skills and understand the basic technological details of the Internet. O'Conner seems very comfortable with the technological issues, while Kelly feels his greatest strength is in finance and planning. With this in mind, they felt that they would most enjoy developing a consulting practice targeted at helping businesses surmount the challenges of E-Commerce. Having done their studies well, they first set about researching the opportunities for success in their native Ireland.

## BACKGROUND

How is the Republic of Ireland positioned to take advantage of this technological revolution and be a leader in e-Europe? Ireland is positioned socially, financially, and politically to take advantage of the shift in technology. Ireland is an island nation slightly larger then West Virginia. Forty percent of the population lives within 97 kilometers of the capital city of Dublin. Greater than forty percent of the population is under the age of 25. In 1997, the GDP of Ireland was IR£67 billion. Ireland exports US$60 billion, which is composed primarily of chemicals, data processing equipment, industrial machinery, and animal products. Sixty-seven percent of these exports go to other EU countries while eleven percent goes to the United States. Ireland imports US$43 billion, which is composed primarily of food, animal feed, data processing equipment, petroleum products, machinery, textiles, and clothing. Fifty five percent of these imports come from other EU countries while fifteen percent comes from the United States. Current unemployment rates are at 4%. Research indicated that this four percent would be unemployable in any vocation requiring third-level education (the American Bachelor's degree). Twenty-one percent of the population is between 0 and 14 years of age. Sixty-seven percent of the population is between 15 and 64 years of age. Twelve percent of the population is greater then 65 years of age. Sixty percent of the population that attends third level education is studying engineering, science, or business studies. As pointed out in a pamphlet from the Irish Industrial Development Agency, "The quality of Ireland's education is exceptionally high. The independent IMD World Competitiveness report ranks Ireland as one of the best in Europe for the quality of education which everyone receives."

Amarach ("tomorrow" in Gaelic) Consulting is Ireland's leading specialist in predictive market research, consumer trend analysis and business forecasting. It produces several reports tracking every aspect of Ireland's usage of the Internet and e-commerce issues and opportunities. Data from Henry and O'Neill (1999) in Figure 1 shows Internet growth in Ireland as a percentage of the total adult population.

Its 6/2000 report indicates that Amarach's market predictions were almost exactly on target. A 2/2000 report indicated that 22% of the adult Irish population was online at that time. The June report indicated that the number of Irish online had grown sharply in the second quarter to reach 25% of Irish people who had access to the Internet. It is important to note that the use of the Internet implies access, primarily at home, work or school. Smaller percentages access the Internet from a friend's house or cyber cafe. The specific point of access is important to note within the context of the market that is being targeted for e-commerce. Obviously, if the consumer is the target audience, an increase in the number of access points

*Figure 1: Internet growth in Ireland*

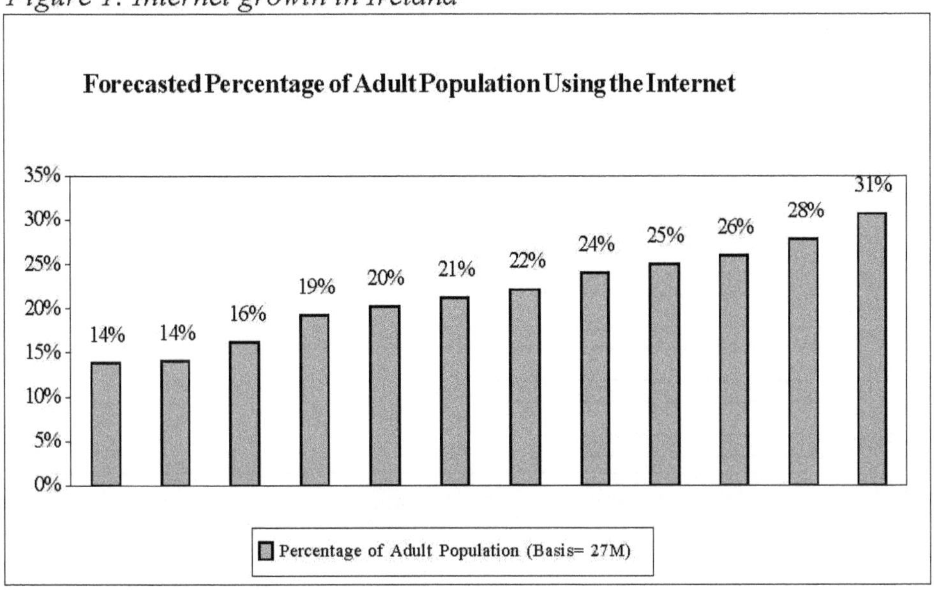

at home is more important. B2B would naturally be looking for an increase in access points at work With an increase in the free Internet Service Provider market, there is a significant increase expected within Ireland.

Another perspective is the percentage of Internet users from home. The implication of the home usage chart is that future growth in Ireland will come from more uses in the home. This will include increasing the number of PCs or other access points to the Internet in the home. These opportunities will proliferate themselves in many ways, including non-traditional access points to the Internet.

Because the largest percentage of adult users is based primarily at home, the most immediate needs for e-commerce would be in the B2C marketplace. With "work" only 2% behind, however, there are more significant, longer-term possibilities within this market of B2B players.

A 3/2000 survey from Amarach indicated that sixty seven percent of people surveyed would prefer to access the Internet through their TV and the remaining thirty three percent said they would prefer mobile phone access if given only the two options.

Cellular phones are also an area that must be monitored with respect the opportunities in the e-commerce marketplace. Welcoming the announcement of the 1,000,000th customer, Stephen Brewer, Eircell Chief Executive, said in a press release, "This is a significant benchmark in the history of Eircell's growth and in the growth of mobile phone usage in Ireland. In 1995 only 3% of the Irish population had a mobile phone, this figure now stands (3/28/2000) at 43%-with over one million of these people having an Eircell phone." Forty-three percent of the population of

*Figure 2: Internet usage in four countries*

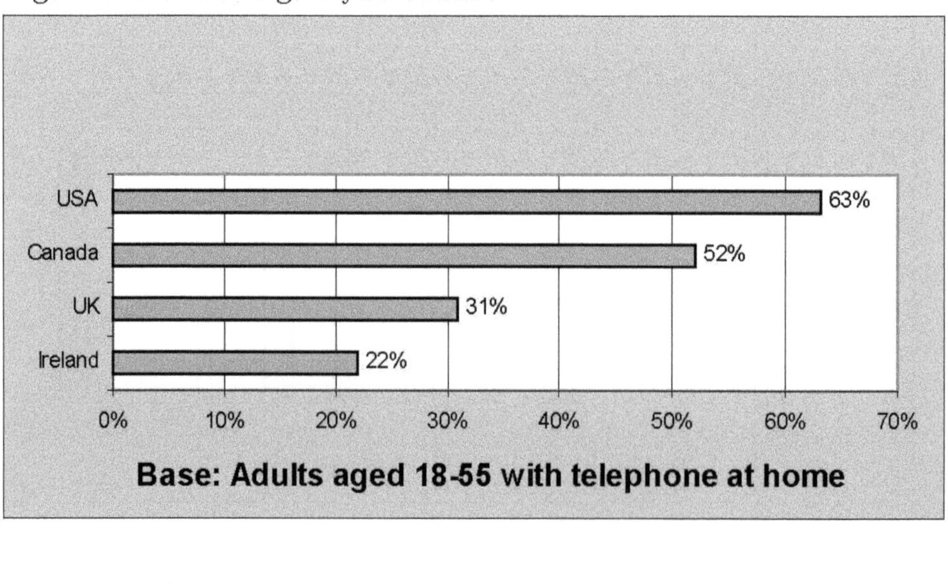

*Figure 3: Home users as a percentage of all Internet users*

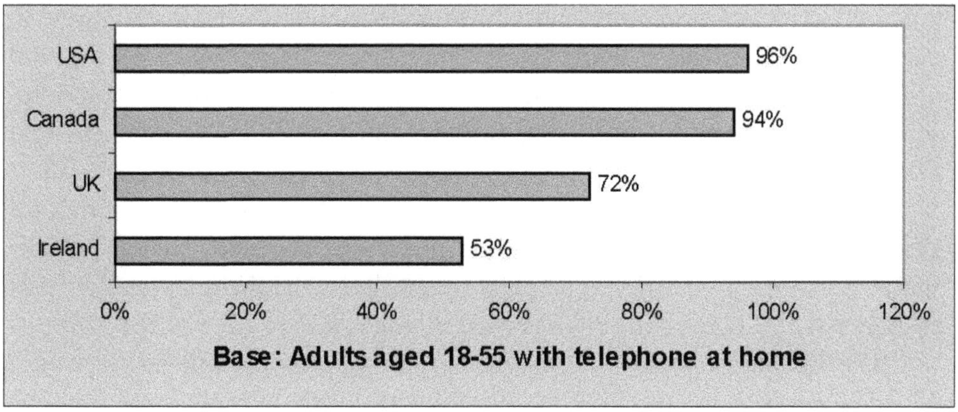


*Figure 4: Internet usage in Ireland*

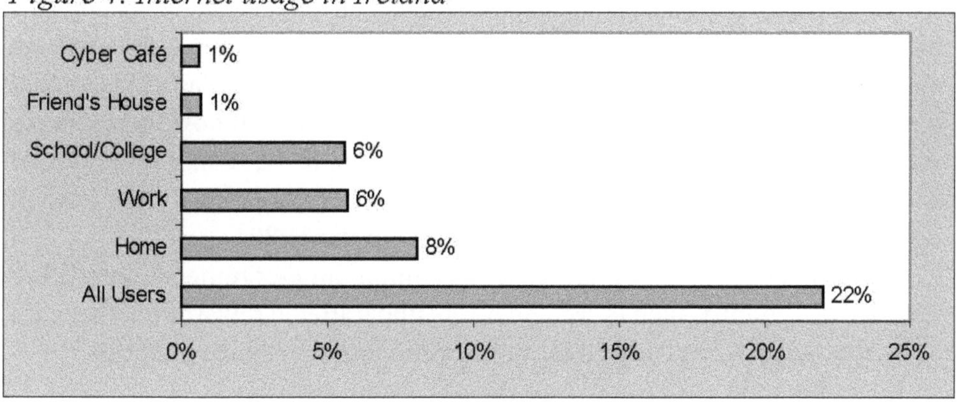


Ireland is approximately 1.55 million potential customers for Wireless Access Protocol (WAP) e-commerce services.

The emergence of WebTV will also foster significant growth in the Irish consumer market in the coming years. Ninety seven percent of all Irish homes have a television with at least the three basic channels located within the country. Also, thirty percent of the population has some type of cable service and twenty two percent are reported to have a satellite dish. Since eighty-five percent of homes have the ability to access cable television, the cable modem market will also be another interesting development to watch and monitor.

An additional social issue that puts Ireland in a key position to leverage the emerging e-commerce trend is that it's the only English speaking country participating in the European Monetary Unit. Ireland will have the ability to trade and cross-market within the other European Union countries. Politically, the government is working closely with industry and academia to anticipate future skill needs. As stated in an Irish Industrial Development (IIDA) pamphlet, "A US$400 million program has been put in place which will double the number of computer science graduates and quadruple the number of software graduates within the next 5 years."

Culturally, the Irish are very customer satisfaction oriented. The Irish are easygoing, lighthearted, good-humored, polite, and cheerful. They are quick-witted and have the ability to laugh at themselves (Brigham Young University, 1999).

The largest negative factor affecting continued growth in the Celtic Tiger's economy relates to the skyrocketing cost of real estate and the potential for inflation to hit the eight percent mark in 2001. Wages are accelerating, spending is increasing and the government is lowering taxes. Government spending is also up. All of these factors in combination with Ireland giving up its monetary policy in favor of the Euro are causing this massive inflation. Basic economic textbooks would say that a government should increase interest rates, raise taxes, or cut spending. Pascal states that the Irish government is doing the opposite. Time will tell how these moves will affect the overall economy though it seems to be a simple matter of economics at this point. If interest rates or taxes are not raised soon Ireland could be facing inflation similar to the USA in the 1980s.

O'Conner and Kelly knew that developing an e-commerce consulting practice in a country such as Ireland offers unparalleled rewards but there are areas that must be considered before leaping into a business situation.

## MARKET SUMMARY

Though the permutations of e-commerce operations are many (Business, Consumer and Government), it appears strategically advantageous in this environment to concentrate on the business-centered areas. Therefore, we suggest concentration in the B2B and B2C. Table 1 below, extracted from Mitchell (2000), represents the size of the potential e-commerce market within the next three years in Ireland.

Estimates by the International Data Group (IDG) have anticipated that in 2001 the entire e-commerce market within Western Europe will be US$30 billion. This research from IDG also indicates that Germany and the UK will be rivals for the spot of the largest Internet market in Western Europe. Germany is expected to drive one third of the e-commerce in 2001. American investors and technology companies fit naturally into the Irish culture. As

Ireland is the only English speaking country participating within the Euro Currency market there are natural advantages to starting this enterprise in Ireland before extending it to other parts of Europe.

B2C is the first area to focus on. Ireland is not that dissimilar to the USA in its purchase patterns while online. The top five products most often purchased online in Ireland include: books, CDs, travel, videos, and software. The typical Irish purchaser is a male, aged 25 to 49 who is employed, owns a credit card, and uses the Internet from home. The ownership of a credit card is where the B2C market in Ireland hits the wall. The strategic guideline here is that the availability of credit distinguishes B2C consumers more clearly from their peers who do not have credit. For B2C e-commerce to become a true powerhouse, new means must be found of putting credit into young hands.

There is a strong need for more Irish companies to have online offerings, and they need to have a higher brand awareness of indigenous online brands to turn this around (Carpenter 2000). Again, significant market opportunities exist to get the smaller companies online.

Where will B2C growth come from within the Irish market? Clearly, the overall growth in people coming online will probably affect the e-commerce bottom line more than any other factor. The push here is to support access from home-based Internet users. Advances in WAP technology products will also stimulate growth within this sector.

B2B is the largest area that we will look at within the scope of market opportunities. This is by far the largest area within the e-commerce sector with respect to the amount of money being spent online. Initial investment is a significant barrier to entry while return is not a short time frame. The rewards will be massive for those who capture the market with the proper branding as it's emerging. The Gartner Group says, "e-business is no longer just an issue of competitiveness–it's a matter of survival," and Kennedy (2000) states, "Global B2B e-commerce will hit US$2.7 trillion in 2004. While Internet trade between business partners will continue to

*Table 1: Size of potential e-commerce market within the next three years in Ireland*

| | **B2C Purchases** | **B2B Purchases** | **Total** |
|---|---|---|---|
| **Usage in 2000** | 15% of Internet Users | 19% of Businesses | **€325M** |
| **Market Value in 2000** | €75M | €250M | |
| **Usage in 2003** | 32% of Internet Users | 64% of Businesses | **€5.3B** |
| **Market Value in 2003** | €1.1B | €4.2B | |
| | **14.7x increase** | **16.8x increase** | |

flourish, e-marketplaces will fuel most of the growth, reaching 53% of all online business trade within 5 years."

A risk/opportunity to be cognizant of in the B2B arena is the diminishment of brand loyalty. Because information about products, including competitors, is available in a few keystrokes, the days of comparison on price will quickly evaporate. Competitive advantage will be based on customer satisfaction, the ability to provide value added services, as well as complementary product lines.

It is vital for any firm or consulting company to recognize that B2B business on the Web is still a business. A writer in the magazine B2B says: "You have to remember the basic rules when it comes to Internet technologies. It's got to make money or save money, and if you can't identify upfront how it's going to do that, in the majority of cases it's going to be a waste of money." The B2B business model is more sustainable than the B2C model because of the necessity for critical mass before becoming successful. A successful consulting company will assist its clients in overcoming these hurdles to add value to the top and bottom lines. Kafka (2000) remarks that "the key determinant in creating value in a B2B hub is increasing the number of participants, that is, creating liquidity."

O'Conner and Kelly wrapped up their marketing research by agreeing that the emphasis in this consulting company would be in the B2B area with consumer and government activities taking less emphasis. There should also be some effort dedicated to R&D of future e-commerce opportunities, including advanced technology investigation. They would focus on mid-sized companies given the likely presence of larger consulting firms handling the major business enterprises.

## OPPORTUNITIES

O'Conner and Kelly continued to investigate the opportunities within the e-commerce market within Ireland.

As they discovered in their B2B research, the first issue that many Irish companies are having is getting started. There is confusion as to what products or services individual companies can or should offer via the Internet. Today, only 12% of Irish companies have e-commerce offerings. In a personal interview, John C. Moynihan, Director of Customer Marketing for Coca-Cola Greater Europe in London, acknowledged that they are attempting to get more into the e-commerce sector in Europe but are having many difficulties. The need created here for a consulting company is in the area of *strategic development and planning*. A team of e-commerce experts will guide management within

Irish businesses to plan, target, acquire, implement, and measure the specific areas within B2B or B2C that are applicable to their environment.

Strategic development and planning does not stop once the Web services are up and running. Many companies go online only to find they do not keep their sites up to date and lose customers to lack of interest. The resulting service offering is to establish a Content Management Office. This office will continually monitor the Web site and confirm that the latest content/information pertaining to clients is visible to the outside world and that the e-commerce offerings are up to date and valid.

Another area that will be key to B2B will be the Web site's integration with back office systems, including legacy and ERP systems. Sales, inventory, and distribution systems must be integrated with the client's Web site so that misleading information is not portrayed to the public. This type of integration is often difficult when the back office systems are not up to date systems or are on alternative platforms including mid-range or mainframe computers. This consulting company will a specialize in solving these types of issues for clients.

Five focus areas: strategic development and planning, infrastructure, hosting, content management, and legacy/ERP integration services will be at the core of their offerings to clients. An additional aspect of service to clients will be continual education of clients on the latest forms of technology specific to their industry. Through the targeted service offerings above, including concentration in B2B and B2C e-commerce, this company can offer a significant value added for clients.

## RESOURCE REQUIREMENTS

Next O'Conner and Kelly addressed the organizational issues of their venture. They weren't sure whether their venture should be a stand-alone business or a component of some larger general consulting enterprise. They did, however, come up with a preliminary organizational structure shown in Figure 5. It wasn't clear who would take which position or how they would fill the other positions and hire additional consultants.

O'Conner and Kelly agreed that Dublin is the appropriate site for their operations given the size of the city and the pace of business activity. They did understand that location might not be all that important given the growing ubiquity of communications technologies. They weren't sure how who should take which position in this organization, but they did know that the General Manager would be key in making the organization work. Would it make sense for one of them to assume this role, or would it make more sense to try to hire someone with more practical experience? This person would have to establish the other staff very quickly. These five people would then be responsible for coordinating the ongoing operations of the office, as well as retaining additional colleagues. Why is there no CIO? Because this is a technology consulting company and the GM is considered the strategic pivot point within the organization for technology decisions.

The GM may elect to initially establish a relationship (outsource) with a local office of Chartered Accounts within Dublin instead of hiring a Controller. All areas of financial responsibilities will fall under this person's area including A/R, A/P, treasury operations, tax, and general accounting.

## RISKS & REWARDS

Are there risks in establishing a new e-commerce company during a time of rapid growth and inflationary instability? A large issue that would seem to be uncontrollable would be the inflationary pressures within Ireland. At a time when interest rates should be tightening, the government should be spending less and increasing taxes. The entrepreneurs felt that none of these factors would be a hindrance to their success.

Another risk within this market place is the talent shortage due to a significant increase in multinational investment in the country and a low unemployment rate. The solution to this issue is offering a superior product that is recognized within the IT industry and providing opportunities for people to be included in the future growth of the Irish economy.

A challenging issue within the Irish economy is the cost of real estate (largely impacted by inflationary pressures). In the past four years it has not been atypical of metropolitan area homes to increase in value by 400%.

*Figure 5: Suggested organizational structure*

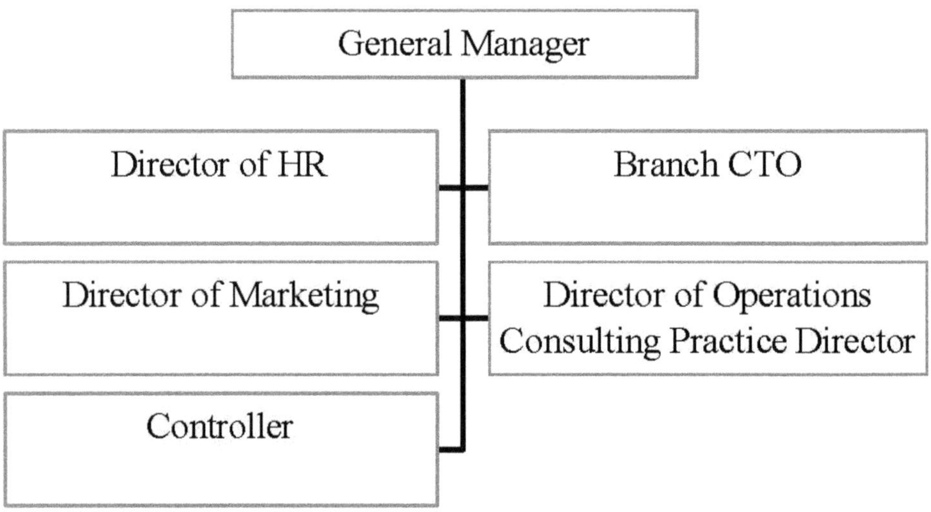

Cultural issues that may impact the business include a lack of trust in the security of technology. Amarach Consulting measured three distinct groups, including "multi-techs, mobile techs, and anti-techs" (MacCarvill, 2000). Unfortunately, the anti-techs seem to make up about twenty-eight percent of the population. The fortunate part is that this group does not represent the target market that is sought within the scope of the client base.

## NEXT STEPS

Having explored the organizational structure and the risks and rewards associated with the venture, O'Conner and Kelly considered their financing. They hoped that funding for this project could come from an already established consulting firm that has a desire to expand into the European market place. Additional funding should be sought from the IIDA and other Irish business incubators.

Pre-Implementation efforts include obtaining office space and recruiting the five top positions within the Irish market place. This should involve advertising in advance in many Dublin-based newspapers and recruiting agencies. Additional funding should be sought at this time to capitalize on the Irish desire to grow multinational influence within their borders. Another pre-implementation effort is partnering with an already established Irish consulting company in a strategically complementary aspect of IT and e-commerce. This partnership will provide insights not simply to the business atmosphere, but to the cultural and legal sensitivities not already outlined in this chapter. "In fact, in an e-business environment where partnerships are everything, relationship management becomes one of the key corporate competencies, and the IS executive who possesses these skills can ride them into new roles as GMs or COOs" (Field 2000).

Once the implementation of this plan was begun, O'Conner and Kelly knew it could be implemented quickly. Three months should be used to establish the office and hire the first wave of people to run the organization. Within six to nine months, the business should be up and operational including placement of consultants at client locations or on client projects working remotely.

O'Conner and Kelly knew that E-commerce is so much more then a technological revolution. It is a shift in thinking that is real, predictable, and appreciable. As they read in an April 2000, B2B article, Nicholas Negreponte, Director of MIT's Media Lab made an interesting comment about the future of business in an electronic world: "When you go out today to buy a car, you are actually buying a piece of metal with four wheels on it. Tomorrow when you buy a car you will be buying an entire transportation service [system]." We think that's a fundamental example of what this chapter has attempted to reveal. Price will not be the competition in the future. Other factors will influence people's decisions to buy, e.g., service, wrap-around features, and customer services.

The Celtic Tiger's influence on e-Europe will not be felt for a few years yet, however the time to capitalize on this European secret treasure chest is now. Ireland is not a market for "day traders" of technology but a safety position for a long-term success story. So many factors are influencing Ireland's ability to be a major player in the European e-conomy of the future that it's almost scary that the IT community isn't jumping up and down yelling and screaming: "Hey look here. Look at the open door to Europe. Look at all this free money!" The world of technology is moving so quickly that people don't have time to pick up their heads to look more then twelve months down the road. For those who do, and see the green of Ireland, they will also see the green of dollars before long. O'Conner and Kelly want to be there.

## STUDY QUESTIONS

The authors of this chapter have suggested a strategy for a pair of entrepreneurs seeking to make their fortune in e-commerce using Ireland as a focus. This strategy is very much based upon standard theories of business strategy, organizational growth and information technology management. However, the world is changing so rapidly and the fundamental issues are so fluid that these theories may be called into question. The following questions should be used to open a dialogue into the theories, options and directions possible.

1. What differences in strategy would be appropriate if the entrepreneurs described in this case were American seeking to enter a new market, or Irish, attempting to start a new local venture?
2. Given the shakeout in the "dot-com" world, what advice would be appropriate for the consultants to give for their Irish clients?
3. Using the dynamic resources of the Internet, what current social, political and economic issues are emerging that would change the strategies described in the case? What mechanisms should be incorporated in the theoretical consulting businesses practices to assure it of continued viability?
4. What details are missing from the suggestions for business operations? How important is a comprehensive business plan in this environment?

## REFERENCES


Amárach Consulting. (2000). About the future: Economic change. *Amárach.* Retrieved June 25, 2000 on the World Wide Web: http://www.amarach.com.

B2B Brief. (2000). *B2B*, April, 10.

Brigham Young University and eMSTAR, Inc. (1999). *Culturgram 2000–Ireland*, 153-156.

Business 2 Consumer. (2000). *B2B*, April, 24-25.

Business.com: Inside the Business of E-Business. (2000). *B2B*, May, 10.

Carpenter, P. (2000). *eBrands, Building an Internet Business at Breakneck Speed.* Boston, MA: Harvard Business School Press.

Case Studies. (2001). *VNU Business Publications.* Retrieved on the World Wide Web: http://www.room.vnu.co.uk/e_com/e_05_01.htm.

Codogno, L. (2000). Euro financial outlook. *Global Markets Group: Euro Economics & Interest Rate Research.* Bank of America, July.

Cross, L. (2000). *Overview of the Pfizer Tablet Plant at Ringaskiddy.* Cork Co., Ireland, August.

Dennehy, J. (2000). The pandora's box of e-business. *B2B*, May, 51.

E-Business is Here and Now. Open Up to it. (2000). Pamphlet, Enterprise Ireland, 1-2.

Eircell Press Release. (2000). *Eircell's Customer Base Exceeds 1 Million–An Increase of 66% in 1 Year.* Retrieved on the World Wide Web: http://www.eircell.ie/htm/oneir/company/word.

Ekoniak, J. (2000). Trends in the b-to-b space. *Upside*, August, 166-168.

The Electronic Commerce Association of Ireland. (2000). *EU E-Commerce Directive.* Retrieved on the World Wide Web: http://www.ecai.ie/.

Enterprise Ireland Press Release. (2000). *Irish E-Business Companies on US Partnership Trail in New York's Silicon Alley*, May. Retrieved on the World Wide Web: http://www.enterprise-ireland.com/english.asp.

Evans, P. and Wurster, T. S. (2000). *Blown to Bits.* Boston, MA: Harvard Business School Press.

E-Work Guide to Company Use. (2000). Pamphlet, Enterprise Ireland, 28.

Field, T. (2000). Special report-IS at the crossroads: A matter of life and death. *CIO*, June, 98-138.

The Future of E-Commerce UK Developments. (2000). *VNU Business Publications.* Retrieved on the World Wide Web: http://www.room.vnu.co.uk/e_com/e_04_e50.

Henry, M. and O'Neill, G. (1999). Facts and strategy for e-commerce in Ireland. *Eir-Commerce*, June, 1-11.

Ince, J. (2000). To B2B, or not to B2B. *Upside*, August, 130-160.

Internet Commerce to Top $30bn in Western Europe by 2001. (2000). *VNU Business Publications*. Retrieved on the World Wide Web: http://www.room.vnu.co.uk/e_com/e_05_50.htm.

"Ireland Becoming Europe's e-com Hub: Government Internet Policies Shaped to Appeal to the Major Players in Cyberspace," Backgrounders, http://www.ebizchronicle.com/backgrounders/january/ireland

Ireland Industrial Development Agency. (1999). Achieve European competitive advantage in Ireland. Pamphlet, 1-10.

Ireland Industrial Development Agency. (2000). "e-commerce," Pamphlet, 1.

Irish E-Commerce to be Worth £3.4bn by 2002. (2000). *B2B*, May, 8.

ITS 2007: Opportunities for Ireland's High-Technology Internationally Traded Services (ITS) Sector to 2007. (2000). Pamphlet, Enterprise Ireland, 2-32.

James, M. (2000). Recruitment's e-volution. *Amárach*. Retrieved August 12, 2000 on the World Wide Web: http://www.amarach.com/.

Joshi, M. and Yermish, I. (2000). The digital economy: A golden opportunity for entrepreneurs? *New England Journal of Entrepreneurship*, 3(1), Spring, 15-22.

Kafka, S. (2000). Why all the buzz about B2B? *Upside*, August, 136-140.

Kanter, R. M. (2001). *Evolve! Succeeding in the Digital Culture of Tomorrow*. Boston, MA: Harvard Business School Press.

Kennedy, J. (2000). B2B exchanges-Chasing new vistas. *B2B*, April, 18-21.

MacCarvill, B. (2000). *New Report Shows Irish Youth Crazy for SMS*. Amárach. Retrieved June 25, 2000 on the World Wide Web: http://www.amarach.com/news/press7.

MacCarvill, B. (2000). *One in Four Now Use the Internet*. Amárach. Retrieved June 25, 2000 on the World Wide Web: http://www.amarach.com/news/press9.htm.

MacCarvill, B. (2000). *The Tech Divide-New Research from Amárach Consulting*. Amárach. Retrieved June 25, 2000 on the World Wide Web: http://www.amarach.com/news/press8.htm.

Molloy, C. (2000). E-procurement by Irish firms to grow 21% in 12 months. *B2B*, April, 4.

Morrison, S. (2000). *Technology Operations at Bank of America European HQ in London*, July.

Moynihan, J. (2000). *Customer Marketing and Overview of Coca-Cola Operations for Greater Europe*, August, .

Mr. Digital. (2000). *B2B*, April, 29.

The Numbers Page: 22% of Irish Adults Online-e-commerce Growth Low. (2000). *B2B*, April, 12.

Porter, M. E. (2001). Strategy and the Internet. *Harvard Business Review*, 79(3), March, 63-78.

Smith, I. (2000). Turn your business inside out. *B2B*, May, 36.

Taking Your Business Online. (2000). Pamphlet, Enterprise Ireland, 2-40.

War Stories of the Web: Why Businesses Fail Online. (2000). *B2B*, May, 22-24.

Wheatley, M. (2000). Mobile technology: High wireless act. *CIO*, July, 126-136.

Zachary, G. P. (2000). Euro limits options for Ireland as Dublin copes with inflation. *Wall Street Journal*, May.

# Chapter 17

# Anxiety and Involvement: Cultural Dimensions of Attitudes Toward Computers in Developing Societies[1]

Roger Harris
Universiti Malaysia Sarawak, Malaysia

Robert Davison
City University of Hong Kong, HK

*Information systems (IS) are implemented within a social context consisting of economic, political, cultural and behavioural factors which differ greatly between societies and countries. Failure to take account of such differences can inhibit adoption of information technology (IT) and increase the risks of failure for system implementations. Developing societies are particularly vulnerable to such risk as their social contexts exhibit considerable differences, not only from the developed nations but also among themselves. This study examined the computer anxiety and involvement with personal computers (PCs) of six groups of computer-using students from China, Hong Kong, Malaysia, New Zealand, Tanzania and Thailand. Differences in comuter anxiety were found to exist between some of the groups, which was probably attributable to demographic factors. Differences were found to exist between the PC involvement of some of the groups which could be attributed to cultural factors. Implications for research and practice are drawn.*

Previously Published in the *Journal of Global Information Management, vol.7, no.1,* 

Developing countries are increasingly deploying IT to solve their developmental problems. Lending for IT by the World Bank has been growing at six times the growth rate of total Bank lending, and is present in 90% of the Bank's lending operations. The extent of IT's impact on World Bank development efforts has been rated great or very great by 79% of projects (Hanna, 1993). However, overall use of IT in developing countries remains at a far lower level than in the developed nations. For example, Tokyo has more telephones than the entire African continent (Harris, 1995).

While the provision of the technology is a necessary condition for achievement of the benefits which IT can bring, there is mounting evidence to suggest that this is not sufficient in itself. Changes in the behaviour of individuals and organisations are also required (Nelson, 1990). When the application of IT fails to account for the behaviour of its users, especially insofar as that behaviour should change, then disappointing results, or even failed applications can result (Harris, 1995). Many factors will affect the behaviour of individuals who have the opportunity to use a computer and the influence which attitudes have on individual behaviour has been well documented (Bagozzi, 1992).

Influences on behaviour may also arise from organisational culture (Geertz, 1973) or from national culture (Hofstede, 1985). The implementation of IT in developing countries is often characterised by a transfer of technology from one culture to another. In aid situations, this transfer is often in the form of a donation from a developed country with a western (European) culture to a developing country with a non-western (African, Asian or Latin-American) culture. In such cases, assumptions regarding the changes in behaviour which are required to make the technology successful may not be as valid in the recipient culture as they may have been in the donor culture. For example, Galliers et al. (1994) describe a national computerised land management system in Pakistan which was rejected by senior management seemingly because they were expected to use it for strategic planning, forecasting and decision support even though they belong to a culture in which meticulous forward planning is not emphasised and in which risk-taking and fatalism are valued.

National cultures are often formed from the aggregation of ethnic cultures which may not coincide with national boundaries. This is especially true in, but not confined to, post-colonial nations, many of which are classified as developing countries. Moreover, many developed nations, whose national cultures might be thought of as fairly homogeneous, in fact contain significant minority populations who often consider themselves ethnically and culturally distinct from both the majority population and from each other. Examples include Australian Aborigines, New Zealand Maoris, and the immigrant communities of several European countries. In all cases, individual behaviour is culturally patterned, and in many cases, some catastrophically so, by ethnic differences.

Among the many aspects of attitudes towards computers which are known to influence computer-related behaviour, two are chosen for examination in the present study. Computer anxiety or technophobia tends to inhibit computer use and has been found to be related to a consistent pattern of response towards computers, viz. lower expectations, poorer performance, more subjective anxiety and a higher frequency of debilitative thoughts (Igbaria and Nachman, 1990). On the other hand, involvement with computers has been shown to have a positive influence on computer-usage behaviour by individuals. Moreover, involvement with a product such as PCs can be simultaneously psychological and behavioural. Product involvement represents a continuum that ranges from total apathy to extreme enthusiasm, and while not being observable, it does produce observable behavioural outcomes (Bloch, 1986). For example, Paré and Elam (1995) found computer-related activities, such as watching TV documentary programmes about computers and consulting computer-oriented magazines, to be among the dominant predictors of PC usage. Bloch (1986) defines product involvement as an unobservable state reflecting the amount of interest, arousal, or emotional attachment a consumer has with a product. Product enthusiasm gives rise to early adoption of new products and desire for the latest technology (Bloch, 1986).

This research measured the computer anxiety and PC involvement of a number of different cultural groups of computer users. Six groups of tertiary-level students, totalling 279 respondents, from China, Hong Kong, Malaysia, New Zealand, Tanzania and Thailand, were asked 20 questions about their feelings towards computers. Hong Kong and New Zealand are not classified as developing nations. Hong Kong was included in the study in order to provide the opportunity for comparing the results with at least one developed nation. The subjects from New Zealand were all of Maori extraction, and as such represent an economically disadvantaged and culturally distinct group which can be regarded as a developing society.

University students were selected for the study as a convenience sample, but also as they represent future professionals who are likely to be well-positioned to make good use of computers in their careers, though major behavioural changes may be required (Panko, 1988). In the next section, we describe the six groups of students and their cultural backgrounds. Following this, we employ analysis of variance and partial least squares to analyse the data collected from the students.

*Table 1: Demographic and Economic Statistics*

| | Population: 000 [1] | GNP per Capita: $US[1] | Telephones per '000[2] |
|---|---|---|---|
| Hong Kong | 5,865 | 17,860 | 510 |
| Thailand | 58,824 | 2,040 | 37 |
| Malaysia | 19,032 | 3,160 | 126 |
| China | 1,175,359 | 490 | 15 |
| Tanzania | 26,743 | 100 | 3 |
| New Zealand | 3,462 | 12,900 | 460 |

[1] The World Bank Atlas, 1995. [2] Connors, 1997

## COUNTRY PROFILES

Table 1 provides summary demographic and economic statistics for the countries included in the study. Subsequent subsections provide a brief description of the political, economic and cultural background of each country.

### Peoples' Republic of China (PRC)

The forty-eight students in this sample were final year undergraduates in the Department of Journalism, at Jinan University, Guangzhou, where Mandarin is the official medium of instruction. Journalism students typically have moderate exposure to PC software, being familiar with word-processing, desktop publishing and presentation graphics but not with programming languages. Particular attention is paid to Chinese computing, since the newspaper industry in China has been computerised for many years and journalists require proficiency in Chinese input methods.

Guangzhou is one of the largest cities in China, with a population of approximately five million. With the "open door" policy now practised in China and the promotion of trade and business in Guangzhou, the financial situation of its citizens has improved considerably in recent years. Generally in China, parents place a high value on education and employ a concept roughly translated as "studying is the most highly appreciated activity." University graduates need to be well equipped with computing skills as computerisation has become very popular in Guangzhou businesses. Computer-related courses have been offered in Guangzhou since the early 1980s and there is an increasing trend for families in Guangzhou to have a PC at home. Computer-related activities, such as playing computer games, and reading computer-related magazines or newspapers that focus on the development of technology in the Chinese context, have also become popular. Guangzhou has eight shopping malls selling computers and peripheral products; as in Hong Kong, much of the software is pirated. With an estimated 500,000 individual Internet users in China (about half through universities), by the end of 1997, China represents a small but growing segment of worldwide Internet users.

### Hong Kong

The 53 students in the study were final year undergraduates in the Department of English, at the City University of Hong Kong. They were all Cantonese speakers with English as a second language. They enjoyed moderate exposure to PC software for the purposes of word-processing, desktop publishing and presentation graphics.

Hong Kong, a British colony for 150 years, recently returned (July 1st, 1997) to Chinese sovereignty. Its population is largely of migrant stock, with many of the present older generation arriving after the end of World War II and the civil war that brought the Chinese Communist Party to power in 1949. As in the PRC, Hong Kong Chinese parents place a high value on education, and students tend to be highly oriented towards academic achievement. A higher level of education increases the chance of joining a more secure and respectable profession, as well as enabling children to support their parents later in life. Chinese culture is also noted for its collectivist characteristic where the paramount concern is for the integrity and advancement of the group (Bond and Wang, 1981). Thus, education traditionally focuses on dependence rather than independence, and this can be manifested in a lack of creativity, with learning by rote.

IT is considered important for education, with all tertiary-education institutions being well-endowed with equipment. Popular technology adoption within Hong Kong extends to pagers, mobile phones and other devices which provide information to an information-hungry population. Hong Kong has the world's highest per capita usage of cellular telephones and radio pagers. In 1996, the government reported a density of radio telephones of 19.6 per 100 people, representing an increase of 76% over 1995. The number of fixed telephone lines is also among the world's highest.

### Sarawak (Malaysia)

The 33 students in this group were drawn from undergraduates in Universiti Malaysia Sarawak (UNIMAS) in Sarawak, Malaysia's largest state. They were 1st to 4th year students attending a variety of courses, but were all required to take two mandatory courses: End User Computing and Power Tools for Knowledge Workers, whereby they were exposed to various PC computing packages as well as the Internet. All the students in the study belonged to Sarawak's indigenous population and are known collectively as Dayaks, who make up about 49% of Sarawak's 1.6 million people. Most of the Dayak indigenous peoples are still only marginally involved in the urban economy (Brookfield et al., 1995). Colonial conditions, culminating in the incorporation of Sarawak into Malaysia, have dramatically altered Dayak societies by placing them in a position of subordination.

In May 1996, the Malaysian Government announced its Seventh Malaysia Plan which is intended to provide a springboard for the ambitious "Vision 2020" by which Malaysia intends to become a fully developed country. One component of this plan is the US$8 billion "cybercity" - Cyberjaya - the world's first "intelligent city," to be built near Kuala Lumpur, in which all homes will have Internet access and all transactions will be conducted through smart cards. A key concern of the government is that the benefits of growth are equitably shared among all ethnic groups and states (Wong, 1996). As it is an avowed government aim that Malaysia must be an information-rich society, computer literacy has been identified as essential for the country to progress and develop (Mohamad, 1991).

The Seventh Plan points out that the number of PCs in the country increased from 160,000 in 1990 to 310,000 in 1995, putting the ratio at around 16 per 1,000 inhabitants. By October 1996, it was reported that Malaysia had about 75,000 Internet subscribers with growth rates of 20% per month (Ponniah, 1996).

### New Zealand - Maoris

Our sample of 41 Maori students were undergraduates at the University of Waikato in New Zealand. They were all enrolled on the Maori language version of a course called The Computing Experience, which provides an introduction to computers and their use. It is intended for students who wish to take only one course in computer science. Eleven percent of New Zealand's 3.5 million inhabitants are Maori, but only a small fraction of these are actively involved with IT, as the Maoris have been socioeconomically disadvantaged in the past. This unfortunate situation is changing now, with computer courses run in the Maori language at Waikato University.

The Maori first came to New Zealand some 1,000 years ago, but European colonisation from the late 18th century introduced disease and firearms, and the Maori population plummeted. Although the Maori population is now increasing and Maoritanga (Maori culture) is experiencing a flourishing renaissance, most Maori are now town and city dwellers, and many have lost touch with their original culture. Maori is now an official language of New Zealand, although outside the Maori community it is rare to hear it spoken except on ceremonial occasions. Maori people have established various programmes for the revival of their language, particularly in kindergartens and primary schools.

The attitude of the Maori population towards education is positive, yet as a disadvantaged group, their access to the education system has left something to be desired. As with many minority groups, education is seen as a potential leveraging point to pull them into a better way of life. A small group of skilled Maori people are deeply involved in writing software, getting IT into schools and their community as a way to increase IT use and awareness. However, it will be many years before those who have gone through the preschool language classes come through the education system to the universities and then into the workforce to make significant differences in the status of the Maori.

## Tanzania

Thirty MBA students from the Faculty of Commerce and Management of the University of Dar es Salaam (UDSM) formed our Tanzanian sample. The students held various first degrees and most were employees of public organisations. All of them are Bantu speakers of different tribal origins with Swahili as their first or second language and English as their second or third language. While Tanzanian culture is a product of African, Arab, European and Indian influences, traditional African tribal values, which persist among the rural population, are being consciously adapted to modern life.

Tanzania offers free education to all its citizens, from primary school to university level. Computing education has been available at tertiary level since 1965, and the first computing degree course was offered in 1974, but after training only 12 Tanzanians, it ceased to operate in 1984 (Mgaya, 1994). The students in the study have taken introductory courses in IT and have had limited exposure to various PC software packages, mostly for word-processing and spreadsheet operations. They have also been able to use a group support system installed in UDSM's Management Decision Centre (Splettstoesser, 1996).

Prior to the mid-1980s, Tanzania's policies of "African socialism" targeted self-sufficiency and limited the use of non-indigenous technologies. In the 1970s, the government banned the import of any computer equipment, and this remained effective until the early 1980s. Since 1985, a policy of liberalisation and structural adjustment has caused the gradual adoption of IT in all sectors of the economy, but IT penetration remains very low. Galilava (1996) reports that 40% of all implemented applications serve financial or accounting functions.

The government has not made any serious attempt to formulate a national IT policy. Despite this, there has been a dramatic increase in microcomputer utilisation in all sectors of the economy during the last ten years. Tanzania has an enormous development workload and a comprehensive and conducive framework for institutional capacity building, which can be strongly supported by IT, is urgently needed.

### Thailand

Seventy-four MBA students from Assumption University in Bangkok formed the Thai sample who responded to our questionnaire. Twenty-five of the students were studying full-time and had recently completed their BBA degrees. The remainder were part-time evening students from a wide variety of backgrounds who are working full-time with between two and six years employment experience. They were typically taking the MBA course in order to win promotion. The students were a mix of native Thai and Chinese-Thai. They all speak Thai as a first or second language and English as a second or third.

Thai-Chinese, who have traditionally been the most economically successful, have been assimilated into Thai culture, partly as a result of government policy which required all children to learn the Thai language in primary school, as well as official pressure to adopt Thai names. Additionally, the lack of religious barriers has allowed free intermarriage between ethnic Thais and Chinese, to the extent that in the cities especially, it is difficult to identify anyone as a "pure" member of either group.

The majority of Thais are Buddhist, and Thai intellectuals have espoused a "Buddhist Road to Development" as an independent, indigenous model which relates development to religious goals. They express a particular concern for the negative effects of capitalist development in rural areas and advocate a more self-reliant agricultural economy, appropriate technology and even the revival of the cooperative institutions of communal culture (Cohen, 1993).

Thailand's dramatic economic transformation in recent years, seen mostly in the development of Bangkok's goods and services and the expansion of the business and industrial centres, has generally not brought an improved quality of life to the rural and urban poor (Ekachai, 1994). Thailand has recently embarked on a government-funded national IT policy which aims to develop a National Information Infrastructure (NII). It is not intended to be a massive technology importation scheme, but rather one where the Thai people must develop applications that suit them best. The government recognises that there is a severe shortage of qualified people to fill the many jobs that will open up as a result of the NII unfolding and they plan to introduce IT into the educational curriculum from primary school onwards (NITC, 1995).

## The Research

A questionnaire (see Appendix) was distributed to the subjects. It asked a number of demographic questions as well as those relating to the instruments for measuring computer anxiety and product involvement (with PCs). The instrument to measure computer anxiety was taken from Igbaria (1990) and has been found to have high internal consistency and reliability in prior empirical studies. The instrument asks individuals to indicate their agreement or disagreement with ten statements reflecting anxiety, apprehension, confusion, hesitation etc. in using computers. The response options, anchored on a five point Likert-type scale, range from 1 = strongly disagree, to 5 = strongly agree, such that the higher the score, the higher the computer anxiety. Product involvement with PCs (PC involvement) was measured using an adapted version of a scale developed by Bloch (1981) for measuring involvement with motor cars. Bloch (ibid.) has pointed out that the product-specificity of the original instrument should not limit its usefulness in efforts to further refine the construct of involvement. Respondents were asked to rate their agreement with a number of statements on a scale of 1 to 5, where; 1 = strongly disagree and 5 = strongly agree. The higher the score, the higher the involvement with PCs.

In order to test for the existence of a difference between the attitudes which the study subjects held towards computers, the research took the form of 16 hypotheses as follows:

H 1: University students will differ significantly with regard to their computer anxiety according to their cultural background.

H 2: University students will differ significantly with regard to their PC involvement according to their cultural background.

As both culture and attitudes are known to influence individual behaviour, these hypotheses test the relative potency of culture and each attitude. The test is suggested by the observation that social receptivity (of innovations) has consistently been more determinative of the directions taken by technological trends than have the properties or developments inherent in the technology itself (McC. Adams, 1996).

H 3: University students will differ significantly with regard to their computer anxiety according to their gender.
H 4: University students will differ significantly with regard to their PC involvement according to their gender.

Gender has been shown to be associated with attitudes and computer anxiety, but the nature of the association is sometimes contradictory (Igbaria, 1990). This test explores further some of the conditions under which the influence of gender might operate.

H 5: University students will differ significantly with regard to their computer anxiety according to their father's occupation.
H 6: University students will differ significantly with regard to their PC involvement according to their father's occupation.

These tests explore the relationship between social status, expressed by father's occupation, and attitudes towards computers. The test is suggested by the observation that computers are often characterised as a tool for minority elites, especially in developing countries (Hamelink, 1997).

H 7: Older university students will be significantly more involved with PCs than younger students.
H 8: Older university students will be significantly less anxious about computers than younger students.
H 9: University students with more computing experience will be significantly more involved with PCs than students with less computing experience.
H 10: University students with more computing experience will be significantly less anxious about computers than students with less computing experience.
H 11: University students to whom computers are more available will be significantly less anxious about computers than students to whom computers are less available.
H 12: University students to whom computers are more available will be significantly more involved with PCs than students to whom computers are less available.

Demographic and situational factors such as age and computer availability and experience have been shown to influence attitudes towards computers (Igbaria,1990). These tests are used to isolate their influence from the other factors in the study.

H 13: University students who are more involved with PCs will be significantly less anxious about computers than students who are less involved with PCs .

This test is introduced in order to examine some of the antecedents of computer anxiety, of which little is known (Torkzadeh and Angulo, 1992).

H14: University students from Hong Kong will be significantly more involved with PCs than students from the developing societies.
H15: University students from Hong Kong will be significantly less anxious about computers than students from the developing societies.
H16: University students from more economically advanced countries will be more likely to feel predisposed towards the use of computers than those from less economically advanced countries.

These tests examine the notion that computer users in a developed society are more likely to be willing to use computers than those in developing societies, given the strong correlation between national GDP and spending on information technology. Connors (1997) illustrates the national differences of the countries represented in the study with an "infostructure index" which ranks countries according to the accessibility of information. Hong Kong ranks fifth out of 147 countries, while Malaysia ranks 67, Thailand 85, China 101 and Tanzania 129 (Connors, 1997).

## DATA ANALYSIS

### Summary Statistics

Subjects were gathered from universities in six countries. Summary statistics are shown in Table 2.

### Scale Reliability

Scale reliability for the measurement instruments was assessed with Cronbach's alpha (Cronbach, 1951), for which a minimum value of 0.7 is generally acceptable. The PC involvement scale achieved an alpha of 0.82, and the computer anxiety scale achieved a value of 0.84.

### Analysis of Variance

The first six hypotheses, were tested using analysis of variance, or ANOVA, which is a method of testing the null hypothesis that several group means are equal in a population, by comparing the sample variance estimated

*Table 2: Summary Statistics*

| Cultural Group | Gender | | | Mean Age | MCA [1] (1-5) | MCE [2] (1-5) | Father's Occupation | | | | | | Computer Anxiety | | PC Involvement | |
|---|---|---|---|---|---|---|---|---|---|---|---|---|---|---|---|---|
| | M | F | Total | | | | Man-ual | Cleri-cal | Profes-sional | Tech-nical | Trade Business | Other | Mean | SD | Mean | SD |
| Hong Kong | 8 | 45 | 53 | 22.4 | 4.4 | 3.7 | 9 | 5 | 1 | 11 | 10 | 17 | 2.80 | .56 | 3.43 | .47 |
| Thailand | 27 | 47 | 74 | 26.3 | 3.9 | 3.3 | 0 | 0 | 5 | 2 | 52 | 12 | 2.43 | .75 | 3.57 | .66 |
| Tanzania | 28 | 2 | 30 | 35.9 | 2.7 | 2.5 | 10 | 1 | 7 | 0 | 3 | 9 | 1.78 | .66 | 4.19 | .46 |
| Sarawak | 15 | 18 | 33 | 21.4 | 3.4 | 3.1 | 8 | 4 | 3 | 2 | 0 | 16 | 2.39 | .69 | 3.97 | .49 |
| PRC | 12 | 36 | 48 | 22.5 | 2.9 | 2.7 | 3 | 16 | 11 | 8 | 8 | 2 | 2.55 | .75 | 3.79 | .63 |
| Maori | 22 | 19 | 41 | 30.6 | 3.1 | 2.6 | 16 | 1 | 13 | 1 | 5 | 0 | 2.56 | .84 | 3.62 | .59 |
| Total | 112 | 167 | 279 | 26.0 | 3.5 | 3.1 | 46 | 27 | 40 | 24 | 78 | 56 | 2.47 | .76 | 3.70 | .61 |

*Table 3: ANOVA for Cultural Groups*

| | | Sum of Squares | df | Mean Square | F | Sig. |
|---|---|---|---|---|---|---|
| Computer Anxiety | Between Groups | 21.230 | 5 | 4.246 | 8.353 | .000 |
| | Within Groups | 138.774 | 273 | .508 | | |
| | Total | 160.004 | 278 | | | |
| PC Involvement | Between Groups | 15.439 | 5 | 30.88 | 9.444 | .000 |
| | Within Groups | 89.257 | 273 | .327 | | |
| | Total | 104.696 | 278 | | | |

*Table 4: ANOVA for Gender*

| | | Sum of Squares | df | Mean Square | F | Sig. |
|---|---|---|---|---|---|---|
| Computer Anxiety | Between Groups | 8.699 | 1 | 8.699 | 15.926 | .000 |
| | Within Groups | 151.304 | 277 | .546 | | |
| | Total | 160.004 | 278 | | | |
| PC Involvement | Between Groups | 4.464 | 1 | 4.464 | 12.335 | .000 |
| | Within Groups | 100.233 | 277 | .362 | | |
| | Total | 104.696 | 278 | | | |

*Table 5: ANOVA for Father's Occupation*

| | | Sum of Squares | df | Mean Square | F | Sig. |
|---|---|---|---|---|---|---|
| Computer Anxiety | Between Groups | 5.167 | 5 | .737 | 1.291 | .255 |
| | Within Groups | 154.841 | 271 | .571 | | |
| | Total | 160.004 | 278 | | | |
| PC Involvement | Between Groups | 3.577 | 5 | .511 | 1.369 | .218 |
| | Within Groups | 101.119 | 271 | .373 | | |
| | Total | 104.696 | 278 | | | |

from the group means to that estimated within the groups. One of the assumptions required for this test is that the variances of the groups in their populations are equal, and this is tested using the Levene test for homogeneity of variances (Norusis/SPSS, 1993).

If the observed significance of the test is small, the null hypothesis, that all variances are equal, can be rejected. For the computer anxiety variable, the Levene statistic was 1.734 with degrees of freedom of 5 and 273, and significance of .127. For PC involvement, it was 1.981, with significance of .082. Both levels of significance are considered large enough that the null hypothesis cannot be rejected and there is therefore insufficient evidence to suspect that the variances are unequal. The results of the ANOVA for testing the differences between the two dependent variables for each of the cultural, gender and fathers' occupation groups are shown in Tables 3, 4 and 5.

The results of the tests for each hypothesis are summarised in table 6.

The results of the tests for hypotheses 1-6 are interpreted by examining the significance levels of the F statistic, which is the ratio of the within-groups mean square and the between-groups mean square. If the means of the groups are similar, the ratio approaches 1. The significance level is based on the F value and the degrees of freedom for the two mean squares. If this is small, say .05 or less, the null hypothesis, that the means are the same, can be rejected.

## Multiple Comparisons

A significant F value indicates only that the means of the populations from which the groups are drawn are probably not equal. It does not indicate which pairs of groups are not equal, and the null hypothesis is rejected if any two means of the groups in the test are unequal. A multiple comparison test indicates which means differ significantly from the others. Tables 7 and 8 show the results of Tukey's honestly significant difference test for computer anxiety and PC involvement respectively according to the cultural background groupings. It is evident that the Tanzanian group displays the most significant differences with the other groups on both these variables, as

*Table 6: Results Summary*

| Hypotheses 1 - 6 | | |
|---|---|---|
| *Test: If the F ratio is > 1and its significance level at the degrees of freedom for the two mean squares is < .05, the null hypothesis, that the means are the same, can be rejected.* | *F* and its Significance | Result |
| H1: University students will differ significantly with regard to their computer anxiety according to their cultural background. | 8.353 (.000) | Supported |
| H2: University students will differ significantly with regard to their PC involvement according to their cultural background. | 9.444 (.000) | Supported |
| H 3: University students will differ significantly with regard to their computer anxiety according to their gender. | 15.926 (.000) | Supported |
| H 4: University students will differ significantly with regard to their PC involvement according to their gender. | 12.335 (.000) | Supported |
| H 5: University students will differ significantly with regard to their computer anxiety according to their father's occupation. | 1.291 (.255) | NotSupported |
| H 6: University students will differ significantly with regard to their PC involvement according to their father's occupation. | 1.369 (.218) | NotSupported |
| Hypotheses 7 - 13 | Coefficient and Significance Level | Result |
| *Test: The path coefficients in the PLS analysis should be both substantive (path coefficient >0.1), significant (at p=.05 or better) and with the correct sign (+/-).* | | |
| H 7: Older university students will be significantly more involved with PCs than younger students. | -.149 (.05) | Supported |
| H 8: Older university students will be significantly less anxious about computers than younger students. | -0.119 (not significant) | NotSupported |
| H 9: University students with more computing experience will be significantly more involved with PCs than students with less computing experience. | .023 | NotSupported |
| H 10: University students with more computing experience will be significantly less anxious about computers than students with less computing experience. | -0.255 (.001) | Supported |
| H 11: University students to whom computers are more available will be significantly less anxious about computers than students to whom computers are less available. | .065 | NotSupported |
| H 12: University students to whom computers are more available will be significantly more involved with PCs than students to whom computers are less available. | .040 | NotSupported |
| H 13: University students who are more involved with PCs will be significantly less anxious about computers than students who are less involved with PCs . | -0.227 (.001) | Supported |

| Hypotheses 14 - 15 | | |
|---|---|---|
| *Test: If the F ratio is > 1and its significance level at the degrees of freedom for the two mean squares is < .05, for all comparisons, the null hypothesis, that the means are the same, can be rejected.* | Significance Levels | Result |
| H14: University students from Hong Kong will be significantly more involved with PCs than students from the developing societies. | .042; .000; .096; .452; .598 | As only 2 of the 5 significance levels are < .05, not supported |
| H15: University students from Hong Kong will be significantly less anxious about computers than students from the developing societies. | .715; .000; .000; .020; .570 | As only 3 of the 5 significance levels are < .05, not supported |

*Table 6: Results Summary (continued)*

| Hypothesis 16 | | | | |
|---|---|---|---|---|
| *Test: The index of predisposition for each group (their mean score for product involvement minus their mean score for computer anxiety) should be in the same rank order as the GNP per capita of each group's country.* | Hong Kong | | Tanzania | |
| | Rank Order by Predisposition Index | Rank Order by GNP | Rank Order by Predisposition Index | Rank Order by GNP |
| H16: University students from more economically advanced countries will be more likely to feel predisposed towards the use of computers than those from less economically advanced countries. | 6th | 1st | 1st | 6th |
| Result: Not supported | | | | |

*Table 7: Tukey's Honestly Significant Difference Test for Computer Anxiety*

| | | Mean Diff | Sig |
|---|---|---|---|
| Hong Kong | Thailand | .3732* | .042 |
| | Tanzania | 1.0257* | .000 |
| | Sarawak | .4117 | .096 |
| | PRC | .2587 | .452 |
| | Maori | .2374 | .598 |
| Thailand | Hong Kong | -.3732* | .042 |
| | Tanzania | .6524* | .000 |
| | Sarawak | .0769 | 1.000 |
| | PRC | -.1146 | .954 |
| | Maori | -.1359 | .925 |
| Tanzania | Hong Kong | -1.0257* | .000 |
| | Thailand | -.6524* | .000 |
| | Sarawak | -.6139* | .008 |
| | PRC | -.7670* | .000 |
| | Maori | -.7883* | .000 |
| Sarawak | Hong Kong | -.4117 | .096 |
| | Thailand | -.077 | 1.000 |
| | Tanzania | .6139* | .008 |
| | PRC | -.1531 | .934 |
| | Maori | -.1744 | .902 |
| PRC | Hong Kong | -.2587 | .452 |
| | Thailand | .1146 | .954 |
| | Tanzania | .7670* | .000 |
| | Sarawak | .1531 | .934 |
| | Maori | .0426 | 1.000 |
| Maori | Hong Kong | -.2374 | .598 |
| | Thailand | .1359 | .925 |
| | Tanzania | .7883* | .000 |
| | Sarawak | .1744 | .902 |
| | PRC | .0426 | 1.000 |
| *The mean difference is significant at the .05 level | | | |

*Table 8: Tukey's Honestly Significant Difference Test for PC Involvement*

| | | Mean Diff | Sig |
|---|---|---|---|
| Hong Kong | Thailand | -1.460 | .715 |
| | Tanzania | -.7650* | .000 |
| | Sarawak | -.5444* | .000 |
| | PRC | -.3592* | .020 |
| | Maori | -.1953 | |
| Thailand | Hong Kong | .1460 | .715 |
| | Tanzania | -.6190* | .000 |
| | Sarawak | -.3984* | .011 |
| | PRC | -.2132 | .335 |
| | Maori | .0992 | .998 |
| Tanzania | Hong Kong | .7650* | .000 |
| | Thailand | .6190* | .000 |
| | Sarawak | .2206 | .645 |
| | PRC | .4058* | .028 |
| | Maori | .5698* | .000 |
| Sarawak | Hong Kong | .5444* | .000 |
| | Thailand | .3984* | .011 |
| | Tanzania | -.2206 | .645 |
| | PRC | .1852 | .707 |
| | Maori | .3492 | .095 |
| PRC | Hong Kong | .3592* | .020 |
| | Thailand | .2132 | .335 |
| | Tanzania | -.4058* | .028 |
| | Sarawak | -.1852 | .707 |
| | Maori | .1639 | .758 |
| Maori | Hong Kong | .1953 | .570 |
| | Thailand | .0985 | .998 |
| | Tanzania | -.5698* | .000 |
| | Sarawak | -.3492 | .095 |
| | PRC | -.1639 | .758 |
| *The mean difference is significant at the .05 level | | | |

suggested by their means. In order, therefore, to test that the overall differences were not due solely to the differences which this group displayed, a further ANOVA test was conducted without the Tanzanian data. The results are shown in Table 9. The low value of both the F significance figures suggest that even after excluding the Tanzanian subjects, the remaining cultural groups still differ significantly on both variables, more so for PC involvement than for computer anxiety. The results of the multiple comparisons for the Hong Kong group were used to test hypotheses 14 and 15, that they would differ from the other groups by being more involved and less anxious.

*Table 9: ANOVA for Cultural Groups Without the Tanzanians*

| | | Sum of Squares | Df | Mean Square | F | Sig. |
|---|---|---|---|---|---|---|
| Computer Anxiety | Between Groups | 5.305 | 4 | 1.326 | 2.562 | .039 |
| | Within Groups | 126.286 | 244 | .518 | | |
| | Total | 131.591 | 248 | | | |
| PC Involvement | Between Groups | 7.396 | 4 | 1.849 | 5.432 | .000 |
| | Within Groups | 83.058 | 244 | .340 | | |
| | Total | 90.455 | 248 | | | |

*Table 10: Comparison of GNP per Capital and Predisposition Index*

| Rank Order by GNP per capita | Country | GNP per Capita $US | Mean Computer Anxiety | Mean PC Involvement | Predisp-osition Index | Rank Order by Predisposition Index |
|---|---|---|---|---|---|---|
| 1 | Hong Kong | 17,860 | 2.8 | 3.43 | 0.63 | 6 |
| 2 | New Zealand | 12,900 | 2.56 | 3.62 | 1.06 | 5 |
| 3 | Malaysia | 3,160 | 2.39 | 3.97 | 1.58 | 2 |
| 4 | Thailand | 2,040 | 2.43 | 3.57 | 1.14 | 4 |
| 5 | China | 490 | 2.55 | 3.79 | 1.24 | 3 |
| 6 | Tanzania | 100 | 1.78 | 4.19 | 2.41 | 1 |

Finally, in order to test hypothesis 16, an index of predisposition was calculated for each group by subtracting their mean score for computer anxiety from their mean score for product involvement. The rationale is that involvement provides a positive inducement for computer use, while anxiety is an inhibitor, or negative inducement. The index was used to rank each group and was then compared to their ranking according to the GNP per capita of each country. The results are shown in Table 10.

The relationship of rank orders is the inverse of what was expected, suggesting that the students from poorer countries feel more predisposed to use computers than those from richer countries. However, caution is advised in this interpretation, as other differences, such as demographics, could account at least in part for the relationship we have found.

## Partial Least Squares Analysis

While ANOVA is suitable to test for differences between several population means, it does not provide information to suggest any causal factors for such differences. Structural Equation Modelling (SEM) is a methodology for specifying, estimating, and testing hypothesised interrelationships among variables in the behavioural and social sciences (Raykov et al., 1991). The technique enables the researcher to estimate direct, indirect and total structural effects that might enhance the understanding of relationships among variables within a specific context (Mueller, 1996).

For relationships to be called causal, they require covariation, i.e. the constructs vary together, temporal asymmetry, i.e., the constructs vary in sequence; and the elimination of other possible causes. The third requirement is impossible to achieve, as it can never be proven that all other possible causes are removed. It is necessary, therefore, either to include them all or to proceed without them. Partial Least Squares (PLS) is a second-generation multi-variate analysis technique for SEM which performs factor analysis and regression analysis simultaneously (Thompson, et al. 1991). Figure 1 shows the results for the tests for hypotheses 7-13.

The PLS analysis only shows paths which are both substantive (path coefficient >0.1) and significant. All paths shown are significant at p=.05 or better.

As PLS allows the researcher to re-specify the model after hypothesis testing, in order to test for additional relationships (Pulos and Rogness, 1995), it was decided to test for an influence from computer availability upon computing experience. A strong and significant relationship was discovered, suggesting that individuals in the study who enjoyed a greater availability of computers claimed higher levels of computing experience than those who did

*Figure 1: PLS Analysis for Hypotheses 7-13*

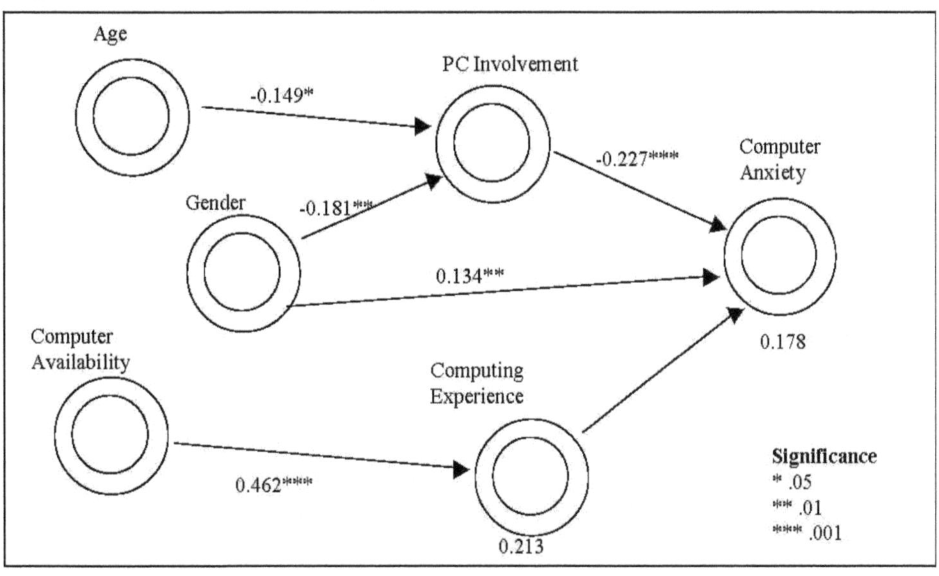

not. Intuitively, one would expect such a result, but the effect of experience in alleviating anxiety suggests that making computers available offers a practical first step towards ultimately reducing users' fears about using them.

## DISCUSSION

The results provide support for seven of the 16 hypotheses. Cultural background appears as a possible influence on both computer anxiety and PC involvement, as does gender, in support of hypotheses 1 to 4, but close examination reveals the interplay of other probable factors. The Tanzanian subjects displayed the most significant differences compared to the other groups on both psychological dimensions. The Thai and Hong Kong groups were the only other pair to exhibit significant differences on the computer anxiety dimension but the levels of PC involvement varied considerably among the groups. Next to the Tanzanians, the Hong Kong group varied the most from the others on PC involvement, notably so from the PRC group.

Differences with the Tanzanian group may be accounted for in part by their demographics. Their means suggest they have the lowest levels of anxiety and the highest levels of involvement of all the groups. However, they registered the lowest levels of both computer availability and computing experience, which presents a paradox as both factors overall appear to attenuate anxiety. However, age and gender induce involvement, thereby indirectly attenuating anxiety. The indirect influence of gender on computer anxiety is 0.041 (-0.181*-0.227), which increases its total influence to 0.175 (0.134+0.041). Although age did not have a significant influence on computer anxiety, there is a slight indirect influence (0.034) through PC involvement. It is probable that this combination of influences works most strongly for the Tanzanian group, which is the oldest and dominated by men.

The picture with the Hong Kong subjects seems to be the converse of that with the Tanzanians. The Hong Kong group is the second youngest, contains a high proportion of females and registered the highest mean level of anxiety and the lowest mean level of involvement. However, they also claimed the highest levels of both computer availability and computing experience. Whilst overall, experience tends to alleviate anxiety, it appears not to do so for the Hong Kong group, and again it is probable that age and gender exhibit the strongest influences for them.

Differences between the levels of computer anxiety among the groups are mainly exhibited by the Tanzanian group, and they appear to be attributable to demographic factors. Reasons for the differences between the groups with regard to their PC involvement are less clear. For the Tanzanians, they are probably also attributable to demographic factors, but the differences between the other five groups are not so easily explained as they are all quite similar demographically.

Overall, the PLS analysis confirms previous research which suggests that women are more likely to be anxious about using computers than men (Igbaria, 1990). However, caution is indicated with regard to this conclusion in that the study subjects were drawn from a variety of courses and levels of study. The analysis explained 17.8% of the variation in computer anxiety, with computing experience and PC involvement (inversely) having the strongest influence. The women in the study also appeared to be less involved with PCs than did the men. Differences between the subjects were sought along the dimension of their father's occupation as an indicator of the social and economic status of the subjects' families. The variable failed to differentiate between the subjects with regard to either their anxiety or their involvement. Age exhibited an influence on PC involvement in the older students who displayed a slightly stronger level of involvement than did the younger students. However, the PLS analysis explains a very small amount of the variation in PC involvement (7.1%), so there are clearly other factors at work which have not been identified. There was insufficient evidence to suggest that the older students were less anxious about computers than the younger students.

Computing experience did not appear to influence PC involvement, but it displayed a significant inverse relationship with computer anxiety, suggesting that students with more experience were less likely to exhibit anxiety about using computers. The availability of computers did not register any influence on either PC involvement or computer anxiety, but the PLS analysis revealed a strong, direct and highly significant influence from availability on computer experience. The model explains 21.3% of the variation in computing experience, which is accounted for by the single variable of availability. Finally, PC involvement displayed an inverse and significant influence on computer anxiety, suggesting that higher levels of involvement alleviate anxiety about using computers.

The inclusion of the Hong Kong students enabled a comparison of the results from a developed society with those from a less developed society. Contrary to expectations, the Hong Kong group were significantly more anxious than two other groups (Thailand and Tanzania) and were less involved than three others (Tanzania, Sarawak and China). The results seem to suggest that such differences are more easily accounted for from the demographic makeup of the groups than from either their cultural or economic background.

## CONCLUSIONS

The purpose of this study was to illustrate the existence of differences between the social, political, cultural and behavioural factors which form the contexts within which IT is used in development scenarios. Data was collected relating to a specific aspect of the behavioural components of computer use in order to illustrate one such difference. The descriptions of the political, economic and cultural backgrounds of the countries from which the study subjects were drawn demonstrate the differences between each of the contexts within which IS implementations take place. In attempting to illustrate a specific example of such differences, the study failed to find differences between the computer anxiety of different cultural groups which might be attributable to their culture, but revealed differences between the PC involvement of different cultural groups which might be so attributable.

It has been shown that computer anxiety can be alleviated by experience of computer use, which, not surprisingly, is induced by having a computer available. In practice, this would imply one PC per potential user for maximum availability, and the study provides empirical evidence for the benefit of this level of provision. However, the absence of an influence of availability on computer anxiety directly, testifies to the importance of PCs actually being used rather than just being provided, and in this respect, it is probable that any type of computer use serves to alleviate computer anxiety, so that computer games and trivial applications can pave the way for more advanced uses by otherwise reluctant users.

Contrary to expectations, the study was unable to expose any differences in the levels of anxiety and involvement which might be accounted for by the economic background of the subjects, as indicated by their father's occupation. However, the inverse relationship between the subjects' apparent predisposition towards computers and the economic status of their country was a surprising result, and presents a challenge for further research to establish the extent to which such a relationship exists among the wider populations and the reasons for it.

The study has some weaknesses. Firstly, a sample of students may not be considered as representative of the user population and generalisations beyond the sample should be made with caution. Other subjects might present different results, for example, if they had not enjoyed the availability of computers which most university students now enjoy. Additionally, the results might have been interpreted with greater confidence had the sample been larger.

The study serves to alert IS practitioners working on development projects to some of the inhibitors to IT adoption which they might encounter as they deal with a variety of cross-cultural interactions. Variations could exist in the rate and level at which different cultures become psychologically involved with PCs, which may inhibit their adoption. The study has identified influences from age and gender, but these factors only explain a small proportion of the variation in PC involvement. As there can be little control over them, it is necessary to identify other factors which are capable of inducing involvement, and which are attuned to the culture in which they are to be applied. Further research is required in order to identify factors that influence PC involvement, as well as finding culturally sensitive methods for inducing it.

The cultural, economic and political differences between the groups suggest that the context against which IS implementations take place differs widely from country to country. Moreover, the factors that may induce involvement create differences at several levels of analysis. First, they exist between the developed societies which create IT and the less-developed societies which use it. Secondly, such differences exist among the less-developed societies themselves. Thirdly, differences exist within both developed and less-developed societies between any culturally distinct groups which exist within them.

Accordingly, information systems in developing societies require two levels of adaptation. First, any assumptions about how IT is to be used, which might have become embedded as a result of being developed within a particular society or culture, usually a developed one, need to be removed. Secondly, information systems need to be adapted to the specific society or culture in which they are intended to be used, because developing countries vary greatly in terms of the social, political, and cultural factors which contribute to the success of IT. We argue that merely "de-culturalising" an IT solution for use in a development scenario denies the existence of enormous cultural variety among the developing societies and therefore risks its rejection if it is not "re-culturalised" for the society in which it is intended to be used.

## ENDNOTE

[1] We are indebted to several colleagues who assisted us with various aspects of data collection and information gathering in early stages of this study, namely: Zhou Duanning, Zhongshan University, Guangzhou, Peoples' Republic of China; Dr. Robert Barbour, University of Waikato, Hamilton, New Zealand; Dr Jack Schleich, Assumption University, Bangkok, Thailand; Dr. Dietrich Splettstoesser, University of Dar Es Salamm, Tanzania; Ada Wong, City University of Hong Kong; Alvin Yeo, Universiti Malay-

sia Sarawak, Malaysia. We also acknowledge the comments by the anonymous reviewers.

[2] MCA = Mean Computer Availability; MCE = Mean Computer Experience

## REFERENCES

Bagozzi, R.P. (1992). The Self-Regulation of Attitudes, Intentions, and Behavior, *Social Psychology Quarterly*, 55(2), 178-204.

Bloch, P.H. (1981). An Exploration into the Scaling of Consumers' Involvement with a Product Class, *Advances in Consumer Research*, 8, 61-65.

Bloch, P.H. (1986). The Product Enthusiast: Implications for Marketing Strategy, *The Journal of Consumer Marketing*, 3(3), 51-62.

Bond, M.H. and Wang, S.H. (1981). Aggressive Behavior in Chinese Society: The Problem of Maintaining Order and Harmony, *Acta Psychologica Taiwanica*, 23(1), 57-73.

Brookfield, H., Potter, L. and Byron, Y. (1995). *In Place of the Forest: Environmental and Socio-economic Transformation in Borneo and the Eastern Malay Peninsular*, The United Nations University, Tokyo.

Cohen, P.T. (1993). Order Under Heaven: Anthropology and the State, in: Evans, G. (Ed.)*Asia's Cultural Mosaic, An Anthropological Introduction,* Simon and Schuster (Asia), Singapore.

Connors, M., (1997). *The Race to the Intelligent State: Charting the Global Information Economy into the 21st Century.* Capstone Publishing , Oxford, United Kingdom.

Cronbach, L.J. (1951). Coefficient Alpha and the Internal Structure of Tests, Psychometrika, 16, 297-334.

Ekachai, S. (1994). *Seeds of Hope: Local Initiatives in Thailand,* Thai Development Support Committee, Bangkok, Thailand: Post Publishing.

Galilava, R. (1996). The Strategic Usefulness of MIS as Perceived by Top Managers in Tanzania, Unpublished MBA dissertation, University of Dar es Salaam.

Galliers R.D., Madon S. and Rashid R. (1994). Information Systems Implementation and Cultural Change in Developing Countries: Linking Theory and Practice, *SEARCC '94, South East Asian Regional Computer Confederation Conference Proceedings,* 93-108, Karachi, Pakistan, November.

Geertz, C. (1973). *The Interpretation of Cultures,* Basic Books: New York.

Hamelink, C.J., (1997). New Information and Communication Technologies, Social Development and Cultural Change, United Nations Research Institute for Social Development, Discussion Paper No. 86, http://www.unrisd.org/engindex/publ/list/dp/dp86/toc.htm#TopOfPage

Hanna, N. (1993). Information Technology in World Bank Lending, World Bank Discussion Paper 206, The World Bank, Washington, U.S.A.

Harris, R.W. (1995). Success with End-User Computing; A Behavioural Perspective for Development, Conference *Proceedings; IT Dev. '95,* Information Technology for Development, 159-167, University of Witwatersrand, Johannesburg, South Africa, July 6-7.

Hofstede, G. (1985). The Interaction Between National and Organisational Value Systems, *Journal of Management Studies,* 22(4), 347-357.

Igbaria, M. (1990). End-User Computing Effectiveness: A Structural Equation Model, *OMEGA: International Journal of Management Science,* 18(6), 637-652.

Igbaria, M. and Nachman, S.A. (1990). Correlates of End User Satisfaction with End User Computing: An Exploratory Study, *Information and Management,* 19(2), 73-82.

McC. Adams, R., (1996). *Paths of Fire: An Anthropologist's Inquiry into Western Technology*, New Jersey: Princeton University Press.

Mgaya, K. (1994). Development of Information Technology in Tanzania, in: Drew, E.P. and Foster, F.G. (Eds) *Information Technology in Selected Countries,* United Nations University, Tokyo, Japan.

Mohamad, M. (1991). Malaysia: The Way Forward, Working Paper presented at the Inaugural Meeting of the Malaysian Business Council, February 28.

Mueller, R.O. (1996). *Basic Principles of Structural Equation Modelling: An Introduction to LISREL and EQS,* New York: Springer-Verlag.

Nelson, D.L. (1990). Individual Adjustment to Information-Driven Technologies: A Critical Review, *MIS Quarterly,* 14(1), 78-98.

NITC-National Information Technology Committee (1995). Social Equity & Prosperity: Thailand IT Policy into the 21st Century, (http://www.nitc.go.th/itplan/itplane.html).

Norusis/SPSS (1993). *SPSS for Windows Base System User's Guide Release 6.0,* SPSS Inc. Chicago.

Panko, R.R. (1988). *End User Computing: Management, Applications and Technology,* John Wiley & Sons, New York.

Paré, G. and Elam, J.J. (1995). Discretionary Use of Personal Computers by Knowledge Workers: Testing of a Social Psychology Theoretical Model, *Behaviour and Information Technology*, 14(4), 215-228.

Ponniah, P. (1996). Telecommunications Industry Sector Analysis: Malaysia. U.S. Department of Commerce, *Internal Trade Administration News Release,* Washington, D.C. 20230, U.S.A., 7th October.

Pulos, S. and Rogness, N. (1995). Soft Modelling and Special Education, *Remedial and Special Education*, May, 16(3), 184-192.

Raykov, T., Tomer, A. and Nesselroade, J.R. (1991). Reporting Structural Equation Modelling in Psychology and Ageing: Some Proposed Guidelines, *Psychology and Aging*, 6(4), 499-503.

Splettstoesser, D. (1996). Development Decision Centers: A Strategy to Improve Development Decision-Making, in: Glasson, B.C., Vogel, D.R. and Bots, P.W.G. Global Inc: An Experiment across Time and Space, *Proceedings of the 2nd International Office of the Future Conference*, IFIP WG 8.4, Tucson, Arizona, April 8-11, 299-315.

Thompson, R.L., Higgins, C.A. and Howell, J.M. (1991). Personal Computing: Toward a Conceptual Model of Utilization, *MIS Quarterly*, 15(1), 125-143.

Torkzadeh, GT., and Angulo, IE., (1992). The concept and correlates of computer anxiety, *Behaviour and Information Technology*, 11(2), 99-108.

Wong, S. (1996). *All Set for the 21st Century*, Star Publications (Malaysia) Bhd, May 7th.

# APPENDIX - THE QUESTIONNAIRE

1. Demographic Questions
1.1 How old are you (to the nearest year) ?
1.2 What gender are you ? Male; Female.
1.3 How much experience of using computers have you had?
1=None, 2=Not much, 3=Some, 4=Quite a lot, 5=A lot.
1.4 When you need to use a computer, is there one available:
1=Never, 2=Not Often, 3=Sometimes, 4=Often, 5= Always ?
1.5 Which of the following categories best describes the occupation of your father ?
1=Manual, 2=Clerical, 3=Professional, 4=Technical, 5=Trade/Business, 6=Other ?

2. Product Involvement
2.1. When I am with friends, we often talk about personal computers.
2.2. I regularly read magazines and newspaper articles about personal computers.
2.3. I am confident about understanding most of what I read or hear about personal computers.
2.4. I would like to own a personal computer of my own.
2.5. I believe it is necessary to keep my knowledge about personal computing up to date.
2.6. I believe personal computers are important in today's world.
2.7. I enjoy using a personal computer.
2.8. I would like to use a PC for more of my regular tasks.
2.9. I would watch a TV programme about new developments in personal computing.
2.10. I often take a close look at PCs, peripherals and software in shops and/or catalogues.

3. Computer Anxiety
3.1. I am not at all confident that I could learn computer skills.
3.2. I am unsure of my ability to learn a computer programming language.
3.3. I will not be able to keep up with important technological advances in computers.
3.4. I feel apprehensive about using a computer.
3.5. If given the opportunity to use a computer, I am afraid that I might damage it in some way.
3.6. I have avoided computers because they are unfamiliar to me.
3.7. I hesitate to use a computer for fear of making mistakes that I cannot correct.
3.8. I am unsure of my ability to interpret a computer printout.
3.9. I have difficulty understanding most technological matters.
3.10. Computer terminology sounds like confusing jargon to me.

## Chapter 18

# The Context of Introducing IT/IS-Based Innovation into Local Government in Colombia

Mahmoud M. Watad
Fairleigh Dickinson University, USA

Information technologies (IT) and information systems (IS) provide opportunities for organizations to improve their productivity and effectiveness. Considerable research has been conducted to study the factors and determinants of the adoption and diffusion of IT/IS into organizations' operations (a sampling: Kwon & Zmud, 1987; Markus & Robey, 1988; Borton & Branchau, 1994; Nidumolu et al., 1996). Most studies support the notion that managerial decisions of IT/IS adoption, and the outcomes of IT/IS introduction in an organization are moderated by the context in which these technologies are introduced (Zmud, 1982; Fichman and Kemere; 1993, Watad and Ospina, 1996). This article adds to this line of research and draws upon theories from the technological innovation literature to explore the contextual factors that shape organizational decisions to adopt IT/IS into Latin American municipalities.

A review of the published literature shows that there has been very little research on the determinants of IT adoption in local government (municipalities), in Latin America. Most studies have tended to focus on the national level and have dealt with transfer and implementation of information technologies or with technology policies (Robey and Rodriquez-Diaz, 1996; Montealegre and Applegate 1994; Correa, 1995; Pers-Nunez, 1994). An-

Previously Published in the *Journal of Global Information Management, vol.7, no.1,* 

other line of research has been growing which focuses on technical applications development, such as databases tools and software engineering (see proceedings, SCI 97).

Overall, this paper provides a model expounding the contextual factors that facilitate the process of adopting IT/IS-based innovation in local government. There are two main constructs in this study: the adoption decision of an innovation and the context of its adoption. This study analyzes adoption cases in which the decision to adopt was positive. While context itself is a difficult construct to operationalize, it is possible to study particular cases under which decisions are made and analyze the contextual factors that affected the dynamic of the adoption decision.

For the purpose of this study, the contextual factors were divided into two groups: external (or environmental such as public dissatisfaction with quality of service) and internal (or organizational such as technological awareness). Furthermore, the study focused only on the triggers rather than on the obstacles (such as innovation cost and complexity), or the challenges (organizational resistance) that hinder the introduction of IT/IS into organizations.

The analytical strategy to identify and explore the relevancy of these contextual factors consists of analyzing two cases of introducing IT/IS projects into two different cities. These projects have in common the introduction of the same type of technology (geographical information systems) and the same type of managerial judgment and justification (the rationale, objectives, and expected benefits were similar). For example, the rationale of introducing IT/IS in both cities was to improve organizational operations. Assuming then that rationale, expected benefits, and technology were relatively similar, the study analyzes the role of contextual factors in triggering the decision to adopt these technologies.

The importance of this study is twofold. First, by identifying the factors that trigger the introduction of IT/IS, the study will provide insights for policy makers to adjust or devise new policies. For example, findings of this study highlight the importance of technological awareness and the availability of external help in facilitating the introduction of technological innovation into organizations. The national government may establish programs which aim at rewarding cities that foster technological awareness, and a willingness to share their experiences. Second, the principles and strategies of effective IT/IS-based innovation are evolving from the experiences of practicing managers. Current experiences need interpretation to provide guidance for new managers (including top management) to help them interact effectively with their environment. The study used in-depth interviews with practicing managers to explore and establish relevancy in the factors that they felt were important to stimulate the introduction of new technologies.

The rest of the paper is structured as follows. The next section contains the theoretical framework followed by the setting and the research methodology; the contextual factors that triggered the adoption decision; and finally, the conclusions and discuses the implications.

## THE CHALLENGE OF INNOVATION AND ORGANIZATIONAL ADOPTION OF IT/IS

This study focuses on the factors that facilitated the adoption of geographical information systems[1] (GIS) as IT/IS-based innovations. GIS introduction is a complex process and it has the potential to radically change organizational operations. Since GIS was adopted for the first time by the organizations studied, it will be considered an innovation. This section will focus first on organizational innovation and then on the challenges of introducing GIS. The research questions of the study are presented at the end of the section.

### Organizational Innovation

Innovation can be defined as inventions or adaptations that result in new ways of doing things. An innovation is an idea, product, practice, or service that is perceived as new by the organization or the unit of adoption (Rogers, 1983). Innovation may also be defined as the process that allows the invention (new idea or new system) to be used in a particular context. Innovation takes place through three overlapping stages: invention, innovation (implementation), and diffusion (Dosi, 1988). The successful introduction of IT/IS into organizational processes constitutes innovation. The widespread and the general acceptance of the IT/IS into organizational operations constitute diffusion. In the context of this paper, since GIS was a new system to the organizations that decided to adopt it; therefore, its development and successful implementation constitutes innovation, and its wide use constitutes diffusion.

In general, an innovation is analyzed along several dimensions (Poutsma et al., 1987). These are: process vs. product, radical vs. incremental, technology-push vs. market pull, and planned vs. incidental innovations. The adoption and implementation of IT/IS into organizational operations would be considered a process-based innovation. The rest of the dimensions are applicable for the adoption of IT. For example, IT may be introduced for streamlining purposes, this will qualify as incremental intervention (enabling minor improvements), while in other cases, it may be introduced as a radical intervention (enabling structural shift in organizational operations).

The innovation literature has identified many factors that may affect an organization decision to adopt an innovation (Gopalakrishnan and Dmanpour, 1997 and 1994; Lai and Guynes, 1994; Fichman,1992; Rogers, 1983). These factors involve political, social, organizational, managerial, and technological characteristics. For example, technological characteristics of an innovation such as relative advantage, complexity, availability, and compatibility; and managerial characteristics such as the existence of local champions, executive commitment, technological awareness; and organizational characteristics such as size structure, and openness; have an affect, to a certain degree, on the adoption decision. Researchers try to isolate specific determinants of innovations. Yet the multi-phase nature of the innovation process adds to the complexity of the problem. For example, organizational conditions that may facilitate idea generation may conflict with the forces that facilitate implementation (Zaltman, et al., 1973).

In summary, there is no unifying theory of innovation adoption, and most researchers agree that innovation adoption and diffusion should be examined in the context of its introduction (Zmud, 1982; Fichman and Kemere; 1993, Watad and Ospina, 1996). Therefore, this research adopts an interpretive approach and examines the adoption of IT\IS-based innovation through the perspective of the participants using their own interpretations of the activities that led to their decisions (Orlikowski and Baroudi, 1991). This approach is useful in studying IS introduction in organizations, primarily due to the influence of the context on the outcome of the adoption.

## The Challenge for Introducing GIS

The pressure for improving performance has prompted local governments to find new ways to conduct their business (Observe and Gambler, 1992; Nedovic and Godshalk, 1996). GISs are useful for local government because of their strategic applications and benefits. These systems provide summarized information to assist in policy making and long term planning. They are instrumental in optimizing resource allocation processes, unifying municipal information, standardizing criteria used for analysis, improving municipal development planning processes, and in providing citizens with accurate data. In general, GIS is helpful in improving policy design, communication, and decision making processes (Somers, 1987; Brown and Brudney, 1993).

Cost and complexity are the two major obstacles to the initial introduction of GIS. The introduction of GIS is very costly, and its purchase and implementation require large investments (Lang, 1990; Newcombe, 1993; Dataquest, Inc., 1994). Moreover, developing GIS requires careful planning and preparation[2]. Given these obstacles, managerial activities aimed at managing the process of introducing IT/IS are crucial for its success. Furthermore, managers face several challenges when they introduce new technologies. First, managers don't have effective tools to measure the benefits (value) of the systems they are introducing. Second, often the unintended outcomes of introducing an IS project exceed its planned objectives. Managers reactions are specifically crucial, when outcomes are negative. Third, they have to respond to competing demands and often are required to manage conflicting objectives (Rainey, 1991).

Due to political scrutiny and the high level of uncertainty of the outcomes of IT/IS-based innovation, public managers tend to be conservative in their efforts to innovate with information technology applications. They tend to react to external pressure more than to initiate or take advantage of opportunities.

Building on the previous discussion of organizational innovation and the challenges of adopting it, the study, therefore, is guided by the following questions:

- What are the factors that helped decision-makers overcome the obstacles (cost, complexity) and the challenges (competing demands, measuring benefits) of introducing new IT/IS?
- Are managerial decisions to adopt an innovation shaped only by internal factors?
- What role does the external environment play in shaping managerial decisions in introducing new (to the organization) IT/IS into their operations?

The underlying logic of these research questions is the proposition that innovation depends on both external and internal factors. Internal factors are not sufficient to trigger the adoption decision, and managers tend to react to external factors more than to internal ones.

## RESEARCH METHODOLOGY AND SETTING

This study is exploratory in nature. The research uses the case study approach. Its aim is to contribute to a broader understanding of the introduction processes of IT. For this purpose, case studies are highly appropriate (Zmud, et al., 1989). According to Yin (1989) case studies are structured and planned to aim at a specific objective. Although no explicit hypotheses were stated, the study is based on a clearly defined research question (see Carstensen et al., 1995, Kautz, 1996). The disadvantage of a case study approach is the generalizability of the results. However, the potential of the results of a case study to contribute to the theories of the field (and to practitioners and policy makers) outweigh the disadvantages. The case study approach was chosen because of the contextual nature of the IT adoption process (Fichman and Kemere, 1993).

The principal methods for collecting data consisted of direct interviews with city and public officials, reviewing documents and city publications, and a direct examination of the systems that have been introduced. The material and documents were collected between February and August 1997. The documents provided information related to the background, structure, and policies of each city; and information related to the characteristics of the GIS projects such as objectives, rationale, hardware, and software. The interviews provided information regarding the contextual factors and the dynamic that led to GIS adoption. Preliminary interviews took place in February 1997 and final interviews took place in August 1997. There were 14 interviews with public officials, city managers, planners and systems developers, and users. The interviews focused on the following question: what are the factors that triggered the municipality's decision to introduce GIS? Managers were asked to answer semi-structured questions about their perception of the factors that triggered the adoption decision.

The cities researched were Cali and Palmira. Cali and Palmira are located in the southwest region of Colombia. Cali's population is approximately 2 million and Palmira's is 300,000. In 1985, there was one personal computer (PC) per 300 employees in Cali's city government. Currently, there is approximately one computer per seven employees, and in some departments such as the planning department, the ratio is one to one. More than one thousand employees (15% of total workforce) received formal training in word-processing, spreadsheets, and Internet applications. In 1987, there were very few computers in the city offices of Palmira. Currently, there are approximately 150 PCs used by 600 employees. In the planning department, for example, there are now 11 computers compared to zero ten years ago. Most workers have attended training sessions in the use of PC technology and software applications.

The process of introducing IT in local government begins with an employee generating an idea. He or she then communicates it to a decision-maker. In most cases, the decision-maker is the immediate supervisor or the director of the department. An adhoc technical team is formed to evaluate the merit of the idea and develop a project proposal (contingent on the director's approval). Upon completion, the proposal is presented in front of El Concejo Municipal and the mayor for approval and resources. In general, the closer the

*Figure 1: Contextual Factors and Adoption Decision*

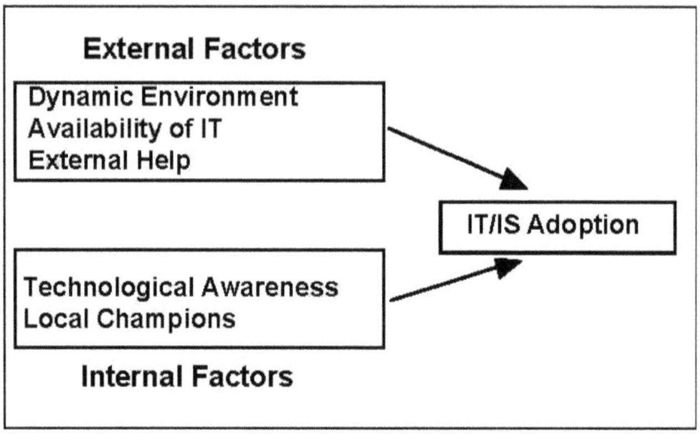


generator of the idea to the decision-maker (or to the mayor) the better the chance for the idea to be approved. It is important to note that a major part of the dynamics of innovation is to convince the director of the department to buy into the project/idea.

It took Cali almost four years and Palmira three years to initiate and implement the main phases of the GIS project. The main phases were completed in 1996 and 1997 respectively. Cali was identified to have a richer experience with IT innovations. Cali introduced GIS before Palmira. It has a larger number of IS personnel, and several entities that deal with IT/IS issues. The initial model of the adoption of IT/IS-based innovation was based on Cali's experience with the introduction of GIS. Palmira, on the other hand. was chosen for validating and refining the model. The model (Figure 1) contains one dependent variable which is the decision to adopt IT\IS, and five contextual variables (two internal and three external). These variables are discussed in the following section.

## THE CONTEXT OF INTRODUCING IT\IS INTO LOCAL GOVERNMENT IN COLOMBIA

Innovation and adoption literature provide many contextual factors that are essential in facilitating the adoption of technological innovation. This study revealed that there are specific internal and external factors that have shaped the context of introducing IT/IS into local government, in Colombia. Internal factors are the level of technological awareness (TA), and the existence of local champions (LC). External factors are: dynamic environment (public dissatisfaction with the quality of service and a new political environment), the availability of IT as a resource, and the availability of external professional help.

## Internal Factors

In general, technology awareness and champions are very important factors for the successful adoption of the IT/IS-based innovation (Huff and Munro, 1985), because, they help unlock obstacles and reduce resistance to change[3].

***Local champions***. According to Curley and Gremillion (1983) champions are those "who make decisive contributions to innovation by actively and enthusiastically promoting its progress through the critical stages of its development and adoption." In both cities (Cali and Palmira), the champions for innovation were middle-level managers and professionals. They have different jobs and skills but share their commitment and drive for finding creative solution to problems facing their organization. These local champions have to convince policy makers to allocate resources for IT/IS-based solutions.

In managing the procurement process, every department has to submit a proposal in order to purchase computers. These proposals are reviewed and authorized by the planning department. It has been difficult to convince the mayor to invest in computers. The director of planning has to pressure the mayor to allocate additional resources for computer purchases. Here the local champions are the driving force between the mayor and technical team or project personnel. Glaser (1994) in a major study on industry transformation through IT, indicates that there are three roles that are key in achieving dominant design: a top-level technology champion; the maestro, a dynamic person with balanced business and technical competence; and the technical team, a group of professionals dedicated to their company. In the case of local governments' adoption of IT, the champions fit Glaser's maestro profile.

In Palmira, the move for privatization required accurate data and information in order to plan strategically. This new environment placed pressure on the planing department to provide answers to new questions. Traditionally, planning bureaus have not been dedicated to planning, but rather to physical control. They were not prepared to handle the new demands for gathering, processing, and distribution of information that was needed to support privatization. The planning department formed the Consejo Municipal de Estadistica (COME)[4]. The director of planning was the main catalyst for COME formation. COME was created with a mandate to collect information and produce reports on privatization issues to the city counsel and the mayor. Local talent and the technical people of the planning department designed software applications to support their efforts of strategic planning. The initial product of their efforts was the development of a geographical information system (GIS).

***Technological awareness.*** Technology awareness by managers determines the relative timing of IT introduction compared with other organizations (Cali introduced GIS before Palmira). Usually high awareness corresponds with high tolerance to uncertainty. Managers with higher awareness tend to try new ideas. Technological awareness and IT knowledge affect the attitude toward IT adoption. Formulating of positive or negative attitude toward an innovation takes place before a decision to adopt (Rogers, 1983). Public managers develop technology awareness by subscribing to professional magazines, attending seminars, visiting other local agencies, and forming relationships with other cities.

In Cali, there are two main entities in the city government that are directly involved in shaping the information infrastructure: the Direccion de Informatica y Telematica (DIT), and the Departamento Administrativo de Planeacion (DAP, 1996). DIT's mission is to unify the municipal information policies and methods to efficiently utilize the municipal network. DIT's responsibility is to design policies and guidelines to ensure compatibility regarding the use of software and hardware. It also advises Central Administration offices and departments on computing and communications issues. Currently, DIT is streamlining the standards for hardware and software throughout city agencies.

DAP has two groups that are involved in IT/IS issues. The telecommunication group within DAP is responsible for promoting scientific and technological research to develop the telecommunication sector in the city. The IS group is responsible for the gathering, processing, and organizing of the municipal information and statistics. For example, the IS group has to develop and coordinate studies that allow the municipal government to forecast population growth. The proposed, developed and successfully implemented GIS was a direct result of the IS group efforts.

## External Factors

External factors, whether they are threats or opportunities, are catalyst for innovation. Public dissatisfaction with the quality of services is a major threat and the availability of IT is a major opportunity. In either case, public managers must respond promptly to environmental changes. The following is a brief description of the factors that facilitated the introduction of IT/IS-based innovation into municipal organizations.

***Dynamic environment: public dissatisfaction and a new political reform.*** In general, an unstable or turbulent environment is a major catalyst for organizational change (Drucker, 1978; Sheldon, 1984). Furthermore, literature supports the notion that substandard performance is a catalyst to stimulate the decision to adopt innovation (Bolton, 1994). In both cities, the infrastructure and services had deteriorated to a point that the public was dissatisfied with governmental policies and solutions. The low quality of service was evident in almost every facet of public service. City managers were aware of these problems and were trying to respond in various ways. For example, Cali is currently decentralizing its service delivery system, and Palmira is in the process of completing the privatization of most of its public services[5].

These initiatives were facilitated by national government reform. In 1991, the Colombian national government issued new laws that decentralized some of its functions. This pushed the locus of authority downward. The laws increased local government control over planning and spending. This new political environment has also allowed cities to consider privatizing some of their public services. The privatization efforts and the desire to improve the quality of service demanded information, which in turn, mobilized cities to introduce GIS.

***The availability of IT as a resource.*** IT resources including software packages, processing technologies, input/output devices, and telecommunications are affordable (IT as a resource is getting cheaper; however, developing and maintaining an IS as a solution is still costly). The rapid advancement of these resources gives managers challenges as well as new tools to conduct their businesses. Managers can build different information infrastructures to deliver organizational services. They can choose either a centralized model, client/server model, and/or an Internet/Intranet model (Tabke, 1996). Cali is improving its information infrastructure to accommodate Internet applications. However, this flexibility is accompanied with an increased level of complexity, which in turn forces managers to increase their level of understanding of the new applications of IT.

Managers must strike a balance between purchasing the software packages or developing them. Findings of this study indicate that on the departmental level (mid-level managers), decisions regarding the type of IT to be purchased were based on the merit of the technology itself. The municipal level (executive) procurement decisions are often based on social networks and personal relationships. It is important to note that there is no formal policy (or legal constraints) that restricts managers on issues related to technology procurement, nor is there a common pool of vendors that managers must use. Therefore, public officials can purchase any type of IT from the appropriate sources.

***The availability of external professional help.*** External assistance was obtained from sister cities, from the national planning department, and from local nonprofit organizations such as the Instituto SER de Investgacion (SER)6. The information systems group at SER developed Sistema de Gernecia Municipal (SGM). SGM supports general municipal administration functions. The national planning department has also sponsored several country-wide IS projects to help cities manage and monitor local projects. For example, the Palmira designers of GIS needed to digitize the city maps. It was estimated that outsourcing this function would cost two million U.S. dollars. The planners consulted with the Instituto Geografico Agustin Codazzi (Bogota). They were able to complete this task for less than $200,000.

Palmira developed a "sister" relationship with a French city. This was made possible through the Colombian Federation of Municipalities. Palmira's personnel went to France several times to learn and examine IT-based innovations. The French experts from the municipal government of Saint-Nazaire visited Palmira and advised its planners on implementation issues. The mechanism of "sister city" proved to be instrumental in helping Palmira to innovate its information management processes. However, the sister city arrangement should be viewed as a mentoring system rather than a strategic partnership.

## THE DYNAMIC OF INNOVATION: DISCUSSION AND CONCLUSIONS

There are several conditions that must exist in order to spark and facilitate the innovation process. Conceptualizing IT/IS-based innovation as an open system model provides a simple tool to explain the dynamics of the innovation process. In this model, the output is the information system and the inputs are the strengths (the existence of champions), weaknesses (lack of technological awareness) opportunities (new technology), and threats (public dissatisfaction) that affect the organization. There is also a feedback loop that informs decision-makers on the usefulness of the information system in reducing threats and weaknesses or enhancing strengths and creating new opportunities to the organization. Opportunities and threats are usually external factors and strengths and weaknesses are internal.

This paper concludes that catalysts for innovation (initiation stage) in these local governments were largely external. They were a mixture of threats and opportunities (public dissatisfaction, national government, IT availability). However, internal factors such as local champions and a high level of technological awareness are keys for successful adoption. They are essential in leading the organization to IT-based innovation and technology-driven solutions. Local champions were instrumental in mobilizing internal resources and reducing organizational resistance. Champions form a bridge between the external and the internal environment (Palmira created COME as a mechanism of partnership) and are helpful in closing the gap between opportunities and problems.

Innovations in Palmira and Cali were realized as a direct response to organizational needs. However, the process and stages that led to the adoption decision of GIS as an organizational solution evolved differently. While the same factors appear to influence managerial decisions to adopt an innovation in both cities, their weight and importance varied between the two cities. In Palmira, the process was facilitated by local champions and they helped align external forces (environmental demands and resources) with internal dynamics (top city officials commitment). In Cali, the process was facilitated by technology awareness. This is a reflection of the higher level of maturity and computing skills among Cali's personnel as opposed to those of Palmira's. The innovation dynamic in Cali could be looked at as an outcome of an incremental progression of computing usage. In this sense, GIS introduction was the highlight of an evolutionary process of using IT. In Palmira, on the other hand, the innovation dynamic was revolutionary. In summary, cities are more likely to adopt IT when they have more technological awareness, local champions, and the environment is more dynamic (new political environment, public dissatisfaction) than stable.

### Internal Implications

Findings show that middle-level managers in both cities were willing to take a measured risk in adopting new technology and innovations. Top management seems to be relaxed or detached from the process. This is a double-edged sword. On the one hand, it gives the champions a wider range for experimentation and mistakes. On the other, the champions have to handle organizational resistance without executive support. It seems that innovations in these local governments originated from middle-managers. This is similar to Japanese organizations which also rely on innovations from middle managers. The key to sustained innovation lies in creating and spreading knowledge among middle managers (Nonaka and Takeuchi, 1995). This is a very important insight and has serious organizational implications. Therefore, fostering mechanisms to enable knowledge sharing among middle-level managers and rewarding those that share their knowledge should be a very high priority for top managers and policy makers.

### External Implications

In dealing with innovations and the adoption of new ways of conducting work, public managers can benefit from the experiences of other public agencies. It is very important to cooperate across the boundaries of their organizations (Cali personnel studied other cities experiences). They can use external resources available through professional, nonprofit organizations (Palmira relied on sister city support and national government). This insight has national-level policy implications. Since innovation is a social process, national politics and reforms (national government intervention) does matter. It is important that the national government initiate policies to encourage: a) building partnerships between local and national government, and b) building partnerships among cities and improving inter-organizational linkages. The national government must continually adjust its policies in favor of supporting local government innovative initiatives.

## ACKNOWLEDGMENTS

The author wishes to thank professors Edgar Varela Barrios and Leonardo Solarte Pazos de la Universidad del Valle, SER officials, professor Sonia Ospina of NYU, the faculty at the college of business of FDU, and Ms. Maria Kocylowsky, head of the business research library at FDU.

## ENDNOTES

[1] GIS and mapping programs range from simple to complex. With a user friendly interface and powerful search-and-retrieve features, GIS summarizes and presents large amounts of information by overlaying data onto maps. Car manufacturers and rental companies are mounting them in cars, and vehicle information and communication systems (VICS) have become widespread in Japan (Pollack, 1997). Businesses use GIS to plan delivery and distribution routes and to plan expansion into new regions (Forrest, 1993; Tweed, 1994). GIS is used in solving problems such as site selection, zone planning, and crime prevention. Police departments use GIS to pinpoint high crime areas to better deploy their resources.

[2] GIS usually have four layers: 1) the required demographic data 2) database management software (DBMS) to manipulate, search and retrieve data, 3) digital maps, and 4) system interface tool such as Mapping software.

[3] Among the obstacles that hinder the innovation process are: limited resources, the lack of computing skills among policy makers, and the favor of the status quo by organizations' members (inertia).

[4] COME (Municipal Committee of Statistics) was formed as a community based committee. The organizers believed that involving the public would facilitate the change process. Fifty percent of its members were government officials. The Planning Department gained recognition through these efforts. Before privatization, it was not well recognized or respected and had a poor public image.

[5] Palmira completed the privatization of the telephone services in six months. In 1995, Cali approved La Reforma Administrative that was aimed at restructuring the municipal administration and decentralizing city services (DAP, 1996). The decentralization of public services is aimed at dividing the city into 20 district service centers, thereby bringing services closer to the residents' homes.

[6] SER is an independent, nonprofit organization, with no political links, dedicated to investigate social development programs.

## REFERENCES


Brown, M. and Brudney, J. (1993). "Modes of Geographic Information System Adoption in Pubic Organizations: Examining the Effects of Different Implementation Structures. "Paper presented at the annual meeting of the American Society for Public Administration. San Francisco, CA, July 17-21.

Carstensen, P., Sorensen, T. and Tuikka T. (1995). "Let's talk about bugs," *Scandinavian Journal of Information Systems*, 7(1), 33-53.

Corra C. M., (1995). "Innovation and technology transfer in Latin America: a review of recent trends and policies," *International Journal of Technology Management,* 10(7), 815.

Curley, K.F. and Gremillion, L.L. (1983). "The role of the Champions in DSS Implementation", *Information and Management,* 6, 203-209.

Damanpour, F. and Evan, W.M. (1984). "Organizational Innovation and Performance: The Problem of 'Organizational Lag'," *Administrative Science Quartile,* v 9, 392-409.

DAP (1996) Cali en cifras.

Dataquest, Inc. (1994). "Government Market Leads Software Revenue." *GIS World,* 7 (5), 12.

Dosi, G. (1988) "The nature of the innovative process," in G. Dosi et al., *Technical Change and Economic Theory,* New York: Pinter Publishers, 221-238.

Drucker, P.F. (1978). *The age of discontinuity.* New York: Harper & Row.

Fichman R.G. (1992). Information Technology Diffusion: a review of empirical research, in *Proceedings of the 13th International Conference on Information Systems*, Dallas, TX,195-206.

Fichman R.G. and Kemerer C.F. (1993). Toward a theory of the adoption and diffusion of software process innovations, in *Proceedings of IFIP Conference on diffusion, Transfer, & Implementation of IT* (ed. L. Levine) Pittsburgh, 21-30.

Forrest, D. (1993). "Data in New Ways," *Computerworld.*

Gopalakrishman S. and Damanpour, F. (1994). "Patterns of generation and adoption of innovation: contingency models of innovation attributes," *Journal of Engineering and Technology Management,* 11, 95-116.

Gopalakrishman S. and Damanpour, F. (1997), "A review of innovation research in economics, sociology and technology management," *Omega,* 25(10.

Huff, S.L. and Muro, M.C. (1985). "Information technology assessment and adoption: A field study," *MIS Quarterly,* December, 327-340.

Kautz K. (1996). Information technology transfer and implementation: The introduction of an electronic mail system in a public service organization" in *Diffusion and Adoption of Information Technology.* Edited by Kautz K. and pries-Heje J. Chapman & Hall

Lai V.S. and Guynes J.L. (1994). A model of ISDN Adoption in U.S. Corporations, *Information & Management,* 26(2).

Lang, L.(1990) "Making a GIS Dream Come True." *Planning,* 56 (7), 14-20.

Montealegre, M. and Applegate, M. (1994). Information technology & organization change: lessons from a less-developed country. In Levine, L. (ed.), *Proceedings of the IFIP TC8.* Pittsburgh, PA, Oct 93, p 99-131.

Nedovic-Budic, Z; and Godschalk, D. (1996) "Human factors in adoption of geographic information systems: a local government case study." *Public Administration Review*, v 56, November/December, 554-67.

Newcombe, T. (1993) "GIS: Beyond Pipes and Plats." *Governing* (December), 25-26, 1993.

Nonaka I. and Takeuchi H. (1995). *The knowledge-creating company: how Japanese companies create the dynamics of innovation.* New York: Oxford University Press.

Osborne, D. and Gaebler, T. (1992). *Reinventing Government.* Reading, MA: Addison-Wesley.

Peres-Nunez, W. "Latin American experience with technology policies: current situation and prospects," *International Journal of Technology Management*, 3(3-4), 332.

Pollack, A. (1997). "Computers let Japan's drivers play in traffic", *NYT,* October 6.

Poutsma E.F., Van Uxem F.W. and Walravens A.H. (1987), *Process Innovation and Automation in Small and Medium Sized Business.* Delft University Press, The Netherlands.

Rainy, H. (1991. *Understanding and Managing Public Organizations,* San Francisco, CA: Jossey-Bass.

Richter, M. (1996). Technology acquisition and implementation: Learning the hard way. *Government Finance Review*,12(1), 7-11.

Robey, D. and Rodriguez-Diaz, A. (1996) "The organizational and cultural context of systems implementation: case experience from Latin America. In Deans C. and Jurison J. (eds.) *Information Technology in a Global Business Environment,* boyed & fraser.

Rogers, E.M. (1983), *The diffusion of Innovation,* 3rd edition, The Free Press, New York.

SCI 97, (1997) World Multiconference on Systemics, Cybernetics and Informatics, *Proceedings,* Caracas, Venezuela.

Sheldon A. (1984). "Organizational paradigms: a theory of organizational change." *Organizational Dynamics,* Winter.

Somers, R. (1991). "GIS in U.S. Local Government." *Cities,* 8 (1), 25-32.

Tabke, B. (1996). "Intranets Boost Productivity." *Net News.* Issue 2, 12.

Tweed, V. (1994). "The Graphic Detail," *Information Week,*" August 29.

Watad M. and Ospina S. (1996). Information technology and organizational change: The role of context in moderating change enabled by technology " in *Diffusion and Adoption of Information Technology.* Edited by Kautz K. and pries-Heje J. Chapman & Hall.

Yin, R. (1989). Research design issues in using the case study method to study MIS. In Cash, J. and Lawrence P. (eds.) *The information systems research challenge: Qualitative research methods.* Vol. 1. Harvard School research Colloquium, 1-6.

Zaltman G., Duncan R. and Holbek J. (1973) *Innovations and Organizations.* Wiley, New York.

Zmud R. W. (1982) Diffusion of Modern software Practices: Influences of centralization and normalization. *Management Science,* 28 (12), 1421-1431.

Zmud, R. Olson, M. and Hauser, R. (1989). Field experimentation in MIS research. In Benbasat, I., ed.' *The information systems research challenge: Experimental research methods,* Vol. 2 Harvard Business School Research Colloquium, 97-112.

## Chapter 19

# The Impact of Information Sharing on Order Fulfillment in Divergent Differentiation Supply Chains

Troy J. Strader
Iowa State University, USA

Fu-Ren Lin
National Sun Yat-sen University, Taiwan, R.O.C.

Michael J. Shaw
University of Illinois at Urbana-Champaign, USA

*Current information technology (IT), such as extended intranets (extranets) utilizing the components of the World Wide Web (WWW), makes information sharing between global supply chain partners feasible. Given this, an important issue is: What impact does efficient information sharing have on supply chain order fulfillment performance (specifically cycle time and inventory costs)? In this study, we focus on divergent differentiation supply chains (commonly associated with industries such as fashion apparel). We identify the impact of information sharing by simulating order fulfillment in this supply chain type and find inventory costs can be reduced while maintaining acceptable order fulfillment cycle times. This is true because information, which provides the basis for enhanced coordination and reduced uncertainty, can substitute for inventory.*

In today's business environment, organizations are finding that efficiency alone is insufficient for maintaining competitiveness. Malone (1991) states that "the revolution of the 1990s is driven not by changes in production and transportation but by changes in coordination" (p. 128), particularly global interorganizational coordination. One example of an attempt to improve coordination and competitiveness is supply chain management. Supply chain management expands the scope of the "organization" being managed beyond the enterprise level to include interorganizational relationships, and they typically involve global operations. Examples include improving coordination between suppliers and manufacturers, as well as between manufacturers and distributors. As improvements in information technology (IT) have enabled the costs of coordination to decrease (Malone, 1987), there has been a general movement toward organizing as partnerships between more specialized firms or business units. Supply chain management is an important topic to study because it is an instance of these partnerships.

Most management information systems (MIS) research related to supply chain management has concentrated on identifying the information requirements of local supply chain node decision making (Billington, 1994; Davis, 1993; Lee, 1992; Lee, 1993a; Lee, 1993b; Swaminathan, 1994). Often this involved development of models of material and information flow through the supply chain network (SCN). The purpose of our research is to analyze the impact of information sharing on one supply chain type. We study divergent differentiation supply chains (commonly seen in industries such as fashion apparel). The term *divergent differentiation* is applied to these supply chains because their structure diverges from a relatively small number of suppliers to a wider range of assemblers and distributors. They also involve products differentiated at the assembly stage by combining a relatively small number of components into a much larger number of final products. It is important to study these supply chains because they are seen in a number of industries, including many types of clothing firms, and involve a large range of product variations, short product life cycles, and a primary objective is responsiveness (Fisher, 1994). Additional characteristics of these supply chains are discussed in more detail later in the paper.

We investigate the characteristics of supply chain management in an environment of electronic commerce. This environment includes (1) centralized, global business and management strategies (e.g. make-to-order, assemble-to-order and make-to-stock), (2) on-line, real-time distributed information processing to the desktop, providing total supply chain information visibility, and (3) the ability to manage information not only within a company but across industries and enterprises (Kalakota, 1996). Within an overall framework for studying electronic commerce, our research is at the application level (e.g., supply chain management) enabled by the information superhighway, multimedia content and network publishing, messaging and information distribution, and common business services infrastructures (Applegate, 1996). This five-level framework was developed to succinctly capture the major elements and features of electronic commerce to enable it to be better understood. Side issues, which are outside the scope of this paper, related to this framework include public policy, legal and privacy issues, as well as technical standards for electronic documents, multimedia and network protocols.

We discuss our analysis in the following sections. We present a brief overview of supply chain management, discuss technologies that enable feasible, efficient, information sharing between supply chain partners, and analyze, based on a set of simulations, how information sharing affects overall supply chain network performance. Finally, we present our conclusions.

## SUPPLY CHAIN MANAGEMENT

We introduce supply chains by presenting (1) a general overview of supply chain management, and (2) a summary description of supply chain management.

### General Overview of Supply Chain Management

A supply chain is a network of facilities that procures raw materials, transforms them into intermediate subassemblies and final products and then delivers the products to customers through a distribution system (Billington, 1994). Supply chains exist in virtually every industry, especially industries that involve product manufacturing. Management of supply chains is a difficult task because of the large amount of activities that must be coordinated across organizational and global boundaries. The most common problems involve coordinating materials inventory and production capacity availability across several organizations to produce products that can satisfy forecasted demand in a highly uncertain environment.

Several factors are making supply chain management an important issue for today's managers. These factors include (1) more instances of multisite manufacturing, where several independent entities are involved in the production and delivery process, (2) increasingly cut-throat marketing channels, (3) the maturation of the world economy, with heightened demand for "local" products, and (4) competitive pressures to provide exceptional customer service, including quick, reliable delivery (Davis, 1993). In the past, management would concentrate on making each node of the supply chain network efficient. What managers are now realizing is that efficiency at each node does not result in the supply chain as a whole operating optimally. Increasingly, the challenges related to improved product quality, customer service and operating efficiency cannot be effectively met by isolated change to specific organizational units, but instead depend critically on the relationships and interdependencies among different organizations (or organizational units) (Swaminathan, 1994).

Supply chain management is a management process that attempts to optimize the operation of the entire supply chain. Different entities in a supply chain typically operate subject to different sets of constraints and objectives. Even when belonging to the same company, supply chain entities often report to different divisions. Supply chain entities are highly interdependent when it comes to improving due date performance, increasing quality or reducing costs. As a result, the welfare of any entity in the system directly depends on the performance of the others and their willingness and ability to coordinate (Swaminathan, 1994). Specifically, supply chain management involves balancing reliable customer delivery with manufacturing and inventory management costs (Billington, 1994). Two metrics commonly used to measure overall supply chain performance include order fulfillment cycle time and inventory level and cost.

One major problem involved in supply chain management is understanding and managing the uncertainties involved in the supply chain. This is especially true in the fashion skiwear industry (where divergent differentiation supply chains are prevalent) where demand is heavily dependent on a variety of factors that are difficult to predict - weather, fashion trends, the economy - and the peak of the retail selling season is only two months long (Fisher, 1994). Different sources of uncertainty exist along a supply chain. They include demand (volume and mix), process (yield, machine downtimes, transportation reliabilities), and supply (part quality, delivery reliabilities) (Billington, 1994; Lee, 1993a; Lee, 1993b). Inventories are often used to protect the chain from these uncertainties.

Another major problem involved in supply chain management is the management of lead-time. This is discussed in more detail in a later section. The role of information technology (IT) in supply chain management is to assist managers in managing uncertainty and lead time through improved collection and sharing of information between supply chain nodes. This will result in better customer service, through better coordination, and improve asset management, by giving decisionmakers the information necessary to optimize inventory and capital asset costs. Many of these improvements occur because IT enables dynamic changes in inventory management and production planning. The difficulty arises when designing an information system that can handle the information needs of each of the supply chain nodes to allow efficient, flexible, and decentralized supply chain management. We discuss these issues throughout this paper.

## Summary Description of Supply Chain Management

The information infrastructure required by supply chain management is, by nature, supported by a distributed information system. Because of this, we feel that a distributed system model is most appropriate to describe a supply chain network. Distributed problem solving is the cooperative solution of problems by a decentralized and loosely coupled collection of knowledge sources (KS's) located in a number of distinct processor nodes (Smith, 1988). Distributed problem solving is often necessary because no one node has sufficient information to solve the entire problem. The components of a distributed, coordination intensive, problem include goals, activities, actors, and interdependencies (Malone, 1990). We feel that a supply chain can be described by identifying its actors, activities, interdependencies, goals and objective. Our supply chain description is summarized in Table 1.

The overall supply chain objective is to balance each of the goals based on their importance to supply chain managers. In some situations costs may be the priority, while in other situations customer service may be the priority.

*Table 1: Supply Chain Management Summary Descriptions*

| | |
|---|---|
| 1. Actors | Suppliers, manufacturers, assemblers, distributors, and customers |
| 2. Activities | Material and information processing |
| 3. Interdependencies | Material shipments and orders, Funds transfer, and Information sharing |
| 4. Goals | Minimize order fulfillment cycle time, & Minimize inventory levels and costs |
| 5. Overall Objective | Balance individual goals based on priorities to produce the best "average" performance, or the best "worst case" performance. |

## TECHNOLOGIES THAT ENABLE INFORMATION SHARING BETWEEN SUPPLY CHAIN PARTNERS

In this section we discuss the issue: Does current information technology make global information sharing between supply chain partners feasible? Several technologies are currently available that potentially make information sharing between supply chain partners feasible. Each partner firm is assumed to have its own information systems that include applications for supporting sales, manufacturing, and so forth. A complete discussion of implementation issues such as the interrelationship between required supply chain business processes and information infrastructure components, as well as the impact that differences in partner's current IT infrastructures, is outside the scope of this paper. We also do not explicitly discuss the costs of implementing the infrastructure. It is assumed that IT costs will continue to decrease to the point where the benefits of using them to share information across a supply chain exceed the costs to do so. Our goal is to show that current technology makes it feasible for supply chain partners to share information in the future. The technologies that we discuss in this section include electronic data interchange (EDI), the Internet-based World Wide Web (WWW), intranets, and expanded intranets (extranets).

EDI is an existing information technology that provides a method of electronic transaction transfer. It is the process of computer-to-computer, business-to-business transaction transfer. EDI involves the direct routing of information from one computer to another without interpretation or transcription by people. To achieve this, the information must be structured according to predefined formats and rules which a computer can use directly (Holland, 1992). One example of where EDI has been shown to improve part of supply chain management is in inventory management, specifically, a just-in-time (JIT) system. EDI technology was shown to facilitate accurate, frequent, and timely exchange of information to coordinate material movements between trading partners. Suppliers receiving JIT schedule information achieved better shipping performance. Similarly, suppliers with the ability to directly map incoming information to internal production control systems were found to enjoy even greater benefits. Moreover, as the supplier handles a higher proportion of customers electronically, it was found that shipment errors continued to diminish (Srinivasan, 1993).

Each year, the use of EDI increases as organizations look for methods to improve enterprise integration and interorganizational coordination. Numerous studies have been done on various aspects of EDI and they all draw the same conclusion. EDI increases the speed and the accuracy of processes compared with non-electronic transfer of information (Snapp, 1990), and it is a potential source of competitive advantage (Johnston, 1988).

When a supplier and a procurer use information technology to create joint, interpenetrating processes at the interface between value-adding stages, they are taking advantage of the electronic integration effect. This effect occurs when information technology is used not just to speed communication, but to change - and lead to tighter coupling of - the processes that create and use information. One simple benefit of this effect is the time saved and the errors avoided by the fact that data need only be entered once (Malone, 1987). This is just one of several benefits derived from supply chain partners using more highly integrated information systems.

A practical problem that must be addressed when designing an EDI process is the lack of a globally recognized standard format for data storage and transfer (Snapp, 1990). Two examples of EDI standards are the American National Standards Institute (ANSI) X.12 and EDIFACT. The ANSI X.12 committee develops standards to facilitate EDI relating to a range of business transactions. EDIFACT is a family of standards similar to X.12 developed by the United Nations (Kalakota, 1996). Because broad-based standards are lacking in many industries, organizations must agree upon the translation software and data format on a project by project basis. Without an agreement upon a standard, the EDI process will not work. The result has been slow deployment and restricted implementation of EDI between companies (Kalakota, 1996). This is one of the reasons why there will be a general movement away from these transaction specific connections to more flexible methods of electronic information transfer. One solution that has been considered by a number of businesses is using the Internet-based WWW and Internet browsers.

The Internet is an example of a global information network composed of an existing set of information technologies that provide a method for electronic information sharing. One component of the Internet is the WWW. At first glance, it appears that messaging-based technologies such as EDI, combined with database and information management services, form the technical foundation for effective electronic commerce solutions. No single one of these technologies can deliver the full potential of electronic commerce, however. What we require is an integrated architecture the likes of which we have never seen before. This integrated architecture is emerging in the form of the WWW (Kalakota, 1996). Although the WWW was not developed specifically for sharing of information among supply chain partners, it provides a model for these types of systems. The WWW was developed to be a pool of human knowledge, which would allow collaborators in remote sites to share their ideas and all aspects of a common project (Berners-Lee, 1994). Because supply chain management is similar to the projects the WWW was designed for (remote sites, shared knowledge, common project) it can serve as a method for sharing of information in a supply chain. The major problem with using the Internet for supply chain management is security. Experts say reports of Internet-related security breaches are rising. Nearly one in four respondents to an *Information Week* survey conducted in February 1996 say fear of Net break-ins is keeping them from using the Net (Violino, 1996). The solution seems to be a more secure version of the Internet, an intranet.

An intranet is essentially any site based on Internet technology but placed on private servers and designed not to allow outsiders in (Miller, 1996). The outsiders in this case would be individuals and companies not directly involved in the management of the supply chain. Intranets use Web-based and Internet technology to inexpensively and easily share [organizational] data across a private network (Carr, 1996). We feel that the "organization" can encompass several separate firms such as in a supply chain. Intranet usage is predicted to overwhelm external Internet usage before the turn of the century. The key enablers of WWW growth are: (1) the proliferation of PCs, LANs, and modems, (2) open standards such as TCP/IP, HTTP, and HTML, (3) cross-platform support, (4) multimedia support and ease of use, and (5) support for secure transactions. [Organizational] intranets can provide information in a way that is immediate, cost-effective, easy to use, rich in format, and versatile (Netscape, 1996). What we have described is an extended intranet (or extranet). This is in line with the third wave of Internet usage identified by Netscape's Marc Andreessen. "We are ready for a new era: the emergence of the extranet, or extended intranet, connecting companies with their suppliers and customers via Web links" (Karpinski, 1997). For example, a distributor (located in North America) in an apparel supply chain can request supply information from an assembler (located in Asia) by sending an HTTP request through their browser to a Web server. The server runs a CGI script to execute programs that access any necessary databases to process the request and return output to the HTTP server. The server translates the output into an HTTP response to the client (Lee, 1993b). To answer the initial question of this section, extranets, utilizing the WWW, its middleware, and browser software, provide a set of existing technologies that make supply chain information sharing feasible. Given that IT costs continue to decline, we also feel that the benefits of sharing information across a supply chain exceed the costs to provide the information infrastructure. The next issue then is, given this, what impact does efficient information sharing have on divergent differentiation supply chain performance?

## PERFORMANCE IMPROVEMENTS ENABLED BY THE INFORMATION INFRASTRUCTURE

We illustrate the usefulness of information sharing by presenting and discussing results from simulations of divergent differentiation supply chain performance under various information- sharing strategies that are enabled by current technology. In this section, we focus on one of the core business processes, the order fulfillment process (OFP), and use the Swarm simulation platform (The Santa Fe Institute, 1996) to simulate the OFP in supply chain networks (Lin, 1996a; Lin, 1996b).

Swarm is a multi-agent simulation platform developed for the study of complex adaptive systems. It was developed at the Santa Fe Institute and aims at providing a general-purpose simulation tool for building simulation models. A detailed description of Swarm is outside the scope of this paper, but can be found in (Lin, 1996b). Our implementation of SCNs in Swarm is described in more detail later.

An order fulfillment process begins with receiving orders from customers and ends with having the finished goods delivered (Lin, 1996a). It consists of several activities (sub-processes), such as order management, manufacturing, and distribution. The main objectives of the OFP can be generalized into two dimensions (Christopher, 1993; Goldman, 1995; Lin, 1996a): delivering qualified products to fulfill customer orders at the right time and right place, and achieving agility to handle uncertainties from internal and external environments.

### Issues in Managing SCNs for Supporting the OFP

Because of the complexity of a SCN, it is a challenge to coordinate the actions of entities within the network to perform in a coherent manner. When orders come into an entity in a SCN, the lead time for delivering products (called the order fulfillment cycle time) is composed of (1) order processing times, including the order transfer time from customers to manufacturers or distributors, and the due date assignment process, (2) material lead times, including material planning and purchase lead time, supplier lead time, transport lead time, receipt and inspection lead time, assembly release time, and material order picking time, (3) assembly lead times, including waiting time, processing times, and transport time to the next stage, (4) distribution lead times, including dispatch preparation time (documents, packages), and transportation time to the customer, and (5) installation lead times. These components of the order fulfillment cycle time distribute across the network, and the variation of lead times at any stage will affect the execution of the other stages and result in uncertainties for the overall order cycle time. This is called the ripple effect. For example, a product that is assembled by component parts from several different suppliers. The cycle time for assembling the product can be affected by the lead-time of material supply from different suppliers. If parts from some of the suppliers come later than the other parts for assembly, the assembly will be delayed due to the unavailability of required parts. This also increases the inventory costs for those available parts. If the product is a component for the downstream manufacturing process, the delay for shipping this product will affect the subsequent stages, and in turn, influence the entire network.

Therefore, the first issue in managing a SCN is how to control the ripple effect of lead-time so that the variability of a SCN can be mitigated. How to coordinate the policies of up and downstream entities in facilitating such variability reduction is the main concern. Demand forecasting is used to estimate demand for each stage, and the inventory between stages of the network is used for protecting against fluctuations in supply and demand across the network such as machine breakdown, extra large demand, etc. Due to the shortening of product life cycles, such protection seems unwise and actually reduces flexibility.

Because of the decentralized control properties of the SCN, control of the ripple effect requires coordination between entities in performing their tasks. The management of interdependencies is the key to smooth material flow within the SCN. The interdependencies between entities of the SCN can be described in the following situations:

1. Producer/consumer dependence can be used to describe the supplier/manufacturer relationship in the SCN. This requires cooperation between suppliers and manufacturers in an efficient and effective way. Efficiency means to reduce material lead times, and effectiveness means to supply only the needed materials. This dependence also implies a constraint satisfaction problem; through the network, it is a constraint propagation issue too.
2. Material flows within the SCN implies a synchronization problem, where related materials for a product are delivered to the manufacturer at a coherent speed which incurs minimal inventory and delay.

Inventory is an unwise approach to dealing with highly changing market demand and short life cycle products. What would be the substitution for inventory? Information can substitute for inventory. The material lead-time information from different suppliers can be used for planning the material arrival, instead of building up inventory. The demand information can be transmitted to the manufacturers on a timely basis, so that orders can

*Table 2: The Properties of Type I, II, and III SCNs*

| Attributes | Type I SCN | Type II SCN | Type III SCN |
|---|---|---|---|
| Manufacturing process | Convergent Assembly | Divergent Assembly | Divergent Differentiation |
| Primary business objectives | Lean production | Customization | Responsiveness |
| Product differentiation | Early | Late | Late |
| Range of product variations | Small | Medium | Large |
| Assembly process | Concentrating at the manufacturing stage | Distributed to the distribution stage | Concentrating at the manufacturing stage |
| Product life cycle | Years | Months to years | Weeks to months |
| Main inventory type | End products | Semi-products | Raw materials |
| Example industries | Automobile and aerospace | Appliance, electronics and computers | Apparel/fashion |

be fulfilled with less inventory costs. The second main issue is how to manage the information flow within a SCN so that decisions made by business entities can take more global factors into consideration. In this way, we can increase SCN visibility. These issues are addressed because of the essential concern: how to make the network respond effectively and efficiently to satisfy customer demand, which leads to the motivation for managing SCNs to support the OFP.

Lin identified three main types of SCNs, Type I, II and III, based on such attributes as manufacturing process, primary business objective, product differentiation, range of product variation, assembly stages, product life cycle, and main inventory type as shown in Table 2 (Lin, 1996a).

The apparel/fashion industry involves Type III SCNs, where acquiring market information to respond to demand, and deferring product differentiation to maintain flexibility to handle constantly changing markets, are two main issues and challenges. In these SCNs, the number of end items is larger than the number of raw materials. There are a small number of suppliers and manufacturers, but a larger number of distributors and retailers.

## The Implementation of Swarm for Simulating Order Fulfillment in SCNs

Figure 1 describes the SCN implementation on the Swarm platform. Theuppermost swarm, the OFP Batch Swarm, is designed to control the whole simulation. It creates two swarms, the OFP Model Swarm and the Statistics Swarm, creates actions, and then activates the simulation process. The OFP Model Swarm is composed of an array of SCN Entities created while building objects. The SCN configuration with each entity's properties and product information are fed in during the entities creation. The OFP model

*Figure 1: The Implementation of SCNs in Swarm*

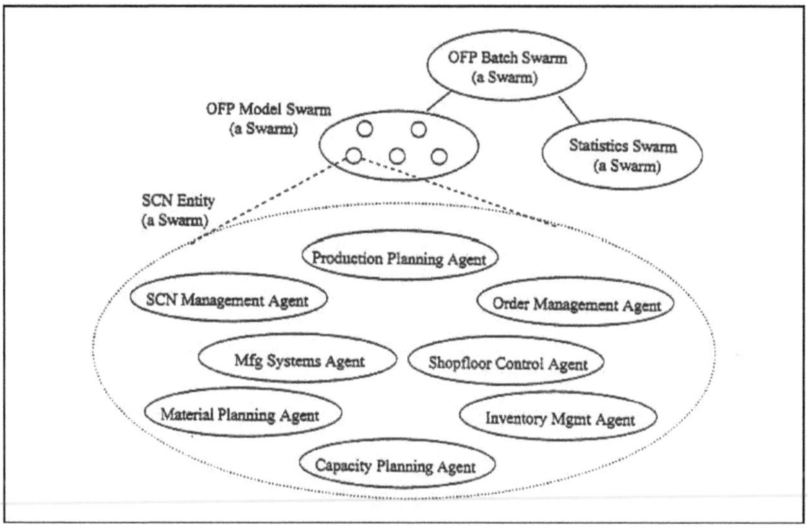


actions are composed of each SCN entity's actions, and are activated when the OFP Model Swarm is activated. A SCN entity is composed of several agents, such as an order management agent, an inventory management agent, and a SCN management agent. An entity with manufacturing capability includes a production planning agent, a capacity planning agent, a materials planning agent, a shop floor control agent, and manufacturing systems agent. A SCN Entity Swarm holds entity level information such as suppliers, customers, order transfer delay time, and product delivery time, which are accessible by internal agents and other entities. The encapsulated agents perform certain functions in enabling the movement of information and material within the entity and between entities. The Statistics Swarm is used to compute the statistics data gathered through the simulation for analysis purposes.

The following scenario describes the interactions among these agents, and Figure 2 summarizes them. For example, an entity ScnESwarm A receives an order from its customer ScnESwarm C. The order flows to the order management agent (OrdM). According to the customer lead times, the inventory availability information (from InvM), the production plan (from PrdP), and the manufacturing capacity (CapP), the order management agent assigns a due date to the order. If the products are in stock, the order is filled by shipping the products from inventory. If the products are in receiving, the due date is set according to the delivery date of the products.

For an entity with manufacturing capability, the order is forwarded to the production planning agent (PrdP) where the schedule for making the products

*Figure 2: SCN Agent Interactions in the Swarm Implementation*

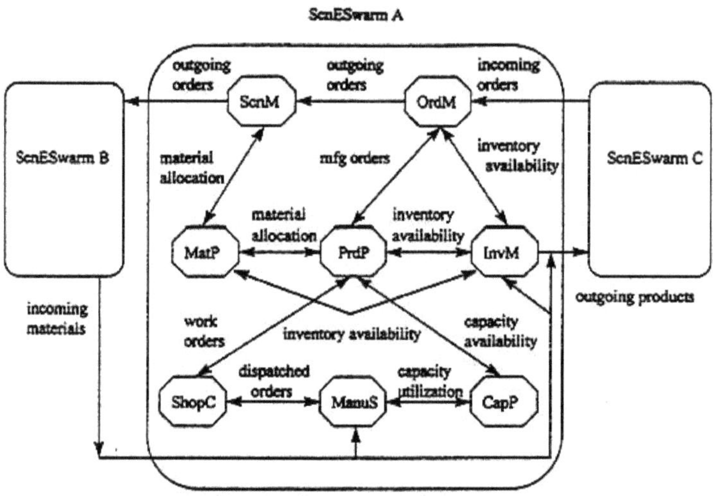


is planned. The capacity planning agent (CapP) and the material planning agent (MatP) are partner agents in generating achievable build plans. The material planning agent obtains build plans from the production planning agent to allocate materials for manufacturing. It also contributes information about material availability to production planning for scheduling. The capacity planning agent (CapP) plans capacity by taking the build plan from PrdP and sends capacity usage information to PrdP for scheduling the build plan. The SCN management agent (ScnM) takes the order information to choose suppliers in allocating material sources. The outgoing orders are transferred through its SCN Entity (ScnESwarm A) to be transferred to other entities (i.e., ScnESwarm B). This describes the information within an entity.

If the entity is a distribution center or a retailer without manufacturing capability, the ordered products are delivered from suppliers as end products to ship to its customers. For an entity with manufacturing capability, the ordered end products are supplied from the shop floor (ManuS) to its customers. The input materials are components for the end products. This represents the material flow with an entity. The interaction of these agents enables the flow of materials and information within an entity, and through the SCN Entity Swarm (ScnESwarm), the information and materials flow across the supply chain network.

In this section, we described the components of our SCN simulation model. The specific business environments that we simulate are described in the next section.

*Figure 3: Simulated SCN Structure—"Scn-III"*

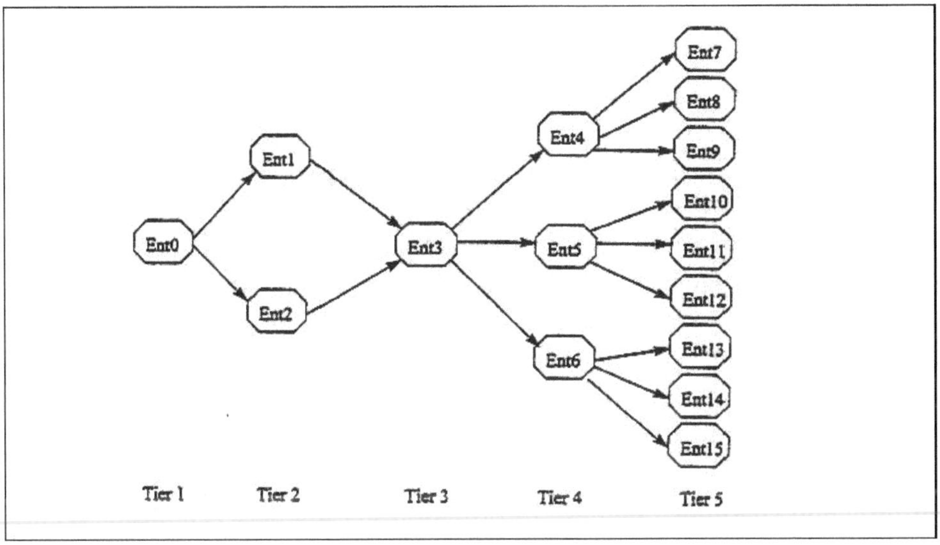


## Simulation Environment

We implemented a supply chain designed to simulate a Type III SCN. "Scn-III" in Figure 3 represents a Type III SCN. It consists of 16 entities aligned into five tiers. The entity at tier 1 is a distribution center without manufacturing capability, entities in tiers 2 and 3 make the products, and entities in tiers 4 and 5 distribute products. The divergent assembly, late product differentiation, many resulting end-items, and manufacturing concentrating at manufacturing stages are common to Type III SCNs.

*Table 3: Simulation Parameter Summary*

| Parameter Type | Simulation Parameter | Value |
|---|---|---|
| Global parameters | Time step | One day |
| | Time period | 240 days |
| | Production planning horizon | 24 days |
| | Forecasting horizon | 48 days |
| | EOQ period | 240 days |
| | Reorder point check | 48 days |
| | Order distribution | U(2,10) units |
| Entity parameters | Manufacturing capability | Yes/No |
| | List of supplier entities | Varies based on SCN structure |
| | List of customer entities | Varies based on SCN structure |
| | Order cost | 2.0 dollars |
| | Inventory holding cost | 0.1 dollars |
| | Production processing time | 0.2 days |
| | Time to forward order | 0.5 days |
| | Time to deliver product | 1.0 days |

Two sets of parameters provide the input to the Swarm system used for these simulations. The parameters and their values are summarized in Table 3.

The global parameters are held constant for each simulation to isolate the cycle time and inventory cost impact of different information sharing and demand management strategies. The interactions between agents are described above in the supply chain structure. Finally, each entity in the structure is characterized by the attributes shown in the entity parameter list. Production costs, such as setup costs, are not included in this study given that they do not directly impact the cycle time and inventory cost performance parameters studied. The values were chosen as reasonably representative of operating supply chains. System documentation and source code is available through the WWW (Lin, 1998).

We conducted experiments to evaluate OFP performance using various information-sharing strategies. Information sharing between business entities considers three issues: (1) the information contents, (2) the depth of information penetration (the number of tiers for which information is accessible), and (3) the information acquisition direction (upward or downward sharing). Agent decision-making processes are held constant to isolate the impact of information sharing.

In the design of the simulation platform, the information acquired by downstream entities is mainly material and capacity availability information from their suppliers. The information acquired by an upstream entity is information about customer demand and orders. The depth of information penetration can be specified in various degrees, e.g., isolated, upward one tier, upward two tiers, downward one tier, downward two tiers, and so forth.

The obtained capacity and material information from suppliers is used to estimate the due dates of incoming orders, which are the basis for generating build plans or reordering schedules. The obtained customer demand information is used to estimate the demand for the next period, so that the production or reordering schedules can adapt to external demand.

Demand management policies, such as make-to-order (MTO), make-to-stock (MTS), and assemble-to-order (ATO) have their characteristics and application situations described in Table 4 (Lin, 1996a; McCutcheon, 1994).

If the amount of customization is low, the firm can usually employ a MTS approach and then use inventories of finished goods to provide short lead times. For products with high customization, the MTS strategy cannot efficiently and effectively match customer preferences. If customers are willing to wait for customized products after submitting orders, the make-to-order strategy can be applied to high-customization firms. When the product

*Table 4: Some Demand Management Policies for the Order Fulfillment Process*

| Policies | Characteristics | Application Situations |
|---|---|---|
| Make-To-Order (MTO) | Production is triggered by customer orders. | High customization pressure but low responsiveness. |
| Assemble-To-Order (ATO) | Final assembly is order-driven, but the component parts are forecast-driven and built to stock. | High customization pressure, high responsiveness, and products with late differentiation |
| Make-to-Stock (MTS) | Production is triggered by inventory replenishment points. | Low customization pressure. |

design allows the product differentiation stage to occur late enough in the production process, the firm can employ an assemble-to-order approach.

## Impact of Information Sharing on Divergent Differentiation Supply Chains

The results from evaluating various information-sharing strategies in divergent differentiation supply chains are shown in Figures 4 through 5. The three information sharing strategies are (1) no information sharing, (2) supply information sharing, and (3) supply and demand information sharing. The

*Figure 4: OFP Improvement in Order Cycle Time Reduction Using Various Information Sharing Strategies in a Type III SCN*

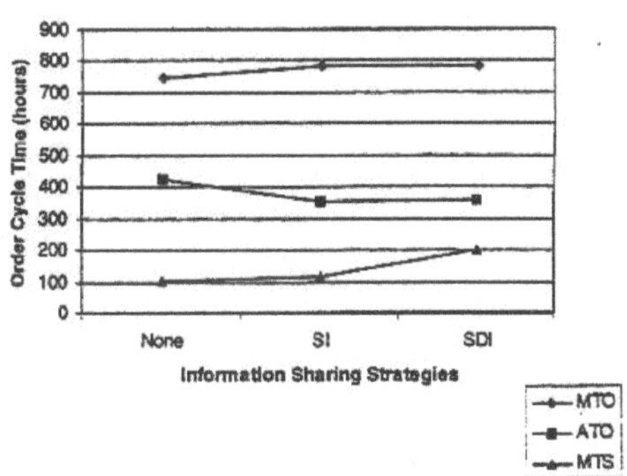


NONE: No information sharing
SI: Supply information is shared
SDI: Supply and demand information is shared
MTO: Make-to-order
ATO: Assemble-to-order
MTS: Make-to-stock

*Figure 5: OFP Improvement in Inventory Cost Reduction Using Various Information Sharing Strategies in a Type III SCN*

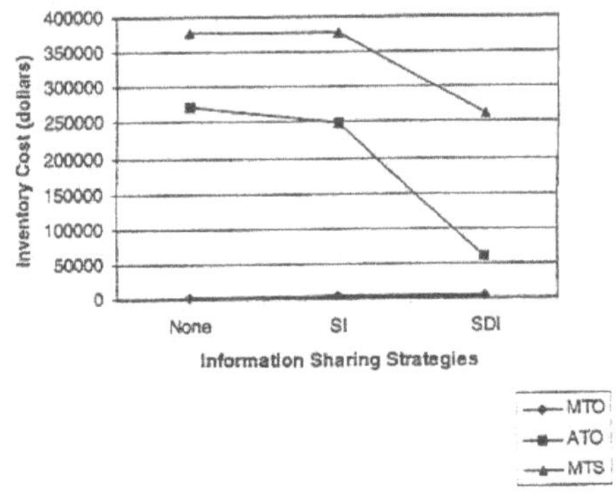


NONE: No information sharing
SI: Supply information is shared
SDI: Supply and demand information is shared
MTO: Make-to-order
ATO: Assemble-to-order
MTS: Make-to-stock

three demand management policies were described earlier. Figures 4 and 5 illustrate order cycle time and inventory cost respectively in a Type III SCN using the three information sharing strategies with the three demand management policies.

In Figure 4 we see that the highest cycle time is for the MTO strategy, while the lowest cycle time is for the MTS strategy. This is the expected result given that the MTO strategy waits for orders to arrive before processing them and the MTS strategy produces products for inventory in anticipation of orders. The most improved cycle time performance (-73 hours or -17%) results from sharing supply information when using an ATO strategy. One interesting result is that the SDI strategy results in an increase (+83 hours or +71%) in the order cycle time when applying the MTS demand management strategy. The demand information provided from downstream is made available to the upstream entities to factor into their component or product reorder decision making instead of just reordering exclusively according to the reorder point calculated by stock level. Therefore, it reduces a significant amount of inventory as shown in Figure 5. However, due to the ripple effect, the demand information fails to penetrate to upper tiers with sufficient lead time, thus the aggregated cycle time increases. Therefore, the information penetration span and lead-time are crucial in implementing the information sharing strategy.

Inventory costs are calculated by accumulating all finished and work-in-process products within the supply chain. In Figure 5, we see that the highest inventory cost is for the MTS strategy, while the lowest inventory cost is for the MTO strategy. This is the expected result given the characteristics of these strategies described earlier. The most improved inventory cost performance (-$188,191 or -76%) results from sharing supply and demand information when using the ATO strategy. Therefore, we feel that the best overall strategy for divergent differentiation supply chains is to apply an ATO strategy while sharing supply and demand information between supply chain partners. Cycle time is held at a reasonable level, but inventory costs are much lower when supply and demand information is shared. Since a Type III SCN faces constantly changing markets, the accessible demand information may allow suppliers to deliver the right products at the right time with less inventory. Our findings support this. The critical component of the information infrastructure is the link between distributors and the market.

Supply information (coupled with a MTS demand management policy) is crucial for OFP improvement when the constituent materials for products are unique and delivered by different suppliers. But as the range of product variations and changes in market demand increases (as commonly seen in fashion/apparel supply chains), the importance of demand information increases. This information, coupled with an ATO demand management policy, results in improved performance in these SCNs.

## CONCLUSIONS

Two sets of conclusions can be drawn from our study. The first set relates to findings specific to divergent differentiation supply chains, while the second set relates to overall findings concerning the relationship between information infrastructure and supply chain management. Management of Type III (divergent differentiation) supply chains is most effective when using an ATO policy coupled with sharing of demand information. Supply information is useful, but less critical. Inventory costs are reduced while cycle times remain relatively stable. These results, related to divergent differentiation supply chains, produce some interesting overall conclusions.

First, supply chain management relies heavily on its information infrastructure. Supply chain performance can be improved through the information sharing and coordination enabled by current technology. Also, an extranet, with its associated components, provides a current technology for the functional requirements required for efficient, electronic, information sharing that is more flexible than EDI, while increasing data and message security relative to the Internet.

Finally, supply chain management involves a fundamental tradeoff between cycle time, inventory and information. In many cases, information can replace inventory while maintaining acceptable cycle times. In the past, when information costs were high, inventory was held to manage uncertainty. Today, when information technology continues to reduce information costs, uncertainty can be reduced resulting in lower inventory requirements. Our results illustrate some of the potential impacts of the electronic integration effect (Malone, 1987). The benefits that we illustrate related to this effect are that supply chain managers may reduce inventory costs because of reduced uncertainty in decision making. This is possible because IT (incorporated into electronic hierarchies) reduces coordination costs. The development of an analytical model to describe this tradeoff is an issue that should be addressed by future research.

Overall, we feel that extranets provide an interorganizational information system that can support the processes and decision making required for effective supply chain management resulting in improved coordination and lower inventory costs. Additional research should involve implementation of current technology for information sharing across supply chains to empirically test our simulation findings.

## REFERENCES

Applegate, L. M., Holsapple, C. W., Kalakota, R., Radermacher, F. J., & Whinston, A. B. (1996). Electronic commerce: building blocks of new business opportunity. *Journal of Organizational Computing and Electronic Commerce,* 6(1), 1-10.

Berners-Lee, T., Cailliau, R., Luotonen, A., Nielsen, H. F., & Secret, A. (1994). The World-Wide Web. *Communications of the ACM*, 37(8), 76-82.

Billington, C. (1994). Strategic supply chain management. *OR/MS Today,* April 1994, 20-27.

Carr, J. (1996). Intranets deliver. *InfoWorld*, 18(8), 61-63.

Christopher, M. (1993). *Logistics and supply chain management*, London, UK: Pitman Publishing.

Davis, T. (1993). Effective supply chain management. *Sloan Management Review,* Summer 1993, 35-73.

Fisher, M. L., Hammond, J. H., Obermeyer, W. R., & Raman, A. (1994). Making supply meet demand in an uncertain world, *Harvard Business Review*, May-June 1994, 83-93.

Goldman, S. L., Nagel, R. N., & Preiss, K. (1995). *Agile Competitors and Virtual Organizations: Strategies for Enriching the Customer*. New York: Van Nostrad Reinhold.

Holland, C., Lockett, G., & Blackman, I. (1992). Planning for electronic data interchange. *Strategic Management Journal,* 13, 539-550.

Johnston, H. R., & Vitale, M. R. (1988). Creating advantage with interorganizational information systems. *MIS Quarterly*, 12(2), 153-166.

Kalakota, R., & Whinston, A. B. (1996). *Frontiers of electronic commerce.* Reading, MA: Addison-Wesley Publishing Company, Inc.

Karpinski, R. (1997). Extranets emerge as next challenge for marketers. *Netmarketing,* April 1997, M-4.

Lee, H. L., & Billington, C. (1992). Managing supply chain inventory: pitfalls and opportunities. *Sloan Management Review*, Spring 1992, 65-73.

Lee, H. L., & Billington, C. (1993a). Material management in decentralized supply chains. *Operations Research,* 41(5), 835-847.

Lee, H. L., Billington, C. & Carter, B. (1993b). Hewlett-Packard gains control of inventory and service through design for localization, *Interfaces,* 23(4), 1-11.

Lin, F. (1996a). Reengineering the order fulfillment process in supply chain networks: a multiagent information systems approach, Ph.D. Thesis, University of Illinois at Urbana-Champaign, August 1996.

Lin, F. (1998). Simulating the Order Fulfillment Process in Supply Chain Networks Using Swarm, http://mais.cm.nsysu.edu.tw/.Malone, T. W., & Crowston, K. (1990). What is coordination theory and how can it help design cooperative work systems?, *Proc. CSCW '90*, 375-388.

Lin, F., Tan, G. W., & Shaw, M. J. (1996b). *Multi-agent enterprise modeling,* University of Illinois at Urbana-Champaign, College of Commerce and Business Administration, Office of Research, October 1996, Working Paper 96-0134.

Malone, T. W., & Rockart, J. F. (1991). Computers, networks, and the corporation, *Scientific American,* 265(3), 128-136.

Malone, T. W., Yates, J., & Benjamin, R. I. (1987). Electronic markets and electronic hierarchies, *Communications of the ACM,* 30(6), 484-497.

McCutcheon, D. M., Amitabh,S. & Meredith, J. R. (1994). The customization-responsiveness squeeze, *Sloan Management Review,* Winter 1994, 89-99.

Miller, M. J. (1996). Your own private Internet, *PC Magazine,* 15(5), 29.

Netscape Communications Corporation. (1996). Intranets redefine corporate information systems, http://home.netscape.com/comprod/at_work/white_paper/indepth.html.

The Santa Fe Institute. (1996). The Swarm Simulation System, URL = "http://www.santafe.edu/projects/swarm/".

Smith, R. G., & Davis, R. (1988). Frameworks for cooperation in distributed problem solving, in *Readings in Distributed Artificial Intelligence*, Ed. Alan H. Bond and Les Gasser, San Mateo, CA: Morgan Kaufmann Publishers, Inc.

Snapp, C. D. (1990). EDI aims high for global growth, *Datamation,* March 1, 77-80.

Srinivasan, K., Kekre, S., & Mukhopadhyay, T. (1993). *Impact of electronic data interchange technology on JIT shipments,* Graduate School of Industrial Administration, Carnegie Mellon University.

Swaminathan, J. M., Smith, S. F., & Sadeh, N. M. (1994). *Modeling the dynamics of supply chains,* The Robotics Institute, Carnegie Mellon University.

Violino, B. (1996). Your worst nightmare, *Information Week,* Feb. 19, 1996, 34-36.

# Chapter 20

# Leapfrogging an IT Labor Force: Multinational and Indigenous Perspectives

Eileen M. Trauth
Northeastern University, USA

*Throughout the world countries recognizing the economic benefits of IT are rapidly developing information sectors. Advanced industrialized nations have been moving gradually from industrial to post-industrial work while newly industrializing nations are leapfrogging directly from agrarian to information-intensive work. At the close of the twentieth century some interesting tensions result from these two types of nations moving in the same direction. An important issue for both advanced industrialized countries and newly industrializing nations is the human resource issue. Managers at multinational firms as well as policy makers in newly industrializing countries recognize the need for developing qualified IT personnel. The Irish IT sector provides a valuable setting for studying human resource issues associated with the rapid development of an information sector because it affords the opportunity to explore both the multinational and the indigenous viewpoints. Two key human resource issues for the global IT sector are explored in this paper: ensuring a supply of qualified IT workers in the host country and managing IT workers in a cross-cultural work environment. This paper is based upon an ethnographic study of the information sector in Ireland. The issues that are raised represent the experiences of American and Irish managers, Irish policy makers and Irish IT workers.*

Previously Published in the *Journal of Global Information Management, vol.7, no.2*, 

As the global economy moves further into the post-industrial era, nations around the world see opportunities for economic development in the information sector of their economies. Advanced industrialized nations have been moving gradually from industrial to post-industrial work over the past several decades and have spawned the multinational IT companies which have enabled the development of a global information sector. Newly industrializing countries, on the other hand, have been leapfrogging in recent years directly from agrarian to information-intensive work. At the close of the twentieth century, some interesting tensions are resulting from both economies moving in the same direction. A critical consideration in the development of a global information sector is the human resource issue: the development and management of appropriately qualified information technology (IT) workers. These are the individuals who fuel the information economy by producing the computers, developing the software and providing the information services that keep it running.[1]

Ireland provides a valuable context for examining IT human resource issues in the information sector for several reasons. Because it developed its information sector through inward investment by multinational firms, the case of Ireland includes two different points of view: 1) that of multinational IT firms from advanced industrialized nations; and 2) that of 'second wave' countries typically, newly industrializing countries. Second wave IT countries such as Ireland are those which are taking an alternate path into the information economy by rapidly moving directly from an agrarian or partially developed industrial economy into an information economy. They have been doing this in the last few decades of the twentieth century. They stand in contrast to 'first wave' countries — typically advanced industrialized nations — which have been moving into the information economy after a gradual transition from an agrarian through an industrial to a post-industrial economy. In addition, since Ireland has been actively engaged in developing its information sector for thirty years, there is ample opportunity to critically analyze the experiences of policy makers, Irish workers and managers, and multinational managers in order to offer some useful lessons applicable to other contexts.

This paper addresses some issues encountered by Irish and American IT professionals in the development of an Irish IT work force in conjunction with an economic development policy of inward investment by multinational IT firms. Specifically, this paper addresses two important sets of human resource issues. The first set relates to ensuring a supply of appropriately qualified IT workers. These issues range from educational policy to creating a national climate and infrastructure supportive of IT work. The second set of human resource issues relates to managing IT workers in a cross-cultural environment. This culturally diverse work environment has resulted from the infusion of multinational IT firms reflecting a variety of national cultures. To the extent that Ireland's experiences are typical of other second wave countries, the lessons learned are extensible to indigenous and multinational managers as well as government policy makers in other countries. As the title of this paper suggests, the lessons apply not just to the host country but to multinational IT management as well.

This paper is situated at the intersection of two important streams of global IT research. One of these streams is producing a body of literature consisting of country-level studies that examine the interaction between IT development, diffusion and use, and societal context.[2] This paper contributes to this body of literature by provided primary-source information about socio-cultural factors in the Irish context and how they are influencing the human resource dimension of IT. The other stream of relevant research is concerned with IT and its role both in developing countries and in economic development. Attention in this body of literature is also given to cross-cultural issues deriving from the interactions between transnational corporations and their foreign subsidiaries.[3] This paper contributes to this body of literature by highlighting the human resource and management issues that arise in a nation desirous of using IT employment for economic advancement.

## METHODOLOGY

This paper is part of an ethnographic study of Ireland's IT sector. The overall objective of this research project was to conduct an interpretive investigation of the interaction between societal context and the evolution of

*Figure 1: Influence-Impact Model of Society-Technology Interaction*

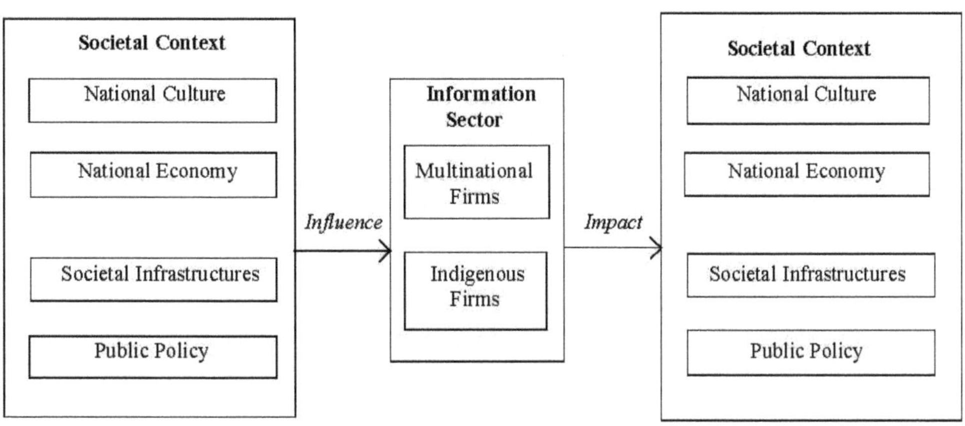

*Table 1: Participating Firms (# Participants)*

| **American Firms (46)** | **Irish Firms (22)** |
|---|---|
| Apple Computer (4) | Computing Workshop AppleCentre (1) |
| Digital Equipment Corporation (9) | Expert Edge (3) |
| IBM (3) | General Controls and Automation (3) |
| Intel (3) | Glockenspiel (3) |
| Prime Computer (8) | HEClissmann (3) |
| Stratus Computer (10) | Mentec (4) |
| Wang Laboratories (9) | System Dynamics (5) |

the primary information sector in Ireland. The general research model used in framing this project is shown in Figure 1.[4]

The primary sources of data for this paper are participant observation and semi-structured interviews conducted at 14 multinational and indigenous IT firms in the hardware, software and information services industries in Ireland. This information is supplemented with public policy documents and interviews with policy makers and cultural observers. Participating firms and the number of participants at each are shown in Table 1.[5] Interviews of one and one-half to two hours in typical duration were conducted with Irish and American IT workers and managers representative of the spectrum of IT work from component assemblers to software engineers to managing directors.[6] Topics covered in the interviews are presented in the Appendix. While these items served as a general guide, the interviews were not restricted to these topics.

Transcripts of the interviews were analyzed using open coding (Strauss, 1987), and a database of respondent characteristics and comments was created to facilitate retrieval and systematic analysis of the data. An in-depth discussion of the methodology can be found in Trauth (1997) and Trauth & O'Connor (1991). The human resource issues that surfaced from this single country case have implications well beyond the Irish shores.

This study was conducted during the period 1989 to 1995 with the bulk of the data collection occurring between September 1989 and August 1990. During that time, I was a Fulbright Scholar living and teaching in Ireland. It was during this time that I conducted the interviews at the IT firms. My posture in this research project was that of an ethnographer from one culture (United States) going to another (Ireland) to learn about a specific facet of it (the information technology sector). During the period of 1991-1995, I spent shorter periods in Ireland collecting documentary information and conducting selected interviews with informants in the IT sector and in government agencies.

## BACKGROUND: DEVELOPING THE IRISH IT SECTOR[7]

Ireland took its first tentative steps to create an information sector in the 1960's, when it began to invite multinational firms to the country and to educate its population to work in them. These were blind steps, as well, in that policy makers were not consciously attempting to create an information sector. Rather, they were attempting to replace dwindling agrarian employment with industrial work. This policy of economic development through inward investment was a direct reversal of the economic policy that had been in place since Ireland achieved its independence from Britain in 1922. In the early years of this country leaders, had endeavored to replace the colonial society with cultural and political sov-ereignty achieved largely through economic isolationism.

When the wisdom of that approach began to falter, Ireland's search for alternatives led it to industrialization-by-invitation: the attempt to rapidly develop an industrial sector by importing both expertise and jobs. In searching for suitable industrial sectors from which to invite multinational firms, Ireland settled upon three: pharmaceuticals, chemicals and electronics. It was from this initial recognition of the economic potential of the electronics sector that the country eventually established its separate information sector.

The expectation of the policy makers who promoted this plan was that multinational firms would bring to Ireland both direct and indirect benefits. Direct and short-term benefits would be the jobs that would keep people in Ireland and off the dole.[8] The anticipated indirect benefit was that foreign investment would provide not only jobs but also a new business climate. It was hoped that these outside influences would help Ireland to more quickly develop an indigenous entrepreneurial capacity. Thus, the long-term benefits would be the spillover effects of the rapid development of technical and business expertise.

The nascent information sector began with firms such as Digital Equipment Corporation, Apple, Amdahl, Analog Devices and Wang which brought employment in computer hardware manufacture when they came to Ireland in the 1970's and early 1980's. The improvements made in the physical and societal infrastructures during the 1980's encouraged these early firms and new ones to consider additional types of work that could be done in Ireland. It was these additional types of work — software development and information processing — which resulted in the recognition of a distinct information sector in Ireland (National Software Directorate, 1992; *Overseas Industry in Ireland*, 1991; Power, 1992). By the 1990s, the information sector was a robust and diversified sector providing employment in hardware manufacture and final assembly, software development and localizing,[9] and information processing and services.[10]

Irish policy makers provided attractive economic incentives for inward investment.[11] These were the necessary conditions for establishing the multinational IT sector in Ireland. But as I learned from both observation and discussions with American informants, the sufficient conditions were an infrastructure supportive of the type of work to be done and a qualified labor force to do it. Multinational firms in the computer industry came to Ireland because they wanted a foothold in Europe and because Ireland provided the most competitive economic incentives. However, once they arrived they found a national infrastructure being developed to keep them there. The crucial ingredient in this mix was the labor force. The young, English speaking, and well-educated workers are one of the main reasons firms that have come to Ireland remain there. By the early 1990s, the quality of the education possessed by the labor force had risen to become a primary selling point of Ireland.

If the reasons for coming to Ireland have to do with financial incentives, the reasons that firms have stayed in Ireland have to do with the culture and society which produces these IT workers. The challenges of ensuring a supply of qualified IT workers and managing them in a cross-cultural work environment are the challenges of maintaining a balance among a variety of socio-cultural factors.

## ENSURING A SUPPLY OF QUALIFIED IT WORKERS

The Irish IT labor force was consistently described by both IT firms and the government agencies promoting the country as young, flexible and well educated in addition to being English speaking. The first and last traits in this list are those upon which insightful policy makers decided to capitalize. The second is related to the cultural context from which this work force emerged. The third trait is both a product of the culture and the result of a concerted effort to align the societal infrastructure with economic needs. Thus, the challenge of ensuring a supply of qualified IT workers has been one of (re)designing societal structures to support this IT work and addressing real or potential barriers to wider participation.

### (Re)designing societal structures to support IT work

Elements of both the societal and the physical infrastructures were adapted to become supportive of IT work and the IT sector. The most important societal aspect has been the educational infrastructure. While the educational system, in a general sense, was initially a contributing factor in Ireland's attractiveness to multinational companies, it also needed to be better aligned with the skill and knowledge demands of IT work in order for this sector to continue to thrive in Ireland. Irish policy makers recognized that the well-educated Irish population was a powerful resource that could be leveraged to support the emerging information sector.

But there were two serious issues to overcome as the country embarked upon its journey into the information sector. The first was establishing equality of access to education. This was accomplished in 1968 when secondary education became State funded. Until that time, the only option was to attend one of the private, religious schools that were geographically and economically unavailable to many people, especially those in rural Ireland.

The other issue was to enable potential IT workers to acquire the specialized skills and knowledge for work in this sector. In the 1960s, the traditional university was not oriented toward vocational education much less vocational education of a technical nature. Consequently in the 1970s and 1980s, two new universities were established and the traditional universities were adapted to incorporate business and IT skills into their curricula. In addition, technical colleges were established around the country. American managers often commented on the high quality of the technicians' education that they considered to be equal or superior to that received by technicians in the States. Finally, evening, adult-oriented programs common in the United States were established. This enabled workers to develop their skills and employment prospects by pursuing undergraduate and graduate studies on a part-time basis. Besides educational programs, the government also sponsored training programs. Those with university degrees but without the proper IT skills as well as those who had been made redundant in other fields were able to acquire the requisite skills for work in the IT sector. In some cases, firms received training grants as an incentive to develop these human resources.

The educational piece is only one side of the equation, however. The other side of creating a national climate that can sustain an IT sector is ensuring a continuing supply of IT jobs. But doing so is not as simple as making sure that multinational firms continue to come to Ireland. There must be an alignment between the particular skill sets being developed in the schools and training programs, and the types of jobs that are becoming available. When there is a misalignment, young Irish graduates have turned to emigration to find jobs non-existent at home. This has become an unintended consequence of the quality of Ireland's educational system: Irish IT workers have become a desired human resource in other countries which have an older and /or less qualified labor force. Consequently, many well-educated Irish graduates are being recruited by foreign firms while still attending university. From an economic development point of view, policy makers would prefer that the fruits of their educational investment remain in Ireland to provide economic payback. As a result of this new wave of emigration, industrial policy assessments have recommended a closer match between the educational plans of the universities and the employment opportunities available in the country.[12]

In addition to redesigning the educational infrastructure, there also was a need to upgrade — and in some cases put in place for the first time — the physical infrastructure needed to support IT work. Transportation, telecommunications and the physical plants to house the IT firms were needed. Thanks to funding from the European Union, Ireland was able to greatly improve its transportation infrastructure. The upgrading of the telecommunications infrastructure, especially in the rural west, had the unintended benefit of furthering Ireland's movement into the information sector. It was the state-of-the-art telecommunications system that encouraged multinational firms to look to Ireland not just for hardware manufacture and final assembly, but also for software development and off-shore data processing.[13]

## Addressing barriers to wider participation in IT work

Three types of barriers to full participation in IT work are in evidence in Ireland. They are barriers related to age, gender and social class. The IT profession in Ireland is viewed as a young person's field. While this is also be the case elsewhere, it is especially noteworthy given the extremely young population of the country.[14] In my interviews, a person over the age of 35 was considered old. I noted a marked difference in the age profiles between the Irish and American sites of the multinational companies I had visited. In the early 1990s when this research began, there was an added reason for this age difference: those over the age of 35 had not had access to free secondary education when they were young. Consequently, they were fewer in number and those who were in IT would have come from families affluent enough have been able to send their children to private, secondary schools. Thus, at the time of my research a conflict between older and younger IT workers had not yet arisen. But looking a few years into the future when the young workers I interviewed would be in middle age and beyond, some challenging future scenarios arise.

The first scenario is that discrimination against "old" people (i.e., those over 35) will persist because they will be seen as not capable of the flexibility or mental alertness needed to cope with technological change. In such a scenario, the country will be depriving itself of the wisdom of experience and the talents of part of its population. Another scenario is a tension between the personal needs of younger and older adults. For example, both young and older respondents described the young workers as more willing to work overtime. Whereas older workers are more settled and have family obligations away from the workplace, the younger workers — especially those who are unmarried — have fewer such obligations and more willingness to work longer hours in order to 'get on' at the firm. Will the IT workplace be able to accommodate workers with a range of motivations and levels of commitment? The final scenario is about career paths. If a mentality persists that only young people can work in IT what will happen when people with university credentials have worked in IT firms for 20 years and want to move up? Where will they go? Will the IT sector be diversified enough to accommodate the career needs of experienced workers or will the IT sector only be able to accommodate young, entry level workers?

These issues are not unique to Ireland. They can be found in the IT sectors of the United States and other countries. However, there are some noteworthy aspects of this finding for Ireland. First, while an age tension may exist in the U.S., older workers in the IT industry have more options because the headquarters are in the U.S. Thus, there is an option to migrate out of a purely technical position into management, for example. Those options are fewer if at all in Ireland. Second, the age differential is more dramatic in Ireland. Both Irish and American respondents noted the difference in the age profile of employees at the American versus the Irish sites of the same multinational firms. Finally, Ireland's dependence upon the IT sector is much greater than in the US. In America, older IT workers could leave the information sector and find employment in other sectors more easily than they could in Ireland.

The second type of barrier that was identified is related to gender. In the course of my interviews and observations, I witnessed the tension between opportunity and restriction with respect to this potential IT labor pool. Women, especially those working in multinational firms, thought it was easier for women to 'get on' in the IT sector than in traditional industries, banking and the civil service. There are three reasons why this is the case. First, because this industry depends upon intellectual not physical strength, women can theoretically compete equally with men. Skill is what counts. Secondly, IT represents a good work setting for women because it is a new industry and, as such, has not had as much time for the old, traditional, patterns to develop. There has been less time for traditions regarding gender-typed work to establish themselves. The final reason for greater opportunity in the IT field is the way in which it is developing in Ireland. Because the IT sector is dominated by American multinational firms, the values about gender that exist in America are being transmitted to Ireland through the medium of corporate culture.

However, these positive sentiments about gender and IT coexist with an acknowledged stereotype of IT work as a male activity. Both men and women recognized that women are not full participants in the IT sector. From there, viewpoints diverged. Men were not sure why, thinking that women must want it that way or that things were changing in the younger generation. Some women believed women *should* be less involved in work outside the home while others felt that the demands of family made women hold themselves back. The status of women in IT emerges from a cultural context of large families and child rearing as a woman's responsibility. Given that reality, some women simply leave the workforce once they begin to have children. Others choose not to progress as far in their careers in order to have time for their family responsibilities. Those women with whom I spoke who had both families and demanding IT jobs acknowledged the inner conflict between career and family.

The final type of barrier relates to social class. This is a complicated barrier to address, in part, because of the reluctance of many Irish people to acknowledge the presence of social class boundaries in Ireland. Part of the country's post-colonial legacy is the desire to reject rigid social class categories. Nevertheless, there is ample evidence of social class boundaries even if they are more permeable, perhaps, than in other countries.[15] The issue at hand is barriers to working class people entering the information sector. To the extent that information sector work is knowledge-based, education is the key. But attitudes are also intertwined with this issue.

Until secondary education became free in 1968, there was a concrete barrier to participation in information sector work. Since the absence of a high school degree is a definite barrier to IT work, there was clear evidence of social class distinctions among workers who would have been of high school age in the late 1960's. But even after secondary education became freely available to all, attitudes about education have continued to serve as a barrier. In family settings without a history of or value placed upon education, there can be pressure on young people to enter the work force as soon as possible in order to add to the family income. Other types of barriers — both educational and attitudinal — come not from the working class people or their families but from middle class people. These types of barriers were in evidence primarily in the indigenous firms. Indigenous firms tend to be smaller and focused on the software and information services subsectors of the information sector. These are also the subsectors having higher educational requirements. Whereas the multinational firms employ workers at a range of levels from technician (with perhaps only the Leaving Certificate[16]) to graduate engineers, the Irish firms tended to require university credentials. Thus, to the extent that there is a barrier to university attendance among working class people, there is a barrier to their participation in indigenous IT firms. In addition, I found evidence of attitudinal barriers in Irish firms. Informants talked about the importance of one's accent and address in securing employment. While they acknowledged this to be much less prevalent in the IT sector than in other sectors in Ireland, I observed and informants agreed that it, nevertheless, exists.

## MANAGING IT WORKERS IN A CROSS-CULTURAL WORK ENVIRONMENT

> Five or seven years ago, if someone asked me what corporate culture was, I'd say, "That's bullshit. You come in, there's a job to do, people got paid, and the product and the process and the information systems, that all dictates what goes on at work." That's not it at all. The culture dictates everything below that. The culture, in one way or another, will dictate the process that you use. And I never would have believed that because it was too 'touchy feely.' At the time, I mean, if it couldn't be measured, if it's not quantifiable, I would have said, "It's not true."

This young American manager shared with me his change in perspective after working for a year in Ireland. He learned from experience that the context has considerable impact on what gets done and how it gets done.[17] The lesson he learned underlies the second set of human resource issues: managing IT workers in a cross-cultural work environment in which the *American culture* is embedded in the *corporate culture* of these firms. Interestingly, the American managers at the multinational firms tended to downplay national culture emphasizing, instead, what they considered to be the unique characteristics of their own corporate cultures. However, as I moved from one American firm to another, I saw considerable similarities in the multinational corporate cultures, all of which reflected American culture. While a firm's corporate culture — its values, management style, method of operations and work environment — are, to a certain extent, unique to each particular firm, it also reflects the national culture in which it developed. Thus, the corporate culture of an American firm brings to Ireland not only its own unique values but also those present in the American culture.

Another interpretation might be that what is at work, here, is a *monolithic IT culture.* While it is true that the culture of firms in the IT sector is different from the culture of firms in the banking sector, for example, it is also true that corporate cultures exist within national cultures. A given IT culture is shaped by real people who come from real countries which have real cultures. One cannot separate the corporate culture of an American IT firm from the American culture from which the entrepreneurs who founded it emerged.

The management question, then, is what happens when these two different cultures interact. In a positive vein, I found that the Irish IT workers generally welcomed the American culture, but for an unexpected reason. They saw an alignment between the open and egalitarian style reflected in the IT culture of the multinationals and what they identified as characteristic of the *essential* Irish culture reflecting their Celtic origins. They contrasted this with those essentially British modes of thought, feeling and behavior that permeated Irish institutions while Ireland was a British colony. Ironically, they identified *indigenous* firms in traditional sectors as reflecting the rigid, hierarchical atmosphere typical of the British management style.

At the same time, I also witnessed a tension between American and Irish managers over how far the cultural influence should go. While it is true that there are two different national cultures interacting, it is also the case that the balance of power is not even. Not surprisingly, the American managers with whom I spoke, viewed the American culture which is embedded in the multinational's corporate culture to be the desired one. Some of them believed that the multinational firms *ought* to have a significant cultural impact. One managing director recommended that multinational firms keep Americans in management control until the corporate culture imported from America is firmly in place. The rationale is that since a firm's success is due, in part, to its corporate culture, that same "successful" corporate culture ought to be in place in the same way at all locations of a multinational firm.

On the other hand, Irish managers — especially Irish human resource managers — argued for tailoring the corporate culture of the multinational firms to the particular national context. They recommended that when multinational companies are setting up operations they should work with managers who understand the local environment. To support his viewpoint, an Irish human resources manager related how his firm learned that lesson. The firm was having difficulty getting the Irish workers to acclimate to the values and processes inherent in its corporate culture. Twice the firm offered a course on those values and twice the outcome was disappointing. The third time, the Irish human resources manager was asked to co-teach the course with American management and the effort was successful.

Recognizing that there are two distinct cultures is only part of the management challenge. The other part is grappling with the direction of cross-cultural influence. American managers tended to think that the direction would be primarily American culture influencing Irish culture. Not surprisingly, this perception did not sit well with all Irish people, as illustrated in one managing director's experience as he discussed cross-cultural effects with local community leaders:

> I said [our firm] would have an impact on the town and the culture. I commented on conflicts that I saw [between the cultures] and said I thought the [firm's] culture would diffuse into the local society. But one man from the residence council said, "The Vikings and the British tried that before and it didn't work!"

One phrase which permeated conversations about cultural differences was "this is Ireland." American managers at all levels encountered this phrase; my own experience with management practices in Ireland bore them out. Responses such as "Well, this is Ireland; things are different in Ireland," or "You don't do that, this is Ireland," were meant to convey that people in Ireland had their own ways of doing things and perhaps did not want to change.

The perceptions on the part of Irish workers and managers was that while the multinationals were bringing certain values and attitudes to the workplace, there was also a significant influence of Irish culture on the multinational IT workplace. One Irish entrepreneur who had strong views on this topic gave an example from a Japanese firm where he had worked.

> I'll tell you this, I saw more Japanese taking an interest in the Irish culture than I saw Irish people taking an interest in the Japanese culture... In fact, it was a problem for some of the Japanese within their own family scenes. The Japanese men didn't like the Japanese women adopting the Irish ways. They didn't like to see their Japanese children running out to do Irish traditional dancing! In fact, some of them wanted to go back to Japan because of that. But also, where the Japanese men tended to adapt or adopt some of the Irish ways, they were often taken back to Japan as well. I think it was nicely summed up by a man that I worked with. After I left [the firm], I went back and met one of the guys three or four years later, and he looked tired and worn. And he was a man in his mid-forties at the time, but literally he looked sixty. He was gray and his skin was dry and he looked very tired. And I said, "John, how are things going?" and he said, "Ach, tough going." I said, "As bad as how it was?" and he said, "Worse." He said, "Just when we had them all nicely educated they sent them all back to Japan!" So the crowd that had learned the Irish ways and had mellowed from some of the Japanese ways were taken out of the situation and put back to Japan.

Using rich description, the point of this man's story is that the Japanese firm was reluctant to leave its expatriate managers in Ireland too long lest they absorb too much of the Irish culture.

But at the same time that there was evidence of resistance to the multinational influence, there was also the recognition that importing another work culture was part of the plan. Like many newly industrializing countries leaping from an agrarian to a post-industrial economy, Ireland did not have much of an industrial work ethic, much less a post-industrial one. Therefore, by bringing in multinational firms in the information industry, Ireland would be able to import a well-established work ethic that would have taken considerably longer to develop if done indigenously. What has helped both Irish workers and multinational managers is that a large percentage of IT workers have no other work experience, no entrenched work habits acquired in other sectors that need to be overcome.

Despite the desire to import an IT work ethic, there has not been unquestioning acceptance of it. Both sides have needed to adapt. For their part, Irish workers have had to cope with the influence of the "live to work" attitude which they attribute to Americans. Quality of life is important in Ireland and an important part of quality of life is time with family. An example repeatedly given was the Christmas holidays. It is traditional in Ireland for workers to have time off during the week between Christmas and New Year's. But for production-oriented computer hardware firms, the end of the quarter and the end of the calendar year is a very important time. While workers understood this, they also regretted the impact on family life.

American managers, on the other hand, have had to integrate important aspects of Irish culture into their management practices. Key among these is the Irish focus on interpersonal interaction and the residual effects on time, communication and an easygoing atmosphere. There was unanimous agreement that Irish IT workers are more social and more communicative than their American counterparts. Observational evidence comes from the time people take before beginning a day's work to discuss experiences from the night before, the common place where people gather for coffee and tea breaks during the day, and the local pub where people congregate after work. While this behavior could be viewed as detrimental to productivity, I was also provided with ample evidence that it contributes to it as well. American managers described Irish workers as more well-rounded, more knowledgeable about the "big picture" at the firm, and not simply their own small piece of it. This, they attributed to the level of social interaction.

There was also unanimity about the Irish attitude toward time. It is a cultural fact that time is a much more fluid concept in Ireland than it is in America (and especially in the American Northeast where most of the firms I visited were headquartered). American management's reaction to the easygoing attitude about life, in general, and time, in particular, represents the range of possible management responses. One approach was to impose strict controls regarding tardiness and implement them with time cards. Another — and more creative — response was explained by an American managing director to illustrate that there is more than one way to achieve a management goal. When he first arrived in Ireland he was advised to implement time cards. But he was also informed that because all employees needed to be seen to be equal, he would have to punch a time card as well. This was unacceptable to him. His creative solution which tapped into the workers' desire for leisure time, was to implement flex time. Workers could arrive between 7:00 and 9:00 in the morning and leave after an eight hour day. To the chagrin of the skeptics, many more workers showed up at 7:00 than at 9:00 so that they would have longer evenings after work!

## CONCLUSION: LESSONS LEARNED

Ireland's experience with leapfrogging the development of an IT labor force by importing expertise and employment has instructive lessons for both policy makers and multinational managers in other countries. These lessons serve to reinforce the research model underlying this project: that the IT sector (including its workforce) is situated within a socio-cultural context which is helping to shape it even as the IT sector, itself, is effecting changes in the society of the host country. Figure 2 shows the Influence-Impact Model applied to the specific issues addressed in this paper.

Because of the specialized skills required for employment in the information sector, education is the key to economic and employment opportunity. Clearly, the educational infrastructure must provide the requisite skills and knowledge. But in order to ensure that all citizens are able to avail themselves of the *economic opportunities* present in the information sector, it is important that they have equal *educational opportunity*. In Ireland and elsewhere economic, gender and class barriers may inhibit full and equal participation in the emerging information society (Clancy, 1988; Trauth, 1993; Trauth, 1995).

*Figure 2: Interaction between Societal Factors and Multinational IT Sector in Ireland*

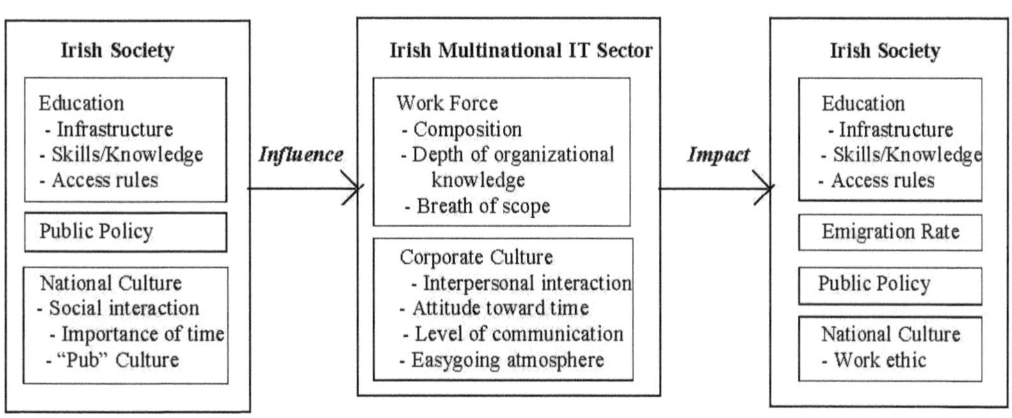

In order to ensure that there is a societal climate which can sustain this IT sector and these workers, the educational infrastructure needs to be aligned with the skills and knowledge required for IT work. Further, there should be coordination between the employment objectives of the policy makers and the educational plans of the universities. There is much irony in the fact that an industrial policy intended to stem the tide of Irish emigration has resulted in some of the best and brightest leaving a country. Clearly, this is an issue for industrial policy makers to address.

Another set of lessons, which are about managing IT workers in a cross-cultural environment, derive from the simple but significant acknowledgement that two different cultures are involved when a multinational IT firm sets up operations in a country. One lesson is that both of these cultures have positive contributions to offer the workplace. There are implications for both multinational and indigenous workers. Multinational managers should strive to understand work patterns and attitudes of the host country culture that, while different, may be able to produce the desired result. Multinational managers should also understand that in introducing a corporate culture they are also introducing a foreign national culture. At the same time, host country managers and policy makers must recognize that when the intent is to import expertise in order to quickly introduce an IT sector, one side effect will be the changes in national culture. One way to cope with these two cultures is by focusing on what the two have in common. In the case of Ireland, for example, managers could leverage the alignment between the egalitarian style of American IT firms and aspects of the essential Irish culture.

By recognizing that there is more than one way to achieve a management goal, multinational managers can develop procedures and management approaches that exploit the best features of the host country. In the case of Ireland, American managers learned that a potentially significant human relations situation could be diffused by employing the mechanism of meeting in the pub to discuss it. Clearly, this management approach fits well with Irish culture but may not suit another cultural context. Likewise, what works best in Japan or in America may not be what works best in Ireland.

In the presence of two distinct cultures, it is also necessary to acknowledge that cultural influence goes in both directions. It is natural that the home culture of the multinational IT firm will influence the society it enters. But it is also natural that the workers will want to influence the corporate culture to make it compatible with their own. By building a permeable wall, by the open exchange of values and norms both cultures can be enriched, something that bodes well for the global IT sector.

## APPENDIX: INTERVIEW CATEGORIES

I. Respondent Background: job description, educational background, work history, family background

II. Socio-cultural Factors - Individual
   - A. Work Life: work ethic, work culture, management style, unions
   - B. Personal Life: quality of life, standard of living, family, religion

III. Socio-cultural Factors - Societal
   - A. Population: age, size, gender, education
   - B. National Traits: authority, personalism, individualism, post-colonial legacy, social interaction, time, island mentality

## ACKNOWLEDGMENTS

This work was initially supported by a Fulbright Scholar Award. I would like to thank my Irish and American informants who made this work possible and my colleagues in the Irish Studies community who have provided critical feedback on this work.

## ENDNOTES

[1] According to Porat (1977) the information economy is divided into the *primary information sector* and the *secondary information sector*. I limited my investigation to the *primary information sector,* which includes those who develop the information technology: the hardware, software and systems. Those working in the secondary information sector, those engaged in information processing activities in the course of doing work related to some other primary activity (such as insurance processing), were not included in the study.

[2] See, for example, Ang and Jiwahhasuchin (1998), Gan (1998), Hernandez, et al. (1996), Hill, et al. (1998), Lally (1994), La Rovere (1996), Odedra-Straub (1993), Trauth, et al. (1998), Van Rychegham (1996), Wilson and Meadows (1998), Wong (1998).

[3] See, for example, Braa (1996), Burn, et al. (1995), Cummings and Guynes (1994), Heeks (1996), Kedia and Bhagat (1988), Lopez (1996), Nelson and Clark (1994), Neo (1991), Walsham, et al. (1988). Books such as Palvia, et al (1992) address both country-specific issues and development/cross-cultural issues.

[4] Previous papers written from this research project have focused on specific dimensions of this study. In contrast, this paper is the result of a meta study of specific findings of the study: those with implications for IT human resources management. Papers from this research project which focus on other, specific dimensions of the study include the following:

Trauth & Pitt, 1992: The impact of telecommunications privatization on later industrializing countries.

Trauth, 1993: The evolution of an educational infrastructure compatible with an information economy.

Trauth, 1995: The influence of the role of women in Irish society on their participation in the information economy.

Trauth, 1996: The cultural and economic effects of importing foreign cultures along with multinational employment.

[5] While multinational firms from other countries such as Japan and Germany are present in Ireland, this research was restricted to American firms only.

[6] On occasion the scheduled time period ended before the respondent had given all of her/his comments and a second interview was held. There were two notable exceptions to the standard interview time and place (at the firm). One was the interview with an American manager which turned into a dinner invitation and an entire evening spent interviewing him and his wife. The other was the six-hour interview with an Irish software engineer, which was conducted in a Dublin pub.

[7] This overview of Ireland's movement into the IT sector has resulted from formal and informal interviews with government officials in Ireland as well as from examination of books and public policy documents. Some of the key books used in this research are the following: Breen, et al. (1990), Daly (1992), Jordan (1992), O'Malley (1989) and Rottman, et al. (1982). Some of the key public policy documents are the following: Whitaker (1958), National Economic and Social Council (1982), *The Way Forward: National Economic Plan 1983-1987, White Paper on Industrial Policy,* 1984, *Review of Industrial Performance,* 1986, *Building on Reality,* 1985-1987, *Industrial Development Act of 1986, Programme for National Recovery,* 1987, National Software Directorate, 1995.

[8] It was the combination of high emigration and high unemployment that signaled the need for a change in the economic policy of isolationism.

[9] In the case of Ireland, localizing software means tailoring commercial packages such as those from Microsoft and Lotus for European consumers.

[10] See, for example, "Europe's Tiger Economy," (1997); "Ireland Shines" (1997); Jordan (1992); and O'Riain (1997).

[11] These financial incentives included both tax relief and financial grants. Depending upon when the firms came to Ireland and where they located, the financial packages varied. But generally the firms were offered zero or low tax rates for a certain period of time. In addition, they were offered grants for establishing and equipping their factories, and training their work forces. In some instances, the Irish government actually built the factories on greenfield sites.

[12] See, for example, Industrial Policy Review Group (1992).

[13] This term refers to data processing activities being carried out in foreign locations. The west of Ireland, for example, has become a remote site for claims processing by American insurance firms, while in Dublin the Financial Services Center provides data processing for European financial services firms. (See, Industrial Development Authority, 1992)

[14] Over the past 30 years, roughly half of the population has remained under the age of 30, oftentimes younger than that.

[15] One signifier of social class that initially puzzled me is the religious order running the secondary school one attended. In my interviews, people would regularly include comments such as: ". . . it was a Jesuit school" or ". . . the Christian Brothers taught me," when giving me their educational history in order to indicate whether they came from a middle class or a working class family.

[16] Attaining the Leaving Certificate is the approximate equivalent of receiving a high school diploma.

[17] See Hall (1994) for an account of how he learned this lesson in working with Native Americans in the American Southwest.

## REFERENCES

Ang, A.Y. and Jiwahhasuchin, S. (1998). "Information Systems Education in Thailand: A Comparison between the Views of Professionals and Academics, *Journal of Global Information Management,* 6(4), 34-42.

Ayiku, M.N.B. (1989). "A Case Study in Process Automation in Ghana," *Information Technology for Development,"* 4(2), 117-156.

Braa, J. and Monteiro, E. (1996). "Infrastructure and Institutions: The Case of Public Health in Mongolia," in *Information Technology Development and Policy: Theoretical Perspectives and Practical Challenges.* E.M. Roche and M.J. Blaine (eds.) Aldershot, UK: Avebury Publishing Ltd, 171-188.

Breen, R., Hannan, D.H., Rottman, D.B. and Whelan, C.T. (1990). *Understanding Contemporary Ireland: State, Class and Development in the Republic of Ireland.* Dublin: Gill and Macmillan.

*Building on Reality: 1985-1987*. Pl.2648. Dublin: The Stationery Office.

Burn, J.M., Ng Tye, E.M.W., and Ma, L.C.K. (1995). "Para-digm Shift — Cultural Implications for Development of IS Professionals," *Journal of Global Information Management,* 3(2), 18-28.

Clancy, P. (1988). *Who Goes to College: A Second National Survey of Participation in Higher Education*. Dublin: Higher Education Authority.

Cummings, M.L. and Guynes, J.L. (1994). "Information System Activities in Transnational Corporations: A Comparison of U.S. and Non-U.S. Subsidiaries," *Journal of Global Information Management,* 2(1), 12-26.

Daly, M.E. (1992). *Industrial Development and Irish National Identity, 1922-1939.* Syracuse, NY: Syracuse University Press.

Europe's Tiger Economy (1997). *The Economist.* (May 17), 21-24.

Gan, S. (1998). "An Overview of Information Technology and Education in Malaysia," *Journal of Global Information Management,* 6(1), 27-32.

Hall, E.T. (1994). *West of the Thirties: Discoveries Among the Navajo and Hopi.* NY: Doubleday.

Heeks, R. (1996). "Promoting Software Production and Export in Developing Countries," in *Information Technology Development and Policy: Theoretical Perspectives and Practical Challenges.* E.M. Roche and M.J. Blaine (eds.) Aldershot, UK: Avebury Publishing Ltd, 77-94.

Hernandez, D., Gibson, R. and McGuire, E.G. (1996). "Informatics in Uruguay: Evolution and Implications," *Journal of Global Information Management,*4(1), 23-31.

Hill, C.E., Loch, K.D., Straub, D.W., and El-Sheshai, K. (1998). "A Qualitative Assessment of Arab Culture and Information Technology Transfer," *Journal of Global Information Management,* 6(3), 29-38.

Industrial Development Act of 1986.

Industrial Development Authority. (1992). *Ireland as a Cost Competitive Location for Back Office Data Processing and Remote Software Development Operations.* Presentation to Taylorix AG.

Industrial Policy Review Group (1992). *A Time for Change: Industrial Policy for the 1990s.* Dublin: Ministry for Industry and Commerce.

Ireland Shines (1997). *The Economist.* (May 17), 16.

Jordan, D. (1992). "The Ireland-Silicon Nexus." *Journal of the West. 31(2),* 79-84.

Kedia, B.L. and Bhagat, R.S. (1988). "Cultural Constraints on Transfer of Technology across Nations: Implications for Research in International and Comparative Management," *Academy of Management Review,* 13(4), 559-571.

Lally, L. (1994). "The Impact of Environment on Information Infrastructure Enhancement: A Comparative Study of Singapore, France and the United States," *Journal of Global Information Management,* 2(3), 5-12.

La Rovere, R.L. (1996). "Diffusion of IT and the Competitiveness of Brazilian Banking," in *Information Technology Development and Policy: Theoretical Perspectives and Practical Challenges.* E.M. Roche and M.J. Blaine (eds.) Aldershot, UK: Avebury Publishing Ltd, 95-112.

Lopez, E.J. and Vilaseca (1996). "IT as a Global Economic Development Tool," in *Information Technology Development and Policy: Theoretical Perspectives and Practical Challenges.* E.M. Roche and M.J. Blaine (eds.) Aldershot, UK: Avebury Publishing Ltd, 57-75.

National Economic and Social Council (1982). *A Review of Industrial Policy: A Report Prepared by the Telesis Consultancy Group.*No.64. Dublin: Author.

National Software Directorate (1992). *The Software Industry in Ireland: A Strategic Review.* Dublin: Industrial Development Authority.

Nelson, K.G. and Clark, T.D. (1994). "Cross-Cultural Issues in Information Systems Research: A Research Program," *Journal of Global Information Management,* 2(4), 19-29.

Neo, B. S. (1991). "Information Technology and Global Competition," *Information & Management,* Volume 20, 151-160.

Odedra-Straub, M. (1993). "Critical Factors Affecting Success of CBIS: Cases from Africa," *Journal of Global Information Management,* 1(3), 16-31.

O'Malley, E. (1989). *Industry and Economic Development: The Challenge of the Latecomer.* Dublin: Gill and Macmillan Ltd.

O'Riain, S. (1997). The Birth of a Celtic Tiger. *Communications of the ACM.* 30(3), 11-16.

*Overseas Industry in Ireland* (1991). A. Foley and D. McAleese (Eds.), Dublin: Gill and Macmillan, Ltd.

Palvia, S., Palvia, P. and Zigli, R. (eds.) (1992). *The Global Issues of Information Technology Management.* Harrisburg, PA: Idea Group Publishing.

Porat, M. (1977). *Information Economy: Definition and Measurement.* Washington, D.C.: Office of Telecommunications.

Power, C. (1990). *1992: A New Dawn for Knowledge-based Industries in Ireland.* Paper presented at the Bolton Trust international seminar, "1992: Its Impact on Technological Education," Dublin, February.

*Programme for National Recovery (*1987). Pl.5213. Dublin: The Stationery Office.

*Review of Industrial Performance* (1986).

Rottman, D.B., Hannan, D.F., Hardiman, N. and Wiley, M.M. (1982). *The Distribution of Income in the Republic of Ireland: A Study of Social Class and Family Cycle Inequalities.* (No. 109). Dublin: Economic and Social Research Council.

Strauss, A.L. (1987). *Qualitative Analysis for Social Scientists.* New York: Cambridge University Press.

Trauth, E.M. (1993). "Educating IT Professionals for Work in Ireland: An Emerging Post-industrial Country," in *Global Information Technology Education: Issues and Trends.* M. Khosrowpour and K. Loch (Eds.) Harrisburg, PA: Idea Group Publishing, 205-233.

Trauth, E.M. (1995)." Women in Ireland's Information Economy: Voices from Inside." *Eire Ireland,* 30(3), 133-150.

Trauth, E.M. (1996). "Impact of an Imported IT Sector: Lessons from Ireland," in *Information Technology Development and Policy: Theoretical Perspectives and Practical Challenges.* E.M. Roche and M.J. Blaine (eds.) Aldershot, UK: Avebury Publishing Ltd, 245-261.

Trauth, E.M. (1997). "Achieving the Research Goal with Qualitative Methods: Lessons Learned along the Way," in *Information Systems and Qualitative Research.* A.S. Lee, J. Liebenau and J.I. DeGross (Eds.) London: Chapman and Hall, 225-245.

Trauth, E.M., Derksen, F.E.J.M., and Mevissen, H.M.J. (1998). "Societal Factors and the Diffusion of EDI," in *EDI and Data Networking in the Public Sector: Governmental Action, Diffusion and Impacts.* K.V. Andersen (Ed.) Boston: Kluwer Academic Publishers, 43-61.

Trauth, E.M. and O'Connor, B. (1991). "A Study of the Interaction between Information Technology and Society: An Illustration of Combined Qualitative Research Methods," in *Information Systems Research: Contemporary Approaches & Emergent Traditions.* H.E. Nissen, H.K. Klein & R. Hirschheim (Eds.) Amsterdam: North-Holland, 131-144.

Trauth, E.M. and Pitt, D.C. (1992). "Competition in the Telecommunications Industry: A New Global Paradigm and Its Limits," *Journal of Information Technology,* 7(1), (March), 3-11.

Van Ryckeghem, D. (1996). "Computers and Culture: Cases from Kenya," in *Information Technology Development and Policy: Theoretical Perspectives and Practical Challenges.* E.M. Roche and M.J. Blaine (eds.) Aldershot, UK: Avebury Publishing Ltd, 154-170.

Walsham, G., Symons, V. and Waema, T. (1988). "Information Systems as Social Systems: Implications for Developing Countries," in *Information Technology for Development,* 3(3), 198-204.

*The Way Forward: National Economic Plan 1983-1987.* (1982). Pl.10161. Dublin: The Stationery Office. October.

Whitaker, T.K. (1958). *Economic Development.*

*White Paper on Industrial Policy (*1984). Pl. 2491. Dublin: The Stationery Office. 12 July.

Wilson, R.E. and Meadows, C.J. (1998). "Tele-Teaching: Australia's Competitive Question," *Journal of Global Information Management,* 6(1), 15-26.

Wong, P. (1998). "Leapfrogging across the Millennium: Information Technology in Singapore Schools," *Journal of Global Information Management,*6(1), 5-13.

# About the Editor

**Dr. Felix B. Tan** is a Senior Lecturer (Associate Professor equivalent in North America) in Information Systems at the University of Auckland, New Zealand. He serves as the Editor-in-Chief of the *Journal of Global Information Management.* He is also the Vice President of Research for the Information Resources Management Association, the Editor of the ISWorld Net's EndNote Resources page, and Editor of the book series *Advanced Topics in Global Information Management.* He had held visiting positions with the Department of Information Systems, National University of Singapore and Richard Ivey School of Business, the University of Western Ontario.

Dr. Tan's current research interests are in business-IT alignment, global information management, management of IT, national information policy issues and cognitive mapping methods. His research has been published in *MIS Quarterly, Journal of Global Information Management,* and *Journal of Information Technology,* as well as other journals and refereed conference proceedings. He has published two other books - one on IT Diffusion in the Asia Pacific region and another presenting Global IT Cases. Dr. Tan has over 20 years experience in information systems management and consulting with large multinationals, as well as university teaching and research in Singapore, Canada and New Zealand.

# Index

## O



## P



## Q



## R



## S



## T

www.ingramcontent.com/pod-product-compliance
Ingram Content Group UK Ltd.
Pitfield, Milton Keynes, MK11 3LW, UK
UKHW050947290926
15470UKWH00005B/1688

9 781931 777117